Doing business in 2004

Doing business *in* 2004

Understanding Regulation

A copublication of the World Bank,
the International Finance Corporation,
and Oxford University Press

1818 H Street NW
Washington, D.C. 20433
Telephone 202-473-1000
Internet www.worldbank.org
E-mail feedback@worldbank.org

1 2 3 4 05 04 03

A copublication of the World Bank and Oxford University Press.

Rights and Permissions

Additional copies of *Doing Business in 2004: Understanding Regulation* may be purchased at http://publications.worldbank.org/ecommerce/catalog/product?item_id=1384804/.

ISBN 0-8213-5341-1

ISSN 1729–2638

Library of Congress Cataloging-in-Publication data has been applied for.

Contents

Acknowledgments

Doing Business in 2004 was prepared by a team led by Simeon Djankov. Caralee McLiesh co-managed development and production of the report. The work was carried out under the general direction of Michael Klein. Simeon Djankov coordinated the work on starting a business and hiring and firing workers. Caralee McLiesh led the work on getting finance. Tatiana Nenova designed and implemented the study on closing a business. Simeon Djankov and Stefka Slavova coordinated the work on enforcing a contract. The team also comprised Ziad Azar, Geronimo Frigerio, Joanna Kata-Blackman, and Lihong Wang and was assisted by Bekhzod Abdurazzakov, Yanni Chen, Marcelo Lu, Totka Naneva, and Tania Yancheva. Zai Fanai and Grace Sorensen provided administrative support.

Andrei Shleifer co-authored the main background studies and provided valuable suggestions throughout the writing of the report. Florencio Lopez-de-Silanes and Rafael La Porta co-authored the background studies on starting a business, hiring and firing workers, and enforcing a contract. Oliver Hart co-authored the background study on closing a business. Bruce Ross-Larson edited the manuscript. Nataliya Mylenko contributed to the research and chapter on getting credit. The survey of credit registries was developed in cooperation with the Credit Reporting Systems Project in the World Bank, and the survey on closing a business was developed with the assistance of Selinda Melnik. Nicola Jentzsch and Fredreich Schneider wrote background papers on the regulation of credit information and the informal economy, respectively. Leszek Balcerowicz, Hernando de Soto, Bradford DeLong, and Andrei Shleifer contributed lectures on the scope of government.

Preparation of the report was made possible by the contributions of more than 2,000 judges, lawyers, accountants, credit registry representatives, business consultants, and government officials from around the world. Many of the contributors are partners in Lex Mundi law firms or are members of the International Bar Association. Their names are listed in the Contributors' section and their contact details are on the *Doing Business* web site.

Individual chapters were refereed by: Elizabeth Adu, Asya Akhlaque, Gordon Betcherman, Harry Broadman, Gerard Byam, Gerard Caprio, Amanda Carlier, Jacqueline Coolidge, Asli Demirguc-Kunt, Julia Devlin, Michael Fuchs, Luke Haggarty, Mary Hallward-Driemeier, Linn Hammergren, Eric Haythorne, Aart Kraay, Peter Kyle, Katarina Mathernova, Richard Messick, Margaret Miller, Claudio Montenegro, Reema Nayar, S. Ramachandran, Jan Rutkowski, Stefano Scarpetta, Peer Stein, Ahmet Soylemezoglu, Andrew Stone, and Stoyan Tenev. A draft report was reviewed by David Dollar, Cheryl Gray, W. Paatii Ofosu-Amaah, Guy Pfeffermann, and Sanjay Pradhan. Axel Peuker, Neil Roger, and Suzanne Smith provided advice and comments throughout the development of the report. Tercan Baysan, Najy Benhassine, Vinay Bhargava, Harry Broadman, Gerard Caprio, Mierta Capaul, David Dollar, Qimiao Fan, Caroline Freund, Alan Gelb, Indermit Gill, Frannie Leautier, Syed Mahmood, Andrei Michnev, John Page, Sanjay Pradhan, Mohammad Zia M. Qureshi, Stoyan Tenev, Cornelius van der Meer, and Gerald West read the penultimate draft and suggested changes. The online service of the *Doing Business* database is sponsored by the Rapid Response Unit of the World Bank Group.

A vibrant private sector—with firms making investments, creating jobs, and improving productivity—promotes growth and expands opportunities for poor people. To create one, governments around the world have implemented wide-ranging reforms, including macro-stabilization programs, price liberalization, privatization, and trade-barrier reductions. In many countries, however, entrepreneurial activity remains limited, poverty high, and growth stagnant. And other countries have spurned orthodox macro reforms and done well. How so?

Although macro policies are unquestionably important, there is a growing consensus that the quality of business regulation and the institutions that enforce it are a major determinant of prosperity. Hong Kong (China)'s economic success, Botswana's stellar growth performance, and Hungary's smooth transition experience have all been stimulated by a good regulatory environment. But little research has measured specific aspects of regulation and analyzed their impact on economic outcomes such as productivity, investment, informality, corruption, unemployment, and poverty. The lack of systematic knowledge prevents policymakers from assessing how good legal and regulatory systems are and determining what to reform.

Doing Business in 2004: Understanding Regulation is the first in a series of annual reports investigating the scope and manner of regulations that enhance business activity and those that constrain it. The present volume compares more than 130 countries—from Albania to Zimbabwe—on the basis of new quantitative indicators of business regulations. The indicators are used to analyze economic outcomes and identify what reforms have worked, where, and why.

What Is New?

Many sources of data help explain the business environment. More than a dozen organizations—such as Freedom House, the Heritage Foundation, and the World Economic Forum—produce and periodically update indicators on country risk, economic freedom, and international competitiveness. As gauges of general economic and policy conditions, these indicators help identify broad priorities for reform. But few indicators focus on the poorest countries, and most of them are designed to inform foreign investors. Yet it is local firms, which are responsible for most economic activity in developing countries, that could benefit the most from reforms. Moreover, many existing indicators rely on perceptions, notoriously difficult to compare across countries or translate into policy recommendations. According to one survey, Belarus and Uzbekistan rank ahead of France, Germany, and Sweden in firms' satisfaction with the efficiency of government. Most important, no indicators assess specific laws and regulations regarding business activity or the public institutions that enforce them. So these indicators provide insufficient detail to guide reform of the scope and efficiency of government regulation.

The indicators in the present volume represent a new approach to measurement. The focus is on domestic, primarily smaller, companies. The analysis is based on assessments of laws and regulations, with input from and verification by local experts who deal with practical situations of the type covered in the report.

This methodology offers several advantages. It is based on factual information concerning laws and regulations in force. It is transparent and easily replicable—allowing broad country coverage, annual updates, and ready extension to new locations. It covers regulatory outcomes, such as the time and cost of meeting regulatory requirements to register a business, as well as measures of actual regulations, such as an index of the rigidity of employment law or the procedures to enforce a contract. It also investigates the efficiency of government institutions, including business registries, courts, and public credit registries. Most important, the methodology builds on extensive and detailed information on regulations—information directly relevant to identifying specific problems and designing reforms.

The *Doing Business* series represents a collaborative effort. The *Doing Business* team works with leading scholars in the development of indicators. This cooperation provides academic rigor and links theory to practice. For this year's report, Professor Andrei Shleifer (Harvard University) served as adviser on all projects. Professor Oliver Hart (Harvard University) advised on the bankruptcy project, and Professor Florencio Lopez-de-Silanes (International Institute of Corporate Governance, Yale School of Management) and Professor Rafael La Porta (Dartmouth) advised on the business registration, contract enforcement, and labor projects.

Each project involves a partnership with an association of practitioners or an international company. For example, the contract enforcement project was conducted with Lex Mundi, the largest international association of private law firms. The project on credit market institutions benefited from collaboration with the law firm of Baker and McKenzie, the International Bar Association Committee on International Financial Law Reform, and Dun and Bradstreet. The bankruptcy project was conducted with the help of the Insolvency Committee of the International Bar Association.

The *Doing Business* project receives the invaluable cooperation of local partners—municipal officials, registrars, tax officers, labor lawyers and labor ministry officials, credit registry managers, financial lawyers, incorporation lawyers in the case of business start-ups, bankruptcy lawyers, and judges. Only those with extensive professional knowledge and experience provide data, and the indicators build on local knowledge.

Once the analysis is completed, the results are subject to a peer-review process in leading academic journals. Simultaneously, the background research is presented at conferences and seminars organized with private-sector partners. For example, preliminary results of the bankruptcy project were discussed with members of the International Bar Association at the association's meetings in Dublin (Ireland), Durban (South Africa), Rome (Italy), and New York (United States). The data are posted on the web (http://rru.worldbank.org/doingbusiness), so anyone can check and challenge their veracity. This continual process of refinement produces indicators that have been scrutinized by the academic community, government officials, and local professionals.

What Does *Doing Business* Aim to Achieve?

Two years ago, the World Bank Group outlined a new strategy for tapping private initiative to reduce poverty. The *Doing Business* project aims to advance the World Bank Group's private sector development agenda:

- *Motivating reforms through country benchmarking.* Around the world, international and local benchmarking has proved to be a powerful force for mobilizing society to demand improved public services, enhanced political accountability, and better economic policy. Transparent scoring on macroeconomic and social indicators has intensified the desire for change—witness the impact of the human development index, developed by the United Nations' Development Programme, on getting countries to emphasize health and education in their development strategies. The *Doing Business* data provide reformers with comparisons on a different dimension: the regulatory environment for business.
- *Informing the design of reforms.* The data analyzed in *Doing Business* highlight specifically what needs

to be changed when reforms are designed, because the indicators are backed by an extensive description of regulations. Reformers can also benefit from reviewing the experience of countries that perform well according to the indicators.

- *Enriching international initiatives on development effectiveness.* Recognizing that aid works best in good institutional environments, international donors are moving toward more extensive monitoring of aid effectiveness and performance-based funding. The U.S. government's Millennium Challenge Account and the International Development Association's performance-based funding allocations are two examples. It is essential that such efforts be based on good-quality data that can be influenced directly by policy reform. This is exactly what *Doing Business* indicators provide.
- *Informing theory.* Regulatory economics is largely theoretical. By producing new indicators that quantify various aspects of regulation, *Doing Business* facilitates tests of existing theories and contributes to the empirical foundation for new theoretical work on the relation between regulation and development.

What to Expect Next

This report summarizes the results of the first year of the *Doing Business* project. The volume is only the first product of an ambitious study of the determinants of private sector development. About a dozen topics in the business environment will be developed over three years. This year, five topics are analyzed. They cover the fundamental aspects of a firm's life cycle: starting a business, hiring and firing workers, enforcing contracts, getting credit, and closing a business. Over the next two years, *Doing Business* will extend the coverage of topics. *Doing Business in 2005* will discuss three new topics—registering property, dealing with government licenses and inspections, and protecting investors. *Doing Business in 2006* will study three other topics: paying taxes, trading across borders, and improving law and order.

The indicators will be updated annually to provide time-series data on progress with reform. Currently the *Doing Business* project does not focus on the political economy of reform. As more data become available, the project will include exploration of political economy issues and measurement of reform impact, as well as the cross-section analysis that this report presents.

The project will also create case studies of reform. It will document past experiences, the forces behind reform, and the features responsible for reforms' ultimate success or failure. This information will help policymakers design and manage reform.

The impact of regulations is measured by their relationship to economic outcomes. Although data on some outcomes such as income growth and employment are readily available, data on others are not. The *Doing Business* project has begun to address this gap by supporting work on the size of the informal business sector and the determinants of entrepreneurship. In future years, other economic outcome variables will be analyzed.

The new data and analysis deepen our understanding of productivity growth and the optimal scope for government in regulating business activity. Under the auspices of the *Doing Business* project, Dr. Leszek Balcerowicz (National Bank of Poland), Professor Bradford DeLong (University of California at Berkeley), Hernando de Soto (Institute of Liberty and Democracy in Lima, Peru), and Professor Andrei Shleifer (Harvard University) have been invited to give lectures on government regulation of business. In coming years other outstanding economic thinkers will be invited to give lectures on *Doing Business* topics.

Updated indicators and analysis of topics, as well as any revisions of or corrections to the printed data, are available on the *Doing Business* Web site: http://rru.worldbank.org/doingbusiness.

Overview

Teuku, an entrepreneur in Jakarta, wants to open a textile factory. He has customers lined up, imported machinery, and a promising business plan. Teuku's first encounter with the government is when registering his business. He gets the standard forms from the Ministry of Justice, and completes and notarizes them. Teuku proves that he is a local resident and does not have a criminal record. He obtains a tax number, applies for a business license, and deposits the minimum capital (three times national income per capita) in the bank. He then publishes the articles of association in the official gazette, pays a stamp fee, registers at the ministry of justice, and waits 90 days before filing for social security. One hundred sixty-eight days after he commences the process, Teuku can legally start operations. In the meantime, his customers have contracted with another business.

In Panama, another entrepreneur, Ina, registers her construction company in only 19 days. Business is booming and Ina wants to hire someone for a two-year appointment. But the employment law only allows fixed-term appointments for specific tasks, and even then requires a maximum term of one year. At the same time, one of her current workers often leaves early, with no excuse, and makes costly mistakes. To replace him, Ina needs to notify and get approval from the union, and pay five months' severance pay. Ina rejects the more qualified applicant she would like to hire and keeps the underperforming worker on staff.

Ali, a trader in the United Arab Emirates, can hire and fire with ease. But one of his customers refuses to pay for equipment delivered three months earlier. It takes 27 procedures and more than 550 days to resolve the payment dispute in court. Almost all procedures must be made in writing, and require extensive legal justification and the use of lawyers. After this experience, Ali decides to deal only with customers he knows well.

Timnit, a young entrepreneur in Ethiopia, wants to expand her successful consulting business by taking a loan. But she has no proof of good credit history because there are no credit information registries. Although her business has substantial assets in accounts receivable, laws restrict her bank from using these as collateral. The bank knows it cannot recover the debt if Timnit defaults, because courts are inefficient and laws give creditors few powers. Credit is denied. The business stays small.

Having registered, hired workers, enforced contracts, and obtained credit, Avik, a businessman in India, cannot make a profit and goes out of business. Faced with a 10-year-long process of going through bankruptcy, Avik absconds, leaving his workers, the bank, and the tax agency with nothing.

Does cumbersome business regulation matter? Yes, and particularly for poor people. In much of Africa, Latin America, and the former Soviet Union, excessive regulation stifles productive activity (figure 1). And government does not focus on what it should—defining and protecting property rights. These are the regions where growth stagnates, few new jobs are created, and poverty has risen. In Africa, poverty rates have increased in the last three decades, with more than 40 percent of the population now living on less than one dollar a day. Two decades of macroeconomic reform in Latin America have not slowed the rise in poverty. And in most former Soviet

Figure 1
Cumbersome Regulation Is Associated with Lower Productivity

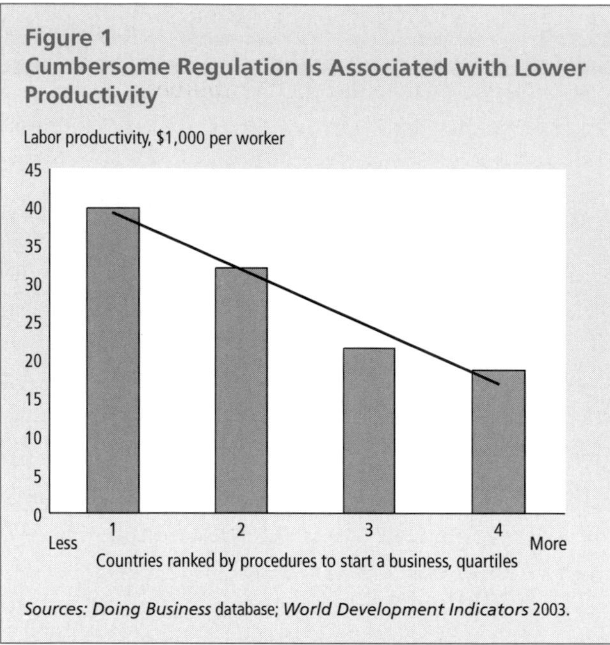

Sources: Doing Business database; *World Development Indicators* 2003.

Figure 2
Heavier Regulation Is Associated with Informality and Corruption

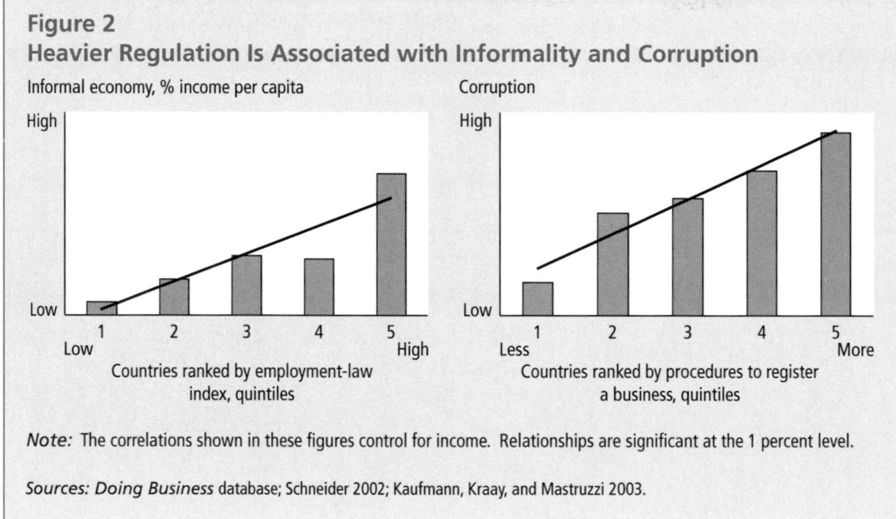

Note: The correlations shown in these figures control for income. Relationships are significant at the 1 percent level.

Sources: Doing Business database; Schneider 2002; Kaufmann, Kraay, and Mastruzzi 2003.

countries, poverty increased in the decade prior to the fall of communism, and even faster thereafter. In 2003, the number of people earning less than a dollar a day remains at 1.2 billion and the number earning less than two dollars a day at 2.8 billion.

"First, I would like to have work of any kind," says an 18-year-old Ecuadorian. The quotation is from *Voices of the Poor*, a World Bank survey capturing the perspectives of poor people around the world. People know how to escape poverty. What they need is to find a decent job. Studies using household survey data confirm this—the vast majority of people who escape from poverty do so by taking up new employment opportunities.

Not any job will lead out of poverty. If it were simply a matter of creating jobs, having the state employ everyone would do the trick. This has been tried in some parts of the world, notably in communist regimes. What is needed is to create productive jobs and new businesses that create wealth. For this, companies need to adjust to new market conditions and seize opportunities for growth. But all too frequently this flexibility is taken away by cumbersome regulation. Productive businesses thrive where government focuses on the definition and protection of property rights. But where the government regulates every aspect of business activity heavily, businesses operate in the informal economy. Regulatory intervention is particularly damaging in countries where its enforcement is subject to abuse and corruption (figure 2).

To document the regulation of business and investigate the effect of regulation on such economic outcomes as productivity, unemployment, growth, poverty, and informality, the *Doing Business* team collected and analyzed data on five topics—starting a business, hiring and firing workers, enforcing a contract, getting credit, and closing a business. The efficiency of the enforcement institutions—commercial registries; municipal offices; tax, fire-and-safety, and labor inspectorates; credit and collateral registries; and courts—has also been assessed.

Doing Business starts by asking five questions. Are there significant differences in business regulation across countries? If so, what explains these differences? What types of regulation lead to improved economic and social outcomes? What are the most successful regulatory models? And, more generally, what is the scope for government in facilitating business activity? As the coverage of topics expands in future editions of *Doing Business*, these questions will be further explored. The analysis in this year's report yields some preliminary answers.

Poor Countries Regulate Business the Most

It takes 2 days to start a business in Australia, but 203 days in Haiti and 215 days in the Democratic Republic of Congo. There are no monetary costs to start a new business in Denmark, but it costs more than 5 times income per capita in Cambodia and over 13 times in Sierra Leone. Hong Kong (China), Singapore, Thailand, and more than three dozen other economies require no minimum capital from start-ups. In contrast, in Syria the capital requirement is equivalent to 56 times income per capita, in Ethiopia and Yemen, 17 times, in Mali, 6 times.

Businesses in the Czech Republic and Denmark can hire workers on part-time or fixed-term contracts for any job, without specifying maximum duration of the contract. Part-time work, exempt from some regulations, is less costly to terminate than full-time employment. In contrast, employment laws in El Salvador allow fixed-term contracts only for specific jobs, and set their duration to be at most one year. Part-time workers receive the benefits of full-time workers, and are subject to the same regulation on procedures for dismissal.

A simple commercial contract is enforced in 7 days in Tunisia and 39 days in the Netherlands, but takes almost 1,500 days in Guatemala. The cost of enforcement is less than 1 percent of the disputed amount in Austria, Canada, and the United Kingdom, but more than 100 percent in Burkina Faso, the Dominican Republic, Indonesia, the Kyrgyz Republic, Madagascar, Malawi, and the Philippines.

Credit bureaus contain credit histories on almost every adult in New Zealand, Norway, and the United States. But the credit registries in Cameroon, Ghana, Pakistan, Nigeria, and Serbia and Montenegro have credit histories for less than 1 percent of adults. In the United Kingdom, laws on collateral and bankruptcy give creditors strong powers to recover their money if a debtor defaults. In Colombia, the Republic of Congo, Mexico, Oman, and Tunisia, a creditor has no such rights.

It takes less than six months to go through bankruptcy proceedings in Ireland and Japan, but more than 10 years in Brazil and India. It costs less than 1 percent of the value of the estate to resolve insolvency in Finland, the Netherlands, Norway, and Singapore—and nearly half the estate value in Chad, Panama, Macedonia, Venezuela, Serbia and Montenegro, and Sierra Leone.

Regulation in poor countries is more cumbersome in all aspects of business activity (figure 3). Across all five sets of indicators, Bolivia, Burkina Faso, Chad,

Figure 3
Poor Countries Regulate Business the Most

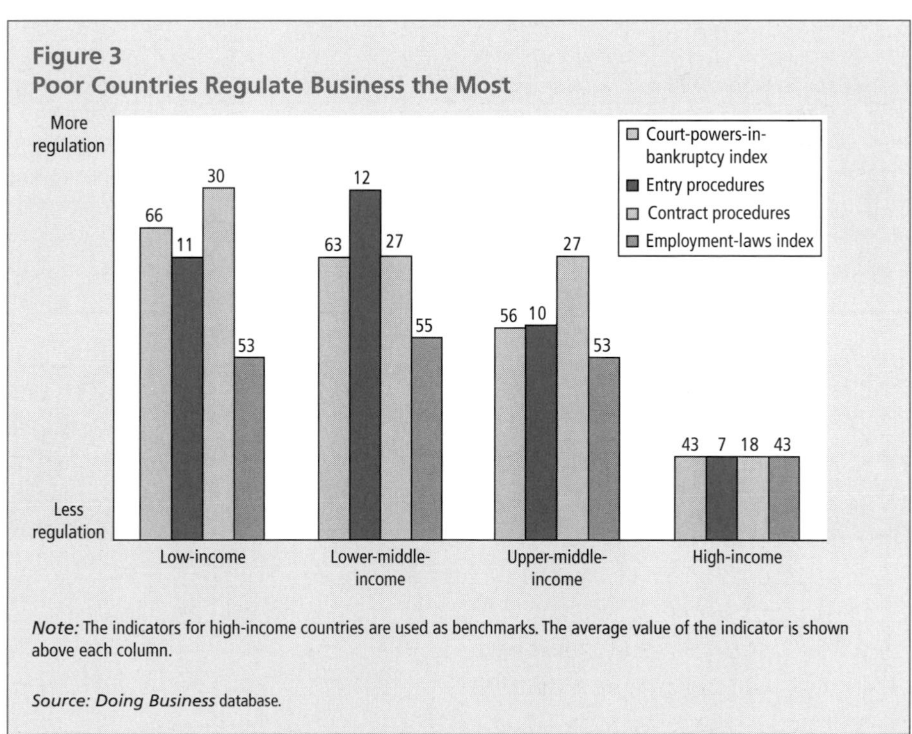

Note: The indicators for high-income countries are used as benchmarks. The average value of the indicator is shown above each column.

Source: Doing Business database.

Costa Rica, Guatemala, Mali, Mozambique, Paraguay, the Philippines, and Venezuela regulate the most. Australia, Canada, Denmark, Hong Kong (China), Jamaica, the Netherlands, New Zealand, Singapore, Sweden, and the United Kingdom regulate the least. There are exceptions. Among the least regulated economies, Jamaica has aggressively adopted best-practice regulation over the last two decades. Contract enforcement, for example, has been improved in line with the latest reforms in the United Kingdom, and bankruptcy law has been revised following the Australian reforms of 1992.

Another important variable in explaining different levels of regulatory intervention is legal origin. Together, income and legal origin account for more than 60 percent of the variation in regulation. While country wealth has long been recognized as a determinant of the quality of institutions (for example, in the writings of Nobel laureate Douglass North), the importance of legal origin has only recently been investigated. The regulatory regimes of most developing countries are not indigenous—they are shaped by their colonial heritage. When the English, French, Spaniards, Dutch, Germans, and Portuguese colonized much of the world, they brought with them their laws and institutions. After independence, many countries revised legislation, but in only a few cases have they strayed far from the original. These channels of transplantation bring about systematic variations in regulation that are not a consequence of either domestic political choice or the pressures toward regulatory efficiency. Common law countries regulate the least. Countries in the French civil law tradition the most.

However, heritage is not destiny. Tunisia, for example, is among the least regulated and most efficient countries in the area of contract enforcement. Uruguay is among the least regulated economies in the hiring and firing of workers. In contrast, Sierra Leone, a common law country, heavily regulates business entry. India, another common law country, has one of the more regulated labor markets and most inefficient insolvency systems.

Heavier Regulation Brings Bad Outcomes

Heavier regulation is generally associated with more inefficiency in public institutions— longer delays and higher cost (figure 4)—and more unemployed people, corruption, less productivity and investment, but not with better quality of private or public goods. The countries that regulate the most—poor

Figure 4
More Regulation Is Associated with Higher Costs and Delays

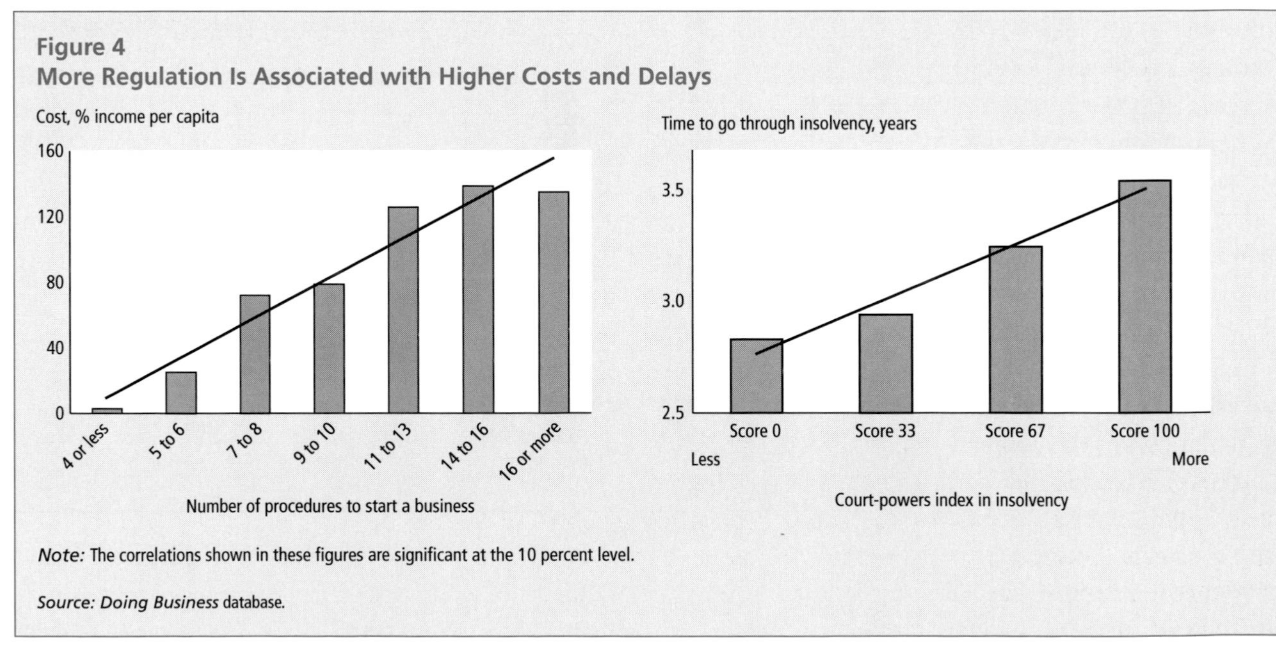

Note: The correlations shown in these figures are significant at the 10 percent level.

Source: Doing Business database.

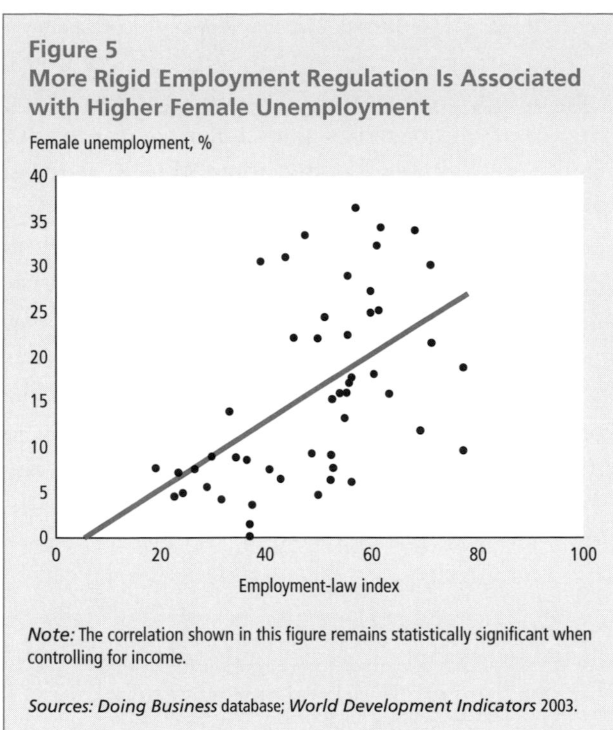

Figure 5
More Rigid Employment Regulation Is Associated with Higher Female Unemployment

Note: The correlation shown in this figure remains statistically significant when controlling for income.

Sources: Doing Business database; *World Development Indicators* 2003.

countries—have the least enforcement capacity and the fewest checks and balances in government to ensure that regulatory discretion is not used to abuse businesses and extract bribes.

Excessive regulation has a perverse effect on the very people it is meant to protect. The rich and connected may be able to avoid cumbersome rules, or even be protected by them. Others are the hardest hit. For example, rigid employment laws are associated especially strongly with fewer job opportunities for women (figure 5). And fewer regulatory restrictions on sharing credit information benefits small firms' access to finance the most. Heavy regulation also encourages entrepreneurs to operate in the informal economy. In Bolivia, one of the most heavily regulated economies in the world, an estimated 82 percent of business activity takes place in the informal sector. There, workers enjoy no social benefits and cannot use pension plans and school funds for their children. Businesses do not pay taxes, reducing the resources for the delivery of basic infrastructure. There is no quality control for products. And entrepreneurs, fearful of inspectors and the police, keep operations below efficient production size.

Critics argue that in developing countries regulation is rarely enforced and plays no role in the conduct of everyday business. Our analysis suggests otherwise. And if it is the case that regulation is irrelevant in poor countries, why not just remove it? A doctor can be hired in place of every government official regulating business activity or compliance with employment laws. A textbook can be printed in place of every batch of paperwork required for this or that license for running a business.

Good regulation does not mean zero regulation. In all countries, the government is involved in various aspects of control of business. The optimal level of regulation is not none, but may be less than what is currently found in most countries, and especially poor ones. For business entry, two procedures—registering for statistical purposes, and for tax and social security—are necessary to fulfill the social functions of the process. Australia limits entry procedures to these two. Sweden has three, including registration with the labor office. New Zealand, the least regulated economy in the world, has 19 procedures to enforce a contract. For employment regulation, Denmark regulates the work week to 37 hours, the premium for overtime pay to 50 percent, the minimum annual paid leave to 27 days, and the severance pay of a worker with 20 or more years of experience to 10 months' wages. It also regulates other aspects of hiring and firing, and the conditions of employment. No one thinks that Danish workers are discriminated against. Yet Denmark is among the countries with the most flexible employment regulation. The Danish example is also an illustration of the difference between rigidity of regulation and social protection. Cumbersome regulation is often an inappropriate tool for protecting weak groups in society.

Instead of spending resources on more regulation, governments are better off defining the property rights of their citizens and protecting them against injury from other citizens and from the state. In *Doing Business*, two examples of such rights are creditor rights—the legal rights of lenders to recover their investment if the borrower defaults—and the efficiency of enforcing property rights through the courts. Countries that protect such rights—rich countries like New Zealand and the United Kingdom, and poor countries like Botswana, Thailand, and South Africa—achieve better economic and social outcomes. In credit markets, assuring lenders of fair returns on investment increases the depth of credit markets and the productivity of investment, even after controlling for income, income growth, inflation, and contract enforcement. Such assurance also increases access to these markets, since lenders are willing to extend credit beyond large and connected firms if they know that their rights to recover loans are secure.

One Size Can Fit All—in the Manner of Business Regulation

Many times what works in developed countries works well in developing countries, too, defying the often-used saying, "one size doesn't fit all." In entry regulations, reducing the number of procedures to only those truly necessary—statistical registration, and tax and social security registration—and using the latest technology to make the registration process electronic, have produced excellent results in Canada and Singapore, Latvia and Mexico—but also in Honduras, Vietnam, Moldova, and Pakistan. Similarly, designing credit information registries has democratized credit markets in Belgium and Taiwan (China), but also in Mozambique, Namibia, Nepal, Nicaragua, and Poland.

Countries like Australia, Denmark, the Netherlands, and Sweden present best practices in business regulation, meaning regulation that fulfills the task of essential controls of business without imposing an unnecessary burden. In these countries, high levels of human capital in the public administration, and the use of modern technology, minimize the regulatory burden on businesses. And where private markets are functioning, competition is a substitute for regulation. By combining simple regulation with good definition and protection of property rights, they achieve what many others strive to do: having government regulators serve as public servants, not public masters.

Aside from how much and what they regulate, good practice countries share common elements in how they regulate. For example, countries with the least time to register a business, such as Canada, have single registration forms accessible over the Internet. Countries that take the least time to enforce a collateral agreement, Germany, Thailand, and the United States, for example, allow out-of-court enforcement. The design of regulation determines the efficiency of economic and social outcomes.

Good practice is not limited to rich countries or countries where comprehensive regulatory reform has taken place. In many instances, reform in some areas of business regulation has been successful. Tunisia has one of the best contract enforcement systems in the world. Latvia is among the most efficient countries in entry regulation. In 2002, Pakistan electronically connected all tax offices in the country, and streamlined business registration. As a result, the time to start a business was reduced from 53 to 22 days. The Slovak Republic recently implemented best-practice laws on collateral. Vietnam revised its Enterprise Law in 1999 to enhance growth in private business activity.

Such partial reforms may lead to a virtuous cycle where the success of one reform emboldens policymakers to pursue further reforms. The Russian Federation simplified business entry in the past year, reducing the number of procedures from 19 to 12, and the associated time from 51 days to 29 days (figure 6). The reforms led to the creation of a large number of new private businesses, which in turn became the constituency for improvements in other regulatory practices. Employment law has since been revised, resulting in more flexibility in hiring and firing workers.

But reform options are not always the same across rich and poor countries. There are cases where good practices in developed countries are difficult to transplant to poor countries. Bankruptcy is one example where the establishment of a sophisticated bankruptcy regime in a developing country generally results in inefficiency and even corruption. Both lenders and businesses suffer. In such instances, developing countries could simplify the models used in rich countries to make them workable with less capacity and fewer resources. In the poorest countries, it is better not to develop a sophisticated bankruptcy system and to rely instead on existing contract-enforcement mechanisms or negotiations between private parties. Similarly, specialized commercial courts

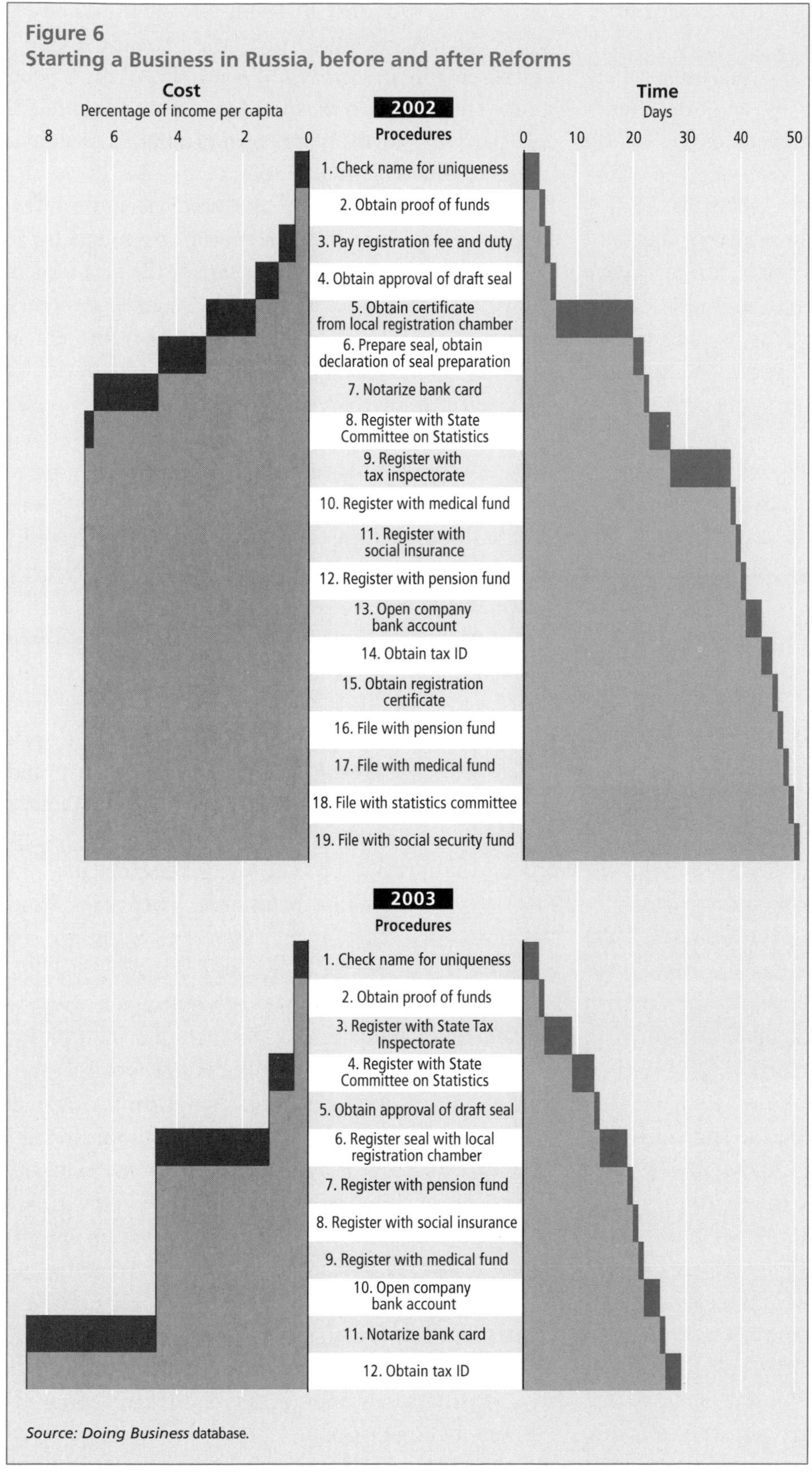

Figure 6
Starting a Business in Russia, before and after Reforms

Source: Doing Business database.

work best in countries with more resources and administrative capacity. Poor countries can implement reforms with the same principle—specialization—but with specialized judges or specialized sections within general jurisdiction courts.

Reform Practice

Regulatory reform has been continuous in most developed countries, improving the environment for doing business.

- Australia has built in regulatory reform by including "sunset" provisions in new regulations, with the regulation automatically expiring after a certain period unless renewed by Parliament. Also, the Office of Regulation Review vets each proposed regulation using a "minimum necessary regulation" principle. In 1996, the office was charged with cutting the regulatory burden on small businesses in half, with annual reviews of progress achieved.
- Denmark revised its business entry regulation in 1996 by removing several procedures, making the process electronic, and eliminating all fees. Since then, a cost-benefit analysis of proposed new regulation is conducted,

resulting in two of every five proposed regulations being shelved.

- In the Netherlands, much of the work on reducing administrative costs is done by an independent agency, ACTAL (Advisory Committee on the Testing of Administrative Burdens). Established in 2000, ACTAL has only nine staff members and is empowered to advise on all proposed laws and regulations. To date, simplification of administrative procedures has been achieved in the areas of corporate taxation, social security, environmental regulation, and statistical requirements. The estimated savings are US$600 million from streamlining the tax requirements alone.
- Sweden has a "guillotine" approach for regulatory reform, in which hundreds of obsolete regulations are cancelled after the government periodically requires regulatory agencies to register all essential regulations.

But there has been much less reform in developing countries, with the result that businesses are sometimes burdened by outdated regulation. For example, the company law regulating business entry dates back to 1884 in the Dominican Republic, to 1901 in Angola, and to 1916 in Burkina Faso. But OECD countries have all revised their laws in the last two decades. Similarly, employment regulation in Africa often dates to colonial times or was revised just after independence. On average, it is over three decades old. This is evidence against the "reform fatigue" in developing countries, often attributed to the work of international aid agencies.

With laws to meet the needs of business developed decades or even a century earlier, it is hardly surprising that those laws often impose unnecessary burdens on business today. But this is also grounds for optimism: outdated regulation is often the result of inertia or a lack of capacity to reform, not of entrenched business or government interests.

There are many reforms where the regulatory burden on business can be reduced, while the government can redirect much-needed resources toward the tasks that really count—such as providing basic social services. Indeed, some countries have recently modernized many aspects of their business regulation, including Jamaica, the Republic of Korea, and Thailand. There is no reason why others should not follow. The benefits can be enormous. So are the costs of not reforming.

Of course, reforms are not always easy. There are also instances where powerful lobbies prevent or reverse regulatory reform. In 1996, the Peruvian government tried to reduce mandatory severance payments by 50 percent. The uproar with unions made the government withdraw the proposal quickly. Instead, severance payments were increased. The German government, in May 2003, proposed far-reaching reforms aimed at making labor markets more flexible. Such proposals have previously been withdrawn after threats of worker strikes. Another ill-fated reform comes from Croatia, where the private notaries' profession has for years undermined the government's efforts to simplify business entry procedures and collateral enforcement. Simplification would mean more competition and a loss of profits for the private notaries. Although *Doing Business* does not address political economy of reform, the report gives other examples of reforms gone awry due to opposing interests.

The analysis presented in this report suggests specific policy reforms (table 1) that illustrate two main themes: first, that poor countries have the furthest to go, and second, that when it comes to the manner of regulation, one size often fits all (in many cases there really is one best practice). The list of reform examples is still incomplete. Future reports aim to enlarge it.

In business entry, reforms that are easy to implement include the adoption of better information and intragovernment communications technology—to inform prospective entrepreneurs and to serve as a virtual one-stop shop for business registration. The introduction of a single registration form and silent consent in approving registration have had enormous success. Reducing the number of procedures to statistical and tax registration and abolishing the minimum capital requirement lighten the burden on entrepreneurs and have been associated with the creation of larger numbers of new businesses. Other reforms that require legislative change include introducing a general-objects clause in the articles of incorporation and

Table 1
Examples of Good Reform Practices

Principles of Regulation	Some Examples
Starting a Business	
• Registration is an administrative, not judicial, process	• China, United States
• Use of single business identification number	• Denmark, Turkey
• Electronic application made possible	• Latvia, Sweden, Singapore
• Statistical and tax registration sufficient to start operations	• Australia, Canada, New Zealand
• No minimum capital requirement	• Chile, Ireland, Jamaica
Hiring and Firing Workers	
• Contracts "at will" between employers and employees	• Denmark, Ireland, Singapore
• No limits on fixed-term contracts	• Australia, Denmark, Israel
• Apprentice wages for young workers	• Chile, Colombia, Poland
• Shift work between slow and peak periods	• Hungary, Poland
Enforcing a Contract	
• Judiciary has a system for tracking cases	• Slovak Republic, Singapore
• Summary procedure in the general court	• Botswana, New Zealand, Netherlands
• Simplified procedure in commercial courts	• Australia, Ireland, Papua New Guinea
• Attorney representation not mandatory	• Lebanon, Tunisia
Getting Credit	
• Strong creditor protection in collateral and bankruptcy laws	• New Zealand, United Kingdom
• No restrictions on assets that may be used as collateral	• Slovak Republic, Hong Kong (China)
• Out of court or summary judgments for enforcing collateral	• Germany, Malaysia, Moldova
• Regulations provide incentives for sharing and proper use of credit information	• Belgium, Singapore, United States
Closing a Business	
• Limited court powers	• Australia, Finland, United Kingdom
• Bankruptcy administrator files report with creditors	• Botswana, Germany, Hungary
• Continued education for bankruptcy administrators	• Argentina, France, Netherlands

Figure 7
Courts and Notaries Are Bottlenecks to Business Start-Up

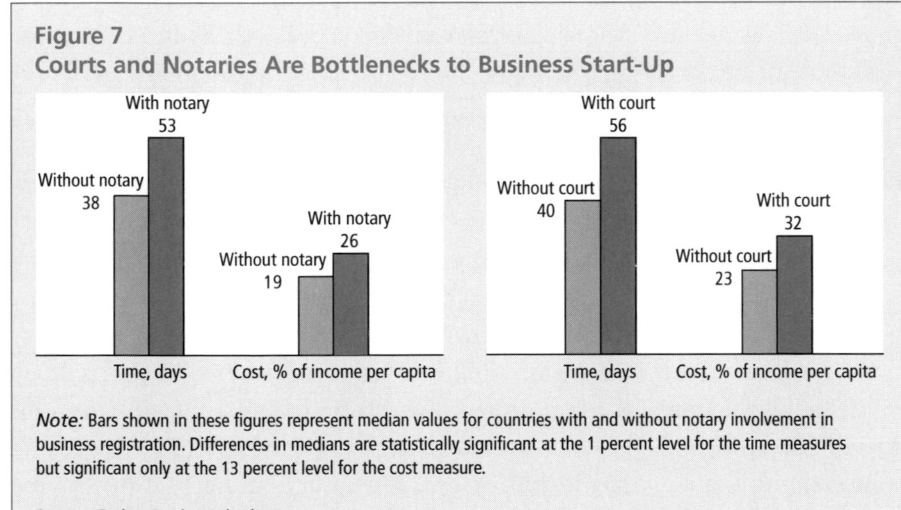

Note: Bars shown in these figures represent median values for countries with and without notary involvement in business registration. Differences in medians are statistically significant at the 1 percent level for the time measures but significant only at the 13 percent level for the cost measure.

Source: Doing Business database.

removing notarial authorizations and court use from the registration process (figure 7). Such reforms may be difficult to implement, as political will in government and the private sector may waver, but they have beneficial effects beyond business entry.

In employment regulation, five types of reform ease the burden on businesses and provide better job opportunities for the poor.

- First, in most developing countries a general reform toward reduction of the scope of employment regulation has yielded positive results. The deregulation experience in Latin America (Chile, Colombia, Guyana, and Uruguay) as well as in transition economies (Estonia) provides many lessons.

- Second, many OECD countries have focused on introducing flexible part-time and fixed-term contracts. These contracts bring groups that are less likely to find jobs (women and youths) into the labor market. Germany has raised the duration of fixed-term contracts to eight years, while Poland does not mandate any duration limit.
- Third, several countries have either reduced the minimum wage (Colombia) or lowered the minimum wage limit for new entrants (Chile).
- Fourth, some countries (Hungary) have made it possible for employers to shift work time between periods of slow demand and peak periods, without the need for overtime payment.
- Fifth, other countries have focused on easing regulation on firing. The most far-reaching reform was recently implemented in Serbia and Montenegro, where the severance payment for a worker with 20 years' tenure was reduced from 36 months to 4 months.

Figure 8
Credit Bureaus Are Associated with More Credit

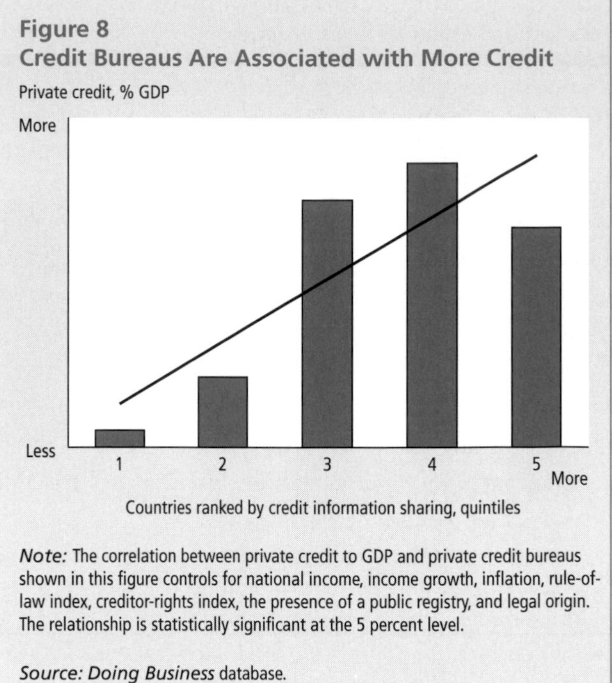

Note: The correlation between private credit to GDP and private credit bureaus shown in this figure controls for national income, income growth, inflation, rule-of-law index, creditor-rights index, the presence of a public registry, and legal origin. The relationship is statistically significant at the 5 percent level.

Source: Doing Business database.

In contract enforcement, establishing information systems on caseload and judicial statistics has had a large payoff. Judiciaries that have established such systems, as in the Slovak Republic, can identify their primary users and the biggest bottlenecks. Simplifying procedures is also often warranted. For example, summary debt collection proceedings of the type recently established in Mexico alleviate court congestion by reducing procedural complexity. When default judgments—automatic judgments if the defendant does not appear in court—are introduced as well, delays are cut significantly.

The structure of the judiciary can also be modified to allow for small claims and specialized commercial courts. Several countries that have small claims courts (Japan, New Zealand, the United Kingdom) have recently increased the maximum claim eligible for hearing at the court. However, the manner of regulation of the judicial process in developing countries may need to be different. Where the judiciary is still in its early stages of development, as in Angola, Mozambique, or Nepal, specialized courts may be premature. There, reformers can establish a specialized section dealing with commercial cases within the general court or train specialized judges.

Simplification of judicial procedures is associated with less time and cost. For example, in some countries, such as Argentina, Bolivia, Morocco, and Spain, businesses are obliged to hire lawyers when resolving commercial disputes. This increases the cost of enforcing contracts, sometimes unnecessarily. In many instances, the manager may simply present to the judge proof of delivery of goods and require payment.

Establishing appropriate regulation and incentives to facilitate private credit bureaus is an essential start to encouraging access to credit (figure 8). In some cases—especially in poor countries where commercial incentives for private bureaus are low—setting up public credit registries has helped remedy the lack of private information sharing, albeit second best to an effective private bureau. The design of credit information regulations influences the impact of bureaus: broader coverage of borrowers and good regulations on collection, distribution, and quality of information (including privacy and data protection) are associated with better functioning credit markets.

Legal creditor protections can be improved by reforming collateral law: introducing out of court or summary enforcement proceedings, eliminating restrictions on which assets may be used as security for loans, and improving the clarity of creditors' liens through collateral registries and clear laws on who has priority in a disputed claim to collateral. Stronger powers for creditors to recover their claims in insolvency are associated with more access to credit.

Three areas of bankruptcy reform give the most promise. The first is choosing the appropriate insolvency law given a country's income and institutional capacity. Ill-functioning judiciaries are better off without pouring resources into sophisticated bankruptcy systems. There is a general misperception that bankruptcy laws are needed to enforce creditor rights. In practice, they often add to legal uncertainty and delays in developing countries. Private negotiations of debt restructuring under contract and secured transactions law and the introduction of summary judgments, like those for simple contract enforcement, will do. The second is increasing the involvement of stakeholders in the insolvency process rather than relying on the court for making business decisions. The third is training judges and bankruptcy administrators in insolvency law and practice

Of course, for governments to undertake reform there needs to be a strong constituency interested in change, so that inertia and the lobbying of entrenched political or business groups can be overcome. By bringing evidence to the debate, *Doing Business* motivates the need for change and informs the design of new regulations and institutions.

1 Building New Indicators of Business Regulation

In 1664, William Petty, an adviser to Cromwell's government and to Charles II after the Restoration, compiled the first known national accounts. He made four entries. On the expense side, "food, housing, clothes and all other necessaries" were estimated at £40 million. National income was split into £8 million from land, £7 million from other personal estates, and £25 million from labor income.[1]

In later centuries, estimates of country income, expenditure, and material inputs and outputs became more abundant. However, it was not until the 1940s that a systematic framework was developed for measuring national income and expenditure, under the direction of John Maynard Keynes.[2] It is hard to underestimate the impact of this new methodology. Complicated transactions data were simplified into an aggregate overview of the economy. Economic performance and structure could be assessed with greater precision than ever before. As the methodology became an international standard, comparisons of countries' financial positions became possible.

Today the macroeconomic indicators in national accounts are standard in every country. Records of overall wealth, production, consumption, wages, trade, and investment across countries are taken for granted. Empirical studies of those data have shed light on new theories of macroeconomic development. But systems for measuring the microeconomic and institutional factors that explain the aggregates are still nascent.

Doing Business addresses the gap by constructing new sets of indicators on the regulatory environment for private sector development. The indicators cover business entry, employment regulation, contract enforcement, creditor rights, credit information sharing systems, and bankruptcy. This is only the beginning of a large agenda of building similar indicators of business licenses, property registries, corporate governance, trade infrastructure, law enforcement, and tax policy.

More than a dozen organizations already produce and periodically update indicators on country risk, economic freedom, and international competitiveness; surveys of firms are now common. New methods are being applied to aggregate indicators, to produce useful gauges of general economic and policy conditions. Surprisingly, none assess the specific laws and regulations that enhance or hinder business activity. Nor do they evaluate the public institutions—courts, credit registries, the company register—that support it. Reformers are left in the dark.

The two types of indicators in *Doing Business* focus on government regulation and its effect on businesses—especially on small and medium-size domestic businesses (which make up the majority of firms, investment, and employment in developing countries). First are measures of actual regulation—such as the number of procedures to register a business or an index of employment law rigidity. Second are measures of regulatory outcomes, such as the time and cost to register a business, enforce a contract, or go through bankruptcy.

Based on readings of laws and regulations, with verification and input from local government officials, lawyers, business consultants, and other professionals administering or advising on legal and regulatory requirements, this methodology has several advantages. It uses factual information and allows multiple interactions with local respondents, ensuring accuracy by

Figure 1.1
Costs of Business Entry in Ethiopia

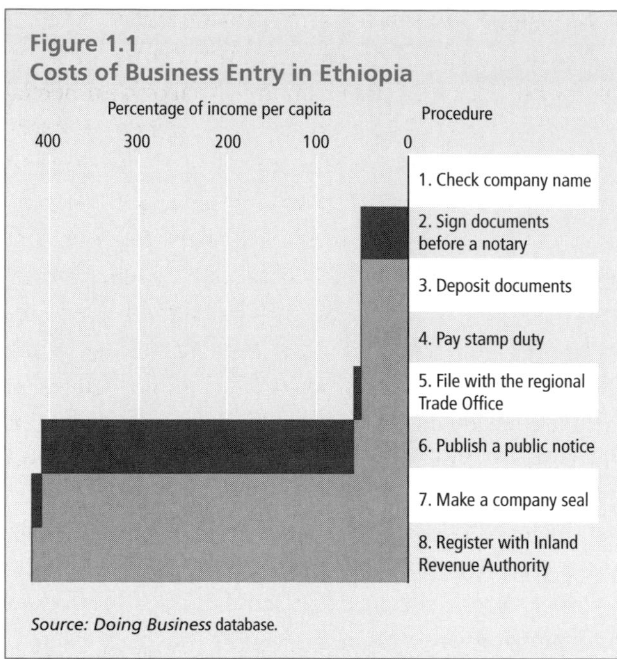

Source: Doing Business database.

clarifying possible misinterpretations of questions. It is inexpensive, so the data can be collected in a large sample of countries. And because the same standard assumptions are applied in data collection, which is transparent and easily replicable, comparisons and benchmarks are valid across countries.

Most important, the analysis has direct relevance for policy reform, which it facilitates in three ways. First, the analysis reveals the relationship between indicators and economic and social outcomes, allowing policymakers to see how particular laws and regulations are associated with poverty, corruption, employment, access to credit, the size of the informal economy, and the entry of new firms. Putting higher administrative burdens on entrepreneurs diminishes business activity—but it also creates more corruption and a larger informal economy, with fewer jobs for the poor.

Second, beyond highlighting the areas for policy reform, the analysis provides guidance on the design of reforms. The data offer a wealth of detail on the specific regulations and institutions that enhance or hinder business activity, the biggest bottlenecks causing bureaucratic delay, and the cost of complying with regulation. A library of current laws, also specifying the regulatory reforms under way, support each indicator set. Governments can thus identify, after reviewing their country's *Doing Business* indicators, where they lag behind and will know what to reform.

For example, in January 2003, Ethiopia was one of the most expensive countries in which to start a new business. The breakdown of the business entry process shows that the cost of entry—more than four times gross national income per capita—is driven mainly by the requirement to publish an official notice in the newspapers (figure 1.1). If the government eliminates the publication fee, the cost plummets to about 50 percent of income per capita, placing Ethiopia below the average in the sample of more than 130 countries. (In June 2003, the Ethiopian government reduced the cost of publishing the notice by 30 percent.)

Another example of how the indicators shed light on policy reforms is the time it takes to enforce a contract in court. Countries that have specialized commercial judges or specialized commercial courts tend to have faster dispute resolution. In countries where commercial sections in general courts or commercial courts were recently established, as in Portugal and Tanzania, the time to recover a debt has been significantly reduced. A reformer can infer that specialization improves efficiency.

Finally, analyses across sets of indicators build the agenda for comprehensive regulatory reform. For example, examination of both entry and labor regulation reveals that a venue to challenge inefficient, unfair, or corrupt regulatory practices is needed. An ombudsman's office or administrative courts in countries with well-functioning public administration, or statutory time limits and a "silence is consent" rule in countries with less administrative capacity would improve entry and labor regulation.

Doing Business Methodology

Features and Assumptions

The methodology followed for each of the topics in *Doing Business* has six standard features:

1. The team, with academic advisers, collects and analyzes the laws and regulations in force.
2. The analysis yields an assessment instrument or questionnaire that is designed for local professionals

experienced in their fields, such as incorporation lawyers and consultants for business entry or litigation lawyers and judges for contract enforcement.

3. The questionnaire is structured around a hypothetical case to ensure comparability across countries and over time.
4. The local experts engage in several rounds of interaction—typically four—with the *Doing Business* team.
5. The preliminary results are presented to both academics and practitioners, prior to refinements in the questionnaire and further rounds of data collection.
6. The data are subjected to numerous tests for robustness, which frequently lead to revisions or expansions of the collected information. For example, following collection and analysis of data on business entry regulation, incorporation lawyers in several countries suggested that the minimum capital requirement be included, because it sometimes constitutes a very large start-up cost. The requirement was included in a follow-up questionnaire. (For another example, the contract enforcement project collected and analyzed data on the recovery of debt in the amount of 50 percent of income per capita, as well as on two other cases—the eviction of nonpaying tenants and the recovery of a smaller debt claim [5 percent of income per capita], which served as robustness checks).[3]

The result is a set of indicators whose construction is easy to replicate. And extending the dataset to obtain other benchmarks is straightforward. For example, *Doing Business* studies a certain type of business—usually a domestic limited-liability company. Analysts can follow the methodology and construct the same measures as benchmarks for sole proprietorships and foreign companies.

The methodology of one project—business entry regulation—is presented in detail below as an illustration of the general approach used in *Doing Business*, before the methodology for the other four sets of indicators is summarized. The data for all sets of indicators are for January 2003.

Starting a business. The project on starting a business records all procedures officially required for an entrepreneur to operate an industrial or commercial business legally. They include obtaining necessary permits and licenses—and completing the required inscriptions, verifications, and notifications—to start operation.[4] The questionnaire calculates the cost and time of fulfilling each procedure under normal circumstances, as well as the minimum capital requirements to operate. The assumption is that such information is readily available to the entrepreneur and that all government and nongovernment entities in the process function efficiently and without corruption.

To make the business comparable across countries, 10 assumptions are employed. The business

- is a limited-liability company (If there is more than one type of limited-liability company in the country, the type most popular among domestic firms is chosen.);
- operates in the country's most populous city;
- is 100 percent domestically owned and has five founders, none of whom is a legal entity;
- has start-up capital of 10 times income per capita, paid in cash;
- performs general industrial or commercial activities, such as the production and sale of products or services to the public;
- leases the commercial plant and offices;
- does not qualify for investment incentives or any special benefits;
- has up to 50 employees one month after the start of operations, all of them nationals;
- has turnover of at least 100 times income per capita; and
- has a company deed 10 pages long.

Obviously, the assumptions enhance comparability at the expense of generality. For example, in many countries, both business regulation and its enforcement are different across different locations within a country. *Doing Business* covers businesses in the largest city. However, one also must be mindful that in many developing countries, inflation data—

one of the staples of macroeconomic analysis—are frequently based on prices of consumer goods in the capital city only. Neither measure is perfect.

To make the procedures comparable across countries, six assumptions are employed:

1. A procedure is defined as any interaction of the business founder with external parties (government agencies, lawyers, auditors, notaries). Interactions between company founders or company officers and employees are not considered separate procedures.
2. The founders complete all procedures themselves, without facilitators, accountants, or lawyers, unless the use of such third parties is required.
3. Procedures not required by law for starting the business are ignored. For example, obtaining exclusive rights over the company name is not counted in a country where businesses are allowed to use a number as identification.
4. Shortcuts are recorded if they fulfill three requirements: they are not illegal, they are available to the general public, and avoiding them causes substantial delays.
5. Only procedures required of all businesses are covered. For example, procedures to comply with environmental regulations are included only if they apply to all businesses.
6. Procedures that the business undergoes to begin electricity, water, gas, and waste disposal services are not included unless they are required for the business to legally start operating.

With those assumptions, four indicators for the requirements to register a business are constructed:

- number of procedures,
- time,
- cost, and
- minimum capital.

The indicators are developed by means of in-house research and expert assessment. The *Doing Business* team starts by studying the laws and regulations on business entry and reviewing publicly available summaries and descriptions of the business registration process. From that research, a detailed list of the procedures, times, costs, and minimum capital requirements is compiled. The list is sent to business registration experts in the country (usually government officials and incorporation lawyers), who are asked to verify the data, identify missing procedures, complete the information about the time required, and make corrections. If there are differences among answers, inquiries are made again until the data can be reconciled.

The texts of the company law, the commercial code, or specific regulations and fee schedules are used as sources for calculating costs. If there are conflicting sources and the laws are not clear, the most authoritative source is used. The constitution supersedes the company law, and the law prevails over regulations and decrees. If disagreeing sources have the same rank, the source indicating the more costly procedure is used, because an entrepreneur never second-guesses a government official. In the absence of fee schedules, a government officer's estimate is taken as an official source. If sources have different estimates, the median reported value is used. If a government officer's estimates are lacking, those of incorporation lawyers are used instead. If several incorporation lawyers have different estimates, the median reported value is used. In all cases, the cost excludes bribes.[5]

Time is recorded in calendar days. It is assumed that the minimum time required to fulfill a procedure is one day. Time captures the median duration that incorporation lawyers say is necessary to complete a procedure. Information is collected on the sequence in which the procedures are to be completed, as well as on procedures that can be carried out simultaneously. If a procedure can be accelerated for an additional cost, the fastest procedure is chosen. It is assumed that the entrepreneur does not waste time and commits to completing each remaining procedure without delay. When calculating the time needed for complying with entry regulations, the time that the entrepreneur spends gathering information is ignored: the entrepreneur is aware of all entry regulations and their sequence from the very beginning.

The minimum capital requirement is the amount an entrepreneur needs to deposit in a bank account to

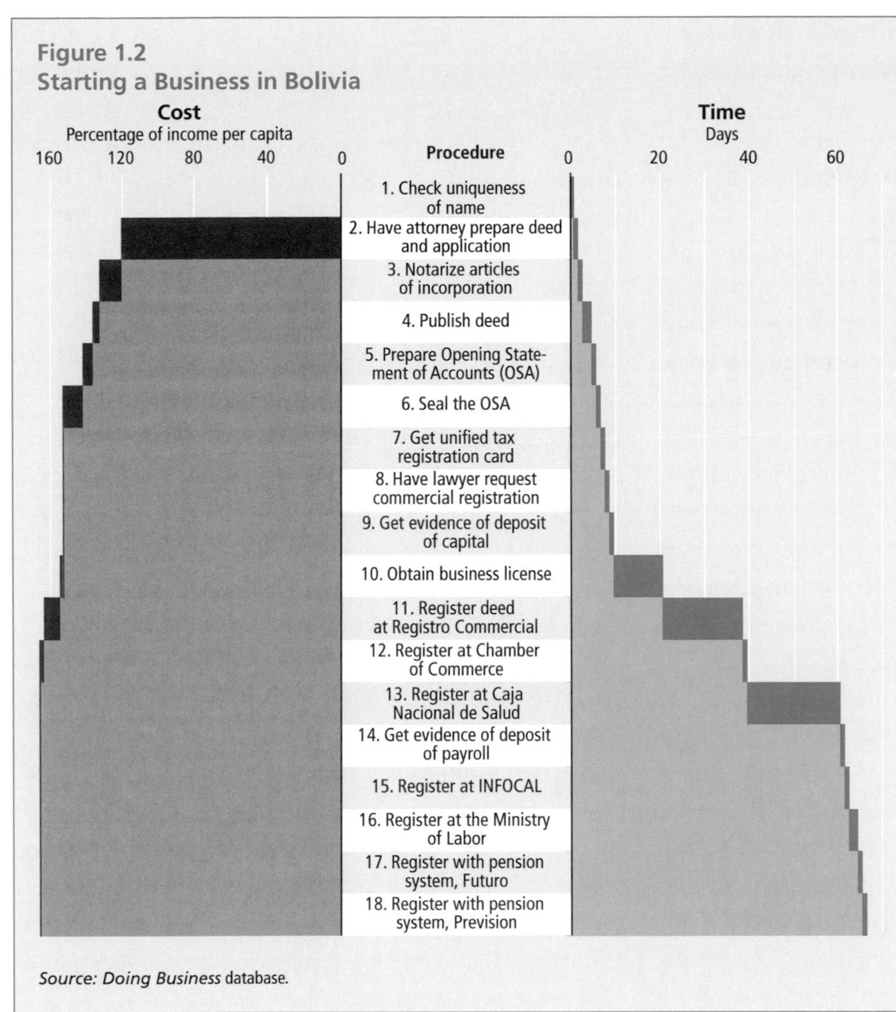

Figure 1.2
Starting a Business in Bolivia

Source: Doing Business database.

obtain a company registration number, as specified in the company law or commercial code.

The data collection results in a file that describes the sequence of procedures—and their time and cost—to start legal operation. Consider the data for Bolivia (figure 1.2). The data represent a good-case scenario because the assumptions necessary to standardize responses across countries remove many possible bottlenecks, such as the entrepreneur's not having correct information about where to go and what documents to submit.

In practice, entrepreneurs may avoid some legally required procedures altogether—say, by not registering for social security or not registering with the chamber of commerce—or they can pay a facilitator for assistance. In both cases, the time would be reduced. So the *Doing Business* time indicator may be either smaller or larger than the average start-up time documented in enterprise surveys. For example, Mozambique's average start-up time in the January 2003 *Doing Business* data was 153 days, but a survey of recently started businesses reported 138 days on average in July 2002. *Doing Business* reported 88 days in India in January 2003, but an enterprise survey conducted in 2002 reported 90 days.[6]

Other Topics

Hiring and firing. The indicators for employment regulation are based on a detailed study of employment laws. Data are also gathered on the specific constitutional provisions related to labor. In most cases, both the actual laws and a secondary source are used to ensure accuracy. Conflicting answers are checked in two additional sources, including a local legal treatise on employment regulation. Legal advice from leading local law firms is solicited to confirm accuracy in all cases.

To make the data comparable across countries, several assumptions about the worker and the company are applied. The worker is a nonexecutive, full-time employee who has worked in the same company for 20 years, has a nonworking wife and two children, and is not a member of a labor union (unless membership is mandatory). The business, a limited-liability manufacturing company that operates in the country's most populous city, is 100 percent domestically owned and has 201 employees.

Three indices of the regulation of labor markets are constructed by examining detailed provisions in the employment laws—flexibility-of-hiring index, conditions-of-employment index, and flexibility-of-firing index, with values between 0 and 100, where a higher value means more regulation. An employment-regulation index averages the values of the three indices.[7]

Enforcing contracts. The indicators on contract enforcement are also constructed by assuming a hypothetical case—a payment dispute of 50 percent of income per capita in the country's most populous city. The data track the procedures to recover debt through the courts. The plaintiff has fully complied with the contract (and is thus 100 percent in the right) and files a lawsuit to recover the debt. The debtor attempts to delay and opposes the complaint. The judge decides every motion for the plaintiff. There are no appeals or postjudgment motions.

The data come from readings of the codes of civil procedures and other court regulations, as well as from administering surveys to local litigation attorneys. Most of the respondents are members of the Lex Mundi association of law firms. At least two association lawyers in each country participated in the survey. The questionnaires were designed with the help of scholars from Harvard and Yale universities and with the advice of practicing attorneys.[8]

On the basis of questionnaire responses, four indicators of the efficiency of commercial contract enforcement are developed:

1. the number of procedures, mandated by law or court regulation, that demand interaction between the parties or between them and the judge or a court officer;
2. the time needed for dispute resolution in calendar days, counted from the moment the plaintiff files the lawsuit in court until the moment of settlement or, when appropriate, payment (this measure includes the days when actions take place and the waiting periods between actions);
3. the official cost of going through court procedures, including court costs and attorney fees; and
4. the procedural complexity of contract enforcement—an index that scores countries on how heavily dispute resolution is regulated.

Figure 1.3
Is the Time to Enforce a Contract Indicator Representative? Yes

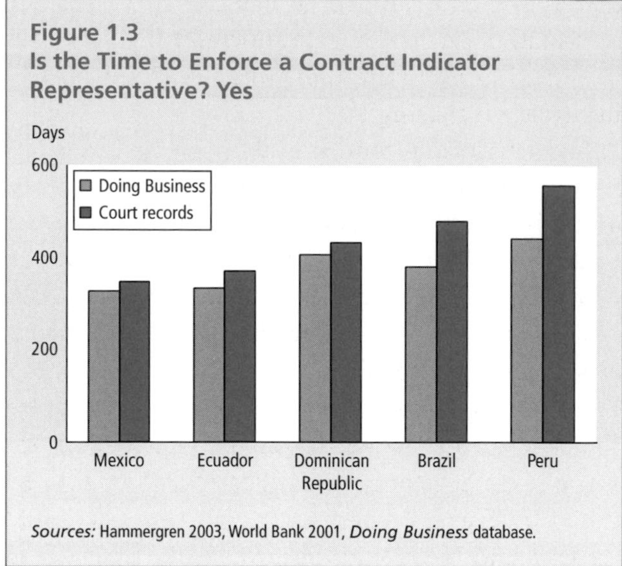

Sources: Hammergren 2003, World Bank 2001, *Doing Business* database.

Are the indicators from a hypothetical case representative of debt recovery practices? Yes. Few countries have done studies on commercial dispute resolution by looking at actual court cases. Where data are available—from Brazil, the Dominican Republic, Ecuador, Mexico, and Peru—the median times are very similar to those reported in *Doing Business* (figure 1.3).[9] For example, a survey of about 500 debt recovery cases in Mexico finds that the median time from filing to service of process is 53 days; from service of process to judgment, 111 days; and from judgment to enforcement, 182 days—a total of 346 days.[10] The respective numbers in *Doing Business* are 55 days, 119 days, and 151 days—a total of 325 days. A study on the Dominican Republic, using more than 2,000 cases, finds that the median duration from filing to judgment is 431 days. *Doing Business* arrives at 405 days. And a study of more than 300 cases in Ecuador finds the duration from filing to resolution to be 369 days.[11] *Doing Business* finds 333 days. Consistent with the good-case scenario of the hypothetical case, our numbers are somewhat lower.

Getting credit. Doing Business constructs two sets of measures on getting financing: sharing credit information and legally protecting creditor rights. The assessment of credit information institutions begins with a survey of banking supervisors. It confirms the presence or absence of public credit registries and private credit bureaus. The survey also collects descriptive data on credit market outcomes and information on related rules in credit markets (collateral, interest rate controls, laws on credit information sharing).

In countries that confirmed the presence of a public registry or a major private bureau, a second survey, on registry structure, laws, and associated rules was conducted. The survey was developed in cooperation with the Credit Reporting Systems Project of the World Bank Group and was reviewed by academic experts on the topic from the University of Salerno. From the responses, measures are constructed for the coverage of the market for credit information, the scope of credit information collected and distributed, the accessibility of the data in the public credit registry, and the quality of information available in the registry.[12] A separate questionnaire on the regulatory framework for sharing credit information is conducted.[13]

The creditor-rights indicator measures four powers of secured creditors in bankruptcy:[14]

1. whether there are restrictions, such as creditors' consent, on entering into reorganization proceedings;
2. whether there is no automatic stay (or "asset freeze") on realizing collateral upon bankruptcy;
3. whether secured creditors are satisfied first on liquidation; and
4. whether management is replaced by a court- or creditor-appointed receiver in reorganization.

A value of 1 is assigned to each variable when a country's laws and regulations provide those powers for secured creditors. The creditor-rights index sums the total score across all four variables. A minimum of 0 represents weak creditor rights; a maximum of 4 represents strong creditor rights. Data for the variables are obtained by reading insolvency laws and legal summaries, then verified by means of a questionnaire submitted to financial lawyers, and then cross-checked against data gathered for the bankruptcy project.

Closing a business. The indicators are derived from questionnaires answered by bankruptcy judges and attorneys at private law firms. The questionnaires were designed with the assistance of scholars from Harvard University and with the advice of practicing attorneys. Most respondents are members of the International Bar Association.

The data track the procedures for a hypothetical business going through bankruptcy. The business is a domestically owned limited-liability company operating a hotel in the most populous city. It has 201 employees, 1 main secured creditor, and 50 unsecured creditors. On the basis of detailed assumptions about the debt structure and future cash flows, it is assumed that the company becomes insolvent on January 1. The case is designed so that the business has a higher value as a going concern—that is, the efficient outcome is either reorganization or sale as a going concern, not piecemeal liquidation.

Six indicators for the bankruptcy process are constructed from responses to the questionnaire:[15]

1. the time to go through bankruptcy;
2. the cost of going through bankruptcy;
3. whether absolute priority for secured lenders is preserved throughout the process;
4. whether the efficient outcome is achieved;
5. an aggregate-goals-of-bankruptcy index, created by averaging the scores for time, cost, priority, and reaching the efficient outcome;
6. an index for court powers in bankruptcy.

Other Indicators in a Crowded Field

Doing Business enters a crowded field of indicators and ratings on various aspects of the environment for doing business (box 1.1). Eight organizations periodically collect such indicators, with a focus on international portfolio investors, global lenders, and executives of multinational companies:

- *Business Environment Risk Intelligence* (BERI),
- *Euromoney Institutional Investor* (EII),
- *International Country Risk Guide* (ICRG), Political Risk Services group,
- *Country Risk Review* (CRR), Global Insight,
- *The Economist* Intelligence Unit (EIU),
- The Heritage Foundation,
- World Markets Research Center, and
- A. T. Kearney.

Box 1.1
Cross-Country Indicators of the Business Environment

World Competitiveness Yearbook

- Published since 1987 by the Institute for Management Development in Lausanne, Switzerland. Until 1996, a joint publication with the World Economic Forum.
- Analyzes the international competitiveness of 49 countries, on the basis of hard data from international organizations and perception surveys of enterprise managers.
- In the 2002 survey, there were 3,532 respondents, or 72 per country on average.
- Hard data cover economic performance, international trade and investment, public finance and fiscal policy, education, productivity, and infrastructure quality. Survey questions cover institutional framework (government efficiency, justice, and security), business legislation (openness, competition regulations, labor regulations, and capital market regulations), management practices, and the impact of globalization.

Source: www.imd.ch.

Global Competitiveness Report

- Published since 1996 by the World Economic Forum in Geneva, Switzerland.
- Analyzes the international competitiveness of 80 countries, on the basis of hard data from international organizations and perception surveys of enterprise managers.
- In the 2002 survey, there were 4,601 respondents, or 58 per country on average.
- Survey questions cover access to credit, public institutions for contract and law enforcement, corruption, domestic competition, labor regulations, corporate governance, environmental policy, and cluster development. Hard data cover economic performance, international trade and investment, public finance and fiscal policy, education, technological innovation, information and communications technology, and infrastructure quality. Starting in 2003, the analysis uses six *Doing Business* indicators on starting a business and enforcing a contract.

Source: www.weforum.org.

Business Environment and Enterprise Performance Survey

- Published in 1999 and 2002 by the EBRD and the World Bank.
- Analyzes government effectiveness, regulatory quality, rule of law, and corruption in 27 transition economies.
- Based on surveys of 6,000 firms in 1999 and 7,500 firms in 2002, with hard data as well as perceptions questions.

Source: www.info.worldbank.org/governance/beeps2002.

Index of Economic Freedom

- Published since 1995 by the Heritage Foundation and the *Wall Street Journal.*
- Analyzes economic freedom in 161 countries.
- Based on assessments by in-house experts, drawing on many public and private sources.
- The index covers 10 areas: trade policy, fiscal burden, government intervention, monetary policy, foreign investment, banking and finance, wages and prices, property rights, business regulation, and black markets.

Source: www.heritage.org.

World Markets Research Center

- Published since 1996 by the World Markets Research Center in London.
- Analyzes the investment climate in 186 countries.
- Based on assessments by 180 in-house experts, drawing on many public and private sources.

Source: www.worldmarketsanalysis.com.

(*contd.*)

Box 1.1
Cross-Country Indicators of the Business Environment (continued)

Economic Freedom of the World

- Published since 1997 by the Fraser Institute.
- Analyzes economic freedom in 123 countries.
- Based on assessments by in-house experts, drawing on many public and private sources. The ratings on the business environment are derivative, based on the *Global Competitiveness Report.*
- The index covers eight areas: size of government, legal structure, security of property rights, access to sound money, freedom to exchange with foreigners, regulation of credit, regulation of labor, and other business regulation.

Source: www.freetheworld.com.

Country Risk Service

- Published quarterly since 1997 by *The Economist* Intelligence Unit.
- Provides international investors with risk ratings for 100 countries.
- Based on assessments by in-house experts, drawing on previous ratings.
- The index covers seven areas of country risk: political, economic policy, economic structure, liquidity, currency, sovereign debt, and banking sector.

Source: www.eiu.com.

International Country Risk Guide

- Published monthly since 1982 by Political Risk Services in Arlington, Virginia.
- Provides international investors with risk ratings for 140 countries.
- Based on assessments by in-house experts, drawing on previous ratings and outside experts.
- The index covers three areas of country risk: political, financial, and economic. Political risk covers law and order, investment profile, and bureaucratic quality.

Source: www.prsgroup.com.

Business Environment Risk Intelligence

- Published by Business Environment Risk Intelligence three times a year since 1966, in Geneva, Switzerland.
- Provides international investors with risk ratings for 50 countries.
- Based on assessments by in-house experts, drawing on previous ratings and outside experts. Their assessments are evaluated by a panel of about 100 external experts.
- The index covers two areas of country risk: political and operational. Operational risk covers the enforceability of contracts, labor costs, bureaucratic delays, short-term credit, and long-term loans.

Source: www.beri.com.

Country Risk Reports

- Published by a U.S. consulting and information company, Global Insight (formerly DRI), since 1996.
- Provides quarterly country risk reviews for 117 countries.
- Based on desk research of 80 in-house experts.
- The index covers 33 immediate risk events and 18 secondary risk events, further classified into policy (tax and nontax) risks and outcome (price and nonprice) risks. Secondary risk events are classified into domestic political, external political, and economic risk.

Source: www.globalinsight.com.

(*contd.*)

Box 1.1
Cross-Country Indicators of the Business Environment (continued)

Country Credit Ratings

- Published every six months since 1979 by *Euromoney Institutional Investor* in New York City.
- Provides international investors with risk ratings for 151 countries.
- Based on assessments by senior economists and sovereign-risk analysts at leading global banks and money management and securities firms.
- The aggregate credit rating is based on nine areas of country risk: political, economic performance, debt indicators, debt in default or rescheduled, credit ratings, access to bank finance, access to short-term finance, access to capital markets, and discount on forfeiting.

Source: www.euromoneyplc.com.

FDI Confidence Index

- Published since 1997 by A.T. Kearney in Chicago, Illinois.
- Provides subjective views on the attractiveness of 60 countries for foreign investment.
- Based on assessments by executive managers of 1,000 global companies.
- Only the aggregate index is published.

Source: www.atkearney.com.

Three others—the World Economic Forum, the Institute for Management Development, and a joint effort between the European Bank for Reconstruction and Development (EBRD) and the World Bank—collect indicators on the general business environment for domestic and foreign companies. The Fraser Institute, in its *Freedom Index*, uses data drawn primarily from the *Global Competitiveness Report* and other indicators to analyze business regulations.

Expert Polls

Services whose primary audience is foreign investors use expert polls to provide frequent updates on global investment risk. New data are released monthly (by Political Risk Services group), quarterly (by BERI, EIU, CRI, EII), or annually (by the Heritage Foundation) for investors allocating global or regional financial portfolios and for multinational corporations deciding which market to enter.

A combination of in-house and outside experts is involved. BERI uses 17 in-house analysts to write initial assessments, which are then provided to a panel of about 100 outside experts. The ratings are constructed by means of the Delphi method, whereby panelists are given their own ratings in previous assessments and the panel's average score on each measure. ICRG also uses a combination of internal analysis of relevant publications and a network of external experts. EII uses outside political analysts and economists at leading global banks and money management and securities firms. CRR indicators are constructed through a similar process, whereby the analysts' reports are first handled by regional risk committees, which revise the scores and submit them to the global risk service committee, all in-house. EIU uses in-house country experts who answer quantitative and qualitative questions about recent and expected political and economic trends.

The expert polls are designed mainly for foreign investors, providing "a means for structuring the composition of global and regional asset deployment that is compatible with executive management's preferences on risk exposure."[16] Foreign investors use such expert advice because they are able to avoid or withdraw from countries with a perceived high level of risk. Local investors who need to operate in

Figure 1.4
Access to Bank Finance and Lending Rates

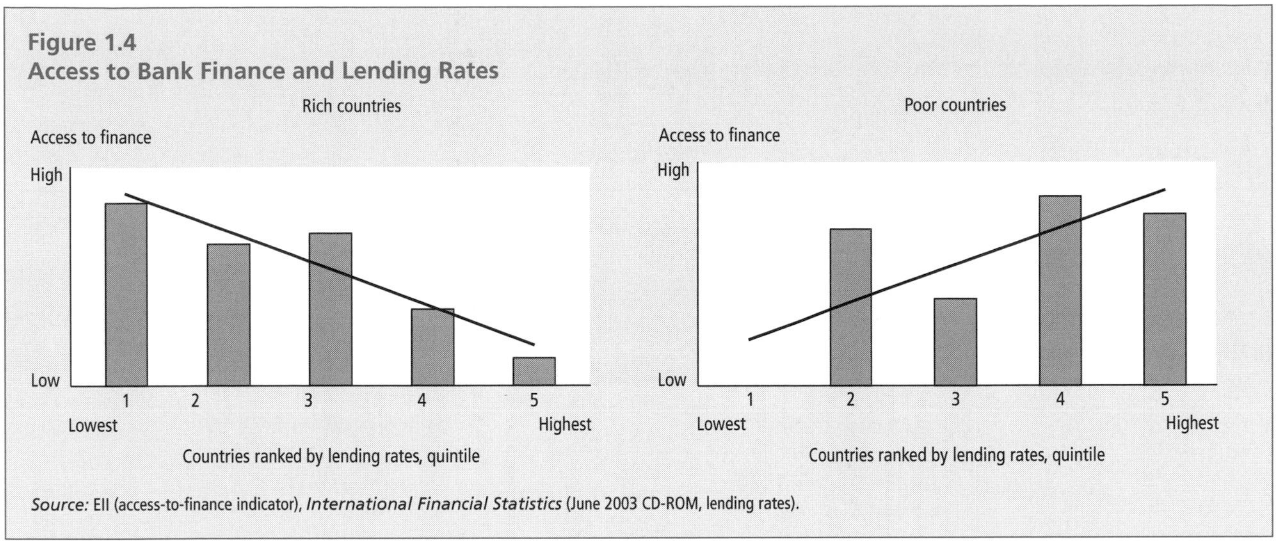

Source: EII (access-to-finance indicator), *International Financial Statistics* (June 2003 CD-ROM, lending rates).

sometimes difficult environments rarely have that choice. Indeed, recent research shows that the indicators generated by experts explain the flow of foreign investment into an economy but not the flow of domestic private investment.[17]

Because foreign investors' interest in many countries is lacking, the experts assessing the less-analyzed countries may not be as well informed about the environment for doing business there. Consider the view of the EII expert panel on access to bank credit. The first graph in figure 1.4 shows a negative relationship between access to bank credit and actual lending rates in the richer half of the *Doing Business* sample. The second graph shows, contrary to expectations, a positive relationship between the two data series in poor countries.

Another example is from a recent study that compares various expert poll ratings in developed and developing countries.[18] The ratings across polls are consistent in developed countries, but not in developing countries (figure 1.5). One conclusion: pollsters pay less attention to countries that do not present large investment opportunities.

Figure 1.5
Polls in Poor Countries Do Not Agree

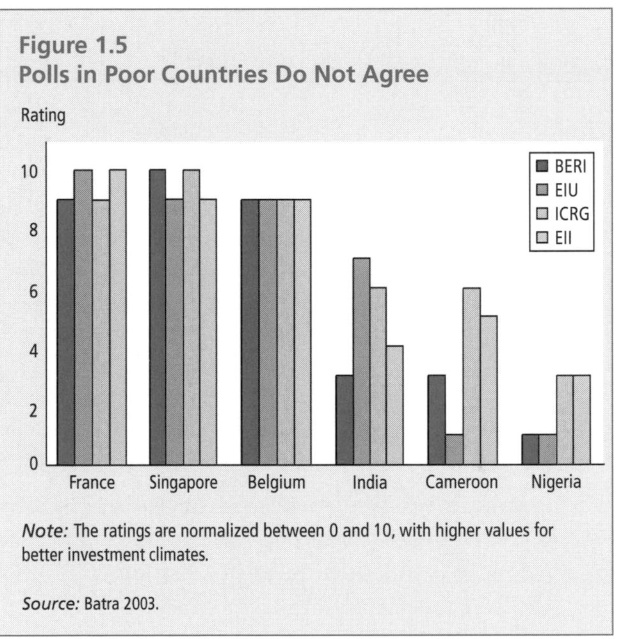

Note: The ratings are normalized between 0 and 10, with higher values for better investment climates.

Source: Batra 2003.

The generality required for making monthly or quarterly updates is adequate for making informed choices about whether to move money in or out of countries but not for guiding policy reform. Take the regulatory component of the *Index of Economic Freedom*, which combines "licensing requirements to operate a business, the ease of obtaining a business license, corruption within the bureaucracy, labor regulations, such as established work weeks, paid vacations, and parental leave, as well as selected labor regulations; environmental, consumer safety, and worker health regulations, and regulations that impose a burden on business."[19] What reforms should the government consider if its country is performing poorly on this indicator? Perhaps reform is needed in all aspects of business regulation, but perhaps it is not.

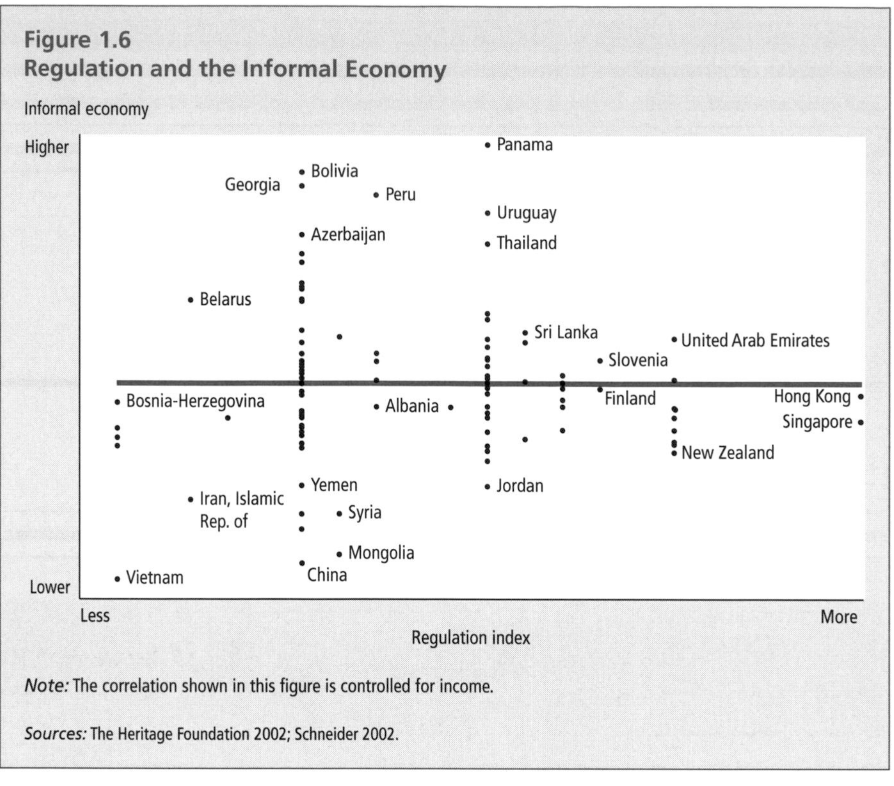

Figure 1.6
Regulation and the Informal Economy

Note: The correlation shown in this figure is controlled for income.

Sources: The Heritage Foundation 2002; Schneider 2002.

for investment. The EBRD–World Bank *Business Environment and Enterprise Performance Survey* uses a mixture of perception and hard-data questions in transition economies.

Enterprise surveys are informative if used appropriately. In many areas, perceptions affect business decisions and thus economic activity. If managers consider the courts to be corrupt and inefficient, they are unlikely to use them. And if managers believe that there is not enough available information on what documents are necessary to apply for a business license, it does not matter that the documents are posted on a government Web site. The information is not easily accessible even if it is available.

As regulatory reform takes place, its effect can be observed in well-designed enterprise surveys. The survey done by the Center for Economic and Financial Research, an independent think tank, covers 2,000 firms in 20 regions of the Russian Federation and asks about actual costs of doing business and general perceptions of the business climate.[21] In August 2001, the Russian Parliament passed a new law limiting the number of inspections of businesses to one per regulatory agency every two years. Before the law took force, many businesses experienced multiple inspections by agencies. With the new law, the average number of inspections in the first half of 2002, compared with the first half of 2001, fell 21 percent. Clearly, there was immediate impact. Such in-depth country surveys can complement the cross-country indicators of the business environment.

But a large body of evidence shows that survey questions on perceptions do not always elicit meaningful responses.[22] Reasons abound—for example, biases in survey design, scaling of responses,

The difficulty in using expert polls for policy reform is seen in the relationship of the burden of regulation to the size of the informal economy. For example, if the measures are adjusted for different country incomes, there is no discernible relationship between the Heritage Foundation's regulatory index and an estimate of informal output (figure 1.6). But a large body of other research shows that excessive business-entry regulation and labor regulation are strong determinants of informality.[20]

Enterprise Surveys

The *Global Competitiveness Report* and the *World Competitiveness Yearbook* report a combination of hard data and perceptions data. The perceptions data come from enterprise surveys on various aspects of the business environment. Managers answer questions on the difficulty of registering a new firm, enforcing contracts through the courts, dealing with labor issues, and so on. A. T. Kearney, in its *FDI Confidence Index*, surveys business executives in the 1,000 largest multinational companies, asking respondents to share their perceptions about the best countries unwillingness of respondents to admit their lack of knowledge or views, lack of a reference point for answering, and sample selection.

Design biases. Simple manipulations of survey design affect the way respondents interpret questions. One bias comes from the ordering of questions. People attempt to provide answers consistent with the answers they have previously given in the survey. In one sociological survey, respondents were asked two questions: "How happy are you with your life in general?" and "How happy are you with your marriage?" When the marriage question came first, the answers to both were highly correlated, but when it came second, they were uncorrelated.[23]

If the survey is long, respondents may exert little effort in answering questions. As a consequence, the ordering of multiple-choice options is important because survey respondents may simply pick the first or last available alternative. Two identical questions in the *Global Competitiveness Report* and the *World Competitiveness Yearbook* ask about the impediments to hiring and firing workers and the ease of creating a new business. Strikingly, the answers to the two questions are highly correlated in the former and unrelated in the latter, in part as a result of the ordering and phrasing of questions.

Response scales. Responses also change according to the scales presented to respondents. In one experiment, some German households were asked how many hours of television they watched each day. Half of the respondents were given a scale that began with a half-hour, then an hour, and proceeded in half-hour increments, ending with four-and-a-half hours. The other respondents were given the same scale, but the first five answers were compressed so that it began with two-and-a-half hours. Twice as many respondents in the second set reported watching television more than two-and-a-half hours a day (37 percent versus 16 percent).[24]

Uninformed answers. Respondents want to avoid embarrassment. In one well-known example, roughly 25 percent of nonvoters report having voted when surveyed immediately after an election. In another example, survey experiments show that respondents answer questions on fictitious issues, such as providing opinions on countries that do not exist, to avoid admitting lack of knowledge.[25]

Figure 1.7
Perceptions Bear No Relation to Actual Tax Rates

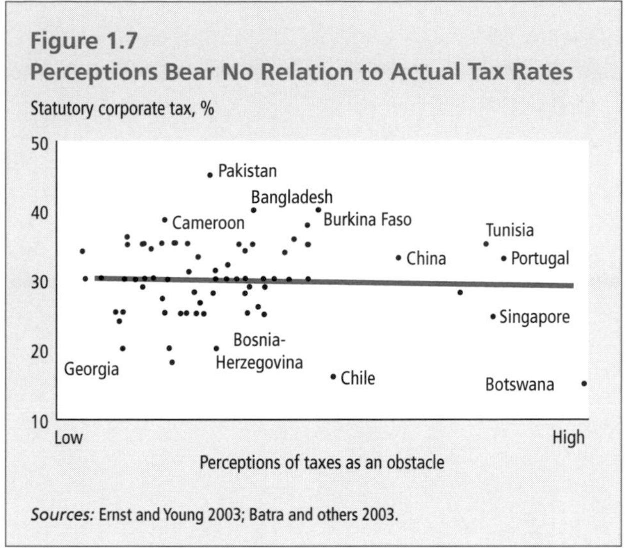

Sources: Ernst and Young 2003; Batra and others 2003.

Lack of a reference point. One example of this defect comes from the United States, where nearly 85 percent of people who need to renew their driver's license report being "better-than-average" drivers. This problem is compounded in cross-country comparisons. One survey asks managers, "Are high taxes a major obstacle to doing business in your country?" When the answers are plotted against the corporate tax rate, the two display no relationship whatsoever (figure 1.7). Managers in every country think tax rates are high.

Sample selection. Nationally representative enterprise surveys are expensive to administer. As a result, almost all firm surveys sample from selected sectors or subsectors within an economy, and many do not cover enough respondents to be statistically representative. Different approaches to sampling can lead to significantly different results, a phenomenon that suggests users should be cautious in generalizing from findings based on a limited pool of firms.

Finally, perceptions measures are often driven by general sentiment but do not provide useful indicators of specific features of the business environment. Consider the 2003 *Global Competitiveness Report.* In the index of the quality of the national business environment, Turkey experiences a dramatic fall in rankings, from 33rd to 52nd (of 75 countries). The report reasons, "Turkey's drop ... is driven by a relative decline in factor quality (university-industry

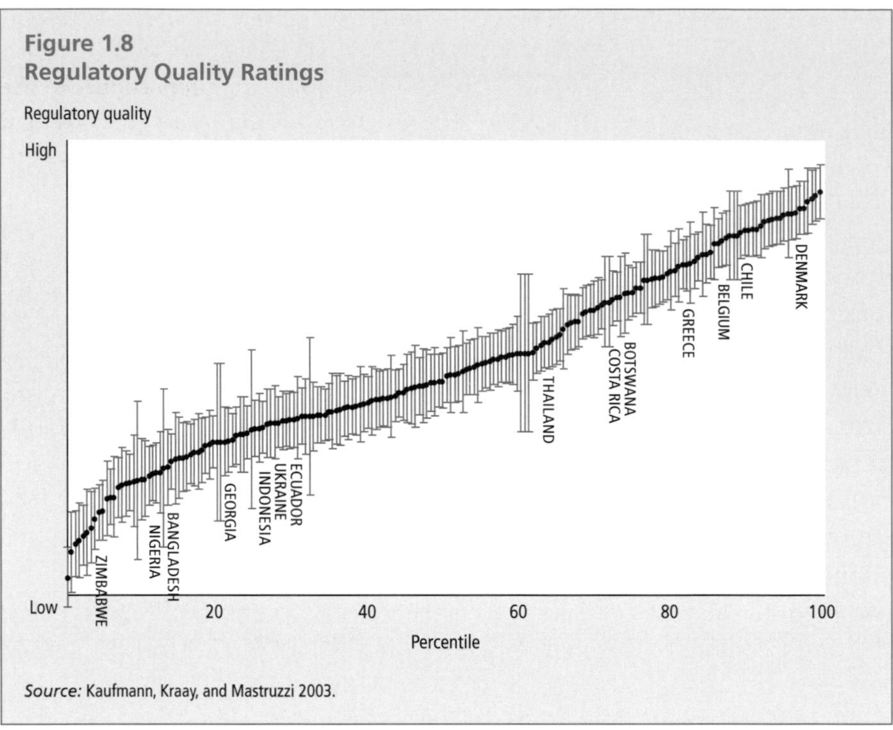

Figure 1.8
Regulatory Quality Ratings

Source: Kaufmann, Kraay, and Mastruzzi 2003.

research collaboration, quality of management schools, administrative burden of start-ups, and others) and context for strategy and rivalry (effectiveness of antitrust policy)."[26] It is hard to imagine how the university-industry research collaboration or the quality of management schools could decline so precipitously in a single year. Also, in 2003, the Turkish government reformed business start-up regulations.[27] Most likely, the change in survey respondents' perceptions was influenced by the financial crisis that started the previous year—that is, it changed the point of reference. Not coincidentally, Argentina, another country in financial crisis in early 2002, also experienced a dramatic fall in business environment rankings.

Aggregate Indicators Are More Robust

The robustness of perceptions indicators is greatly enhanced if they are aggregated. Aggregation brings three benefits: it improves the precision of estimating indicators; it quantifies the explanatory power, giving policymakers the ability to choose which indicators and analyses to rely on; and it increases coverage because some surveys study countries that other surveys do not. However, despite the benefits, aggregated indicators cannot provide detail on the design of underlying regulations and how to reform them.

Using aggregation methodology to study regulatory quality, the World Bank Institute's 2002 regulatory quality indicator measures the incidence of market-unfriendly policies, such as price controls, and perceptions of the regulatory burden on businesses.[28] It uses 60 individual indicators from about a dozen sources. Countries are ranked by using point estimates, with standard deviations informing users about the precision of the ranking (figure 1.8).

The benefits are readily apparent. First, the point estimates have better explanatory power than individual perception surveys do. For example, the aggregate indicators have much greater power in predicting the share of informal activity across countries than the individual indicators do (compare figure 1.9 with figure 1.6). Second, the aggregates also show which of the underlying indicators are most closely related to the composite measure: for example, the regulatory-quality index shows that the World Markets Research Center and the EIU indicators are closest to the underlying aggregate measure that relates more closely to government policies and economic outcomes. Third, almost every country can be covered (the regulatory-quality index covers 199 countries).

An aggregate index of the investment climate—which includes regulatory quality, infrastructure quality, competition, and macroeconomic stability—has recently been constructed at the World Bank Group, by using indicators from 21 databases.[29] As with the previous example, an unobserved-components approach is used to capture the information

Figure 1.9
Regulatory Quality Is Associated with Less Informal Activity

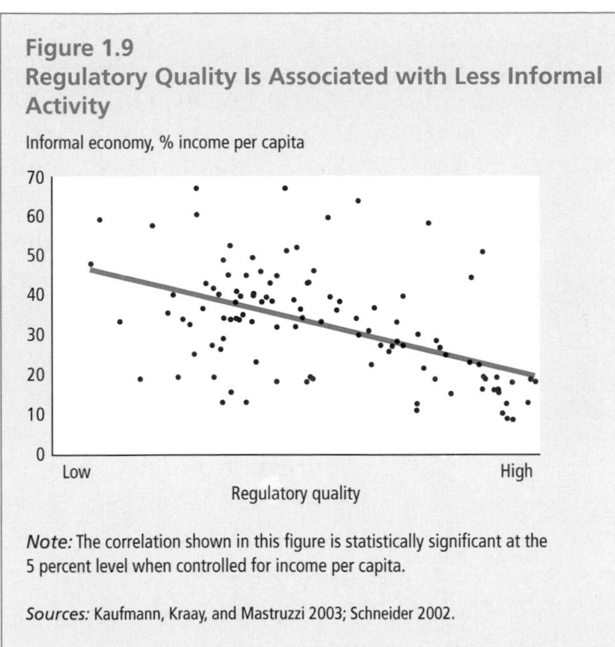

Note: The correlation shown in this figure is statistically significant at the 5 percent level when controlled for income per capita.

Sources: Kaufmann, Kraay, and Mastruzzi 2003; Schneider 2002.

common to a set of indicators and eliminate the idiosyncratic part of each indicator. The index rates the United States, Singapore, Switzerland, Canada, and the Netherlands as the top five economies for doing business. Bangladesh, Haiti, and Mozambique vie for the lowest rating.

Notes

1. Petty 1691.
2. Meade and Stone 1941. Although presented to the British Parliament as a one-off measure, the national accounts quickly became an annual production.
3. For instance, one question is whether the number of procedures in debt recovery is correlated across countries with the number of procedures in resolving a (commercial) tenancy dispute. The answer is yes. For the countries in the *Doing Business* sample, the simple correlation is 0.86. The simple correlation between the number of procedures in debt recovery equivalent to 5 percent and 50 percent of income per capita is 0.94. The high correlations imply that the specific case that was chosen is generally representative for other types of commercial resolution.
4. The methodology was developed by Djankov and others (2002) and adopted with minor changes here.
5. Informal payments are subject to greater measurement error. Moreover, theoretical models in public economics show that bribes are proportional to the severity of regulatory burden—that is, informal payments are an outcome of cumbersome regulations rather than a regulatory obstacle in their own right.
6. World Bank 2002a.
7. The methodology was developed by Botero and others (2003) and adopted with minor changes in this report.
8. The methodology was developed by Djankov and others (2003) and adopted with minor changes in this report. The original study used two cases: a bounced check of 5 percent of GNI per capita, and a landlord-tenant dispute.
9. The work on Latin America is summarized in Hammergren (2003).
10. World Bank 2002b, p. 40.
11. World Bank 2003.
12. Djankov, McLiesh, and Shleifer 2003.
13. Jentzsch 2003.
14. The methodology was developed by La Porta and others (1998) and was adopted with minor changes here.
15. Djankov, Hart, and others 2003.
16. BERI 2002. *User Guide*, p. 1.
17. Batra 2003.
18. Batra 2003.
19. The Heritage Foundation 2002, p. 74.
20. Schneider 2002; Friedman and others 2000; Djankov and others 2002.
21. The survey results are available at www.cefir.ru.
22. Bertrand and Mullainathan 2002.
23. Schwarz, Strack, and Mai 1991.
24. Schwarz and others 1985.
25. Bishop, Oldendick, and Tuchfarber 1986.
26. Cornelius, Porter, and Schwab 2003, p. 38.
27. World Bank 2002c.
28. Kaufmann, Kraay, and Mastruzzi (2003) use the unobserved-component methodology, which expresses survey data as a linear function of the unobserved common component, and a disturbance term capturing perception errors. The assumptions of the model ensure that the distribution of the aggregate indicator is normal and that the means and standard deviations for each country have a natural interpretation. In particular, one can construct a 90 percent probability range around the point estimate where the "true" level of the indicator lies.
29. See Batra 2003 for a detailed description.

2 Starting a Business

In *The Other Path*, Hernando de Soto shows that the prohibitively high cost of establishing a business in Peru denies economic opportunity to the poor. In 1983, de Soto's research team followed all necessary bureaucratic procedures in setting up a one-employee garment factory in the outskirts of Lima. Two hundred and eighty-nine days and $1,231 later, the factory could legally start operation.[1] The cost amounted to three years of wages—not the kind of money the average Peruvian entrepreneur has at his or her disposal. "When legality is a privilege available only to those with political and economic power, those excluded—the poor—have no alternative but illegality," writes Mario Vargas Llosa in the foreword to de Soto's book.

This sentiment is not new. Well into the 19th century, European companies required a state charter or a concession from the state to be registered, and only the rich could afford such.[2] In France, free registration for private companies was proclaimed in 1791, in the aftermath of the revolution. In England, free incorporation was allowed in 1844, a consequence of expanding the franchise to the middle classes.[3]

When European corporate law was transplanted to other parts of the world, whether through willing appropriation or through colonization, it affected the formation of business entities. The 1865 Commercial Code in Chile, following the 1848 Spanish Code, required two separate presidential decrees for company incorporation. In contrast, the first Commercial Code of Colombia, adopted in 1853, did not contain the requirement to obtain a concession from the state. This departure from the Spanish Code was made in the belief that free business incorporation is a right.[4] The 19th century saw a boom in incorporation in the United States, with the passage of general corporate laws—in 1811 in New York, 1839 in Massachusetts, 1844 in England, 1849 in California, and 1883 in Delaware. The main reasons for the rapid expansion were the competition among states in liberalizing their corporate laws and the advent of the railroads. By the late 19th century, the United States had more limited-liability companies than all of Europe.[5]

The incorporation of business is beneficial for four reasons. First, legal entities can outlive their founders. Second, resources are pulled together, as shareholders join forces in establishing the company's capital. Third, the formal introduction of limited liability—starting with the enactment of the Code de Commerce in France in 1807—reduces the risks of doing business. In *The Wealth of Nations*, Adam Smith notes: "These [incorporated] companies have been useful for the first introduction of some branches of commerce by making, at their own expense, an experiment which the state might not think it prudent to make."[6] Limited liability gives one the freedom to innovate and experiment without large negative consequences. Fourth, registered businesses have access to services—provided by public courts or private commercial banks—that are not available to unregistered firms. In short, the establishment of a legal entity makes every business venture less risky and increases its longevity and its likelihood of success.

Two procedures—notification of existence and tax and social security registration—are sufficient for business registration. In reality, all countries impose additional requirements. Further, the regulation of business entry varies systematically across countries.

Richer countries regulate less. So do countries in the common-law tradition.

In poorer countries, market failures may be more severe, and therefore may increase the desire to correct the failures by regulating entry. The temptation should be resisted, for the costs of government inefficiency may outweigh the benefits of stricter regulation. Cumbersome entry regulation is associated with less private investment, higher consumer prices, greater administrative corruption, and a larger informal economy. There are no discernible benefits in improving product quality or in reducing undesirable externalities such as pollution.

Governments can go a long way with simple reforms. These include adopting better information and intragovernment communications technology—to inform prospective entrepreneurs and to serve as a virtual one-stop shop for business registration. Cutting unnecessary steps from the entry process, such as notarial certification of all incorporation documents or registration with the local chamber of commerce, introducing single registration forms, a single company identification number, and silent consent in approving registration (a nonresponse implies approval) have had enormous success. In the Russian Federation, a 2002 reform transferred all registration powers to the State Tax Inspectorate, thereby cutting the number of business entry procedures from 19 to 12. Thanks to a single registration form, separate notification to the local registration chamber, the pension fund, the health fund, the statistical committee, and the social security fund, and application of making a seal are no longer necessary. Moreover, the registration of the new legal entity and tax registration are merged into one procedure.

Reforms that require new legislation include introducing a general-objects clause in the articles of incorporation (which allows a firm to change lines of business without reregistering), eliminating the capital requirement, and removing notarial authorizations and court use from the registration process. Such reforms may be difficult to implement, as they may face stiff opposition from both judges and the legal and notarial professions, but their beneficial effects go far beyond business entry.

How Easy Is Business Entry?

It takes two procedures, two days, and less than 1 percent of annual income per capita to register a private limited-liability company in Australia. It costs nothing to do the same in Denmark, and almost nothing (about 1 percent of annual income per capita) in Canada, New Zealand, Singapore, Sweden, the United Kingdom, and the United States. But it takes 18 procedures to start a business in Algeria, Bolivia, and Paraguay, and 19 procedures in Belarus, Chad, and Colombia. It takes 152 days to do so in Brazil, 168 days in Indonesia, 198 days in the Lao PDR, 215 days in the Democratic Republic of Congo, and 203 days in Haiti. And it costs more than three times per capita income to start a business in Burkina Faso and Nicaragua, four times in Ethiopia and Niger, and more than five times in Cambodia. (In June 2003, the Ethiopian government reduced the cost of business registration by a quarter.) Business entry costs $5,531 in Angola (838 percent of per capita income), $785 in the Democratic Republic of Congo (872 percent of per capita income), and $1,817 in Sierra Leone (13 times per capita income). Contrast this with $28 in New Zealand, $210 in the United States, $264 in the United Kingdom, and $249 in Singapore.

In Mexico—a country with an income per capita of $5,910—the entrepreneur needs to deposit at least $5,180 to start registration. High capital requirements are the norm in the Middle East—at 17 times the income per capita in Yemen, 16 times in Saudi Arabia, and 24 times in Jordan. Some African countries also have high capital requirements: 7 times income per capita in Burkina Faso, 8 times in Niger, 9 times in Mauritania, and 18 times in Ethiopia. In a third of the sample, there are no capital requirements at all.

These numbers show the vast differences in the treatment of new firms across countries. Four measures—the necessary procedures, the associated time and cost, and the minimum capital requirements—capture various aspects of the registration process.

- The number of procedures describes the external parties that the would-be entrepreneur faces. One can think of them as tollbooths—at each procedure, the entrepreneur may be stopped. In many countries, at each procedure involving government officials, a bribe may change hands.

- The number of days and the official costs associated with each procedure are easy to interpret: the higher those numbers, the more cumbersome and costly the registration process and the less likely it is that many entrepreneurs will register businesses.
- The minimum capital requirement is the amount of capital that the entrepreneur needs to put into a bank account before registration starts. The account is frozen during business entry and in many countries remains so until the dissolution of the legal entity.

All entry indicators constructed in this chapter describe a limited-liability company—not a sole proprietorship, a partnership, a cooperative, a joint stock company, or a corporation. Why? Because private limited-liability companies are the most prevalent business form around the world. They are also desirable for economic reasons. Investors are encouraged to venture into business when the potential losses are limited to their capital participation.[7]

Indeed, evidence from 19th-century England, Ireland, and the United States suggests that the introduction of limited liability dramatically increased the number of companies seeking registration.[8] A study of German companies also shows that limited liability is associated with larger firm size.[9] Today, limited-liability companies account for more than 55 percent of registered businesses and 90 percent of output in OECD countries.[10] Even in a transition economy, such as Latvia's, limited-liability companies account for 62 percent of all registered businesses and 93 percent of output.[11] Similarly, limited-liability companies account for 57 percent of private enterprises in Vietnam and more than 70 percent of output.

Sometimes, as part of registration, new businesses have to acquire zoning permits or licenses, so they are included in the entry procedures list of the respective countries.

- In Indonesia, every business needs to apply for a trading license, a procedure that takes two weeks and, in the event the entrepreneur has moved from out of town, requires a "good conduct" note from the police.
- In Ghana, companies are required to obtain an environmental certificate. The company submits an application describing the location, current zoning classification, processes to be used, and likely environmental impact. Environmental officials visit the site once the application is submitted and file a detailed report. It takes at least 90 days to fulfill this procedure.
- In Jordan, all entrepreneurs must apply for a municipal vocational license. The application needs to be accompanied by the certificate of company registration, the membership certificate in the chamber of commerce, an activity approval by the appropriate ministry, a notarized rental contract or ownership title of facilities, and a location map.

On top of these procedures, some businesses need permits and construction approvals, utility connections, and product and process licenses before they can commence operations. Although these procedures are not covered in the data here (because they are not general requirements), they can be significant obstacles for entrepreneurs. In Tanzania, it takes 25 separate procedures to acquire all of the necessary permits and licenses for land and factory use. In the best case, these procedures take 795 days to fulfill, and the official cost is $508, or about twice the income per capita. In Mozambique, it takes 34 procedures, 625 days, and $11,045, or about 50 times the income per capita, to fulfill all requirements for entry of a new manufacturing firm.[12]

Enterprise surveys reveal that entrepreneurs' perceptions of the efficiency of the registration process differ across regions in a country, sometimes dramatically. One reason is that local regulations can affect starting a business, as in Botswana, Brazil, Chile, Colombia, Ecuador, Jordan, Kenya, Malawi, Philippines, Romania, Tanzania, Uganda, Venezuela, and Zimbabwe. The local enforcement of the national company law and regulations explains most differences, as in Vietnam, Russia, and Bulgaria.[13] In Vietnam, the 2000 Enterprise

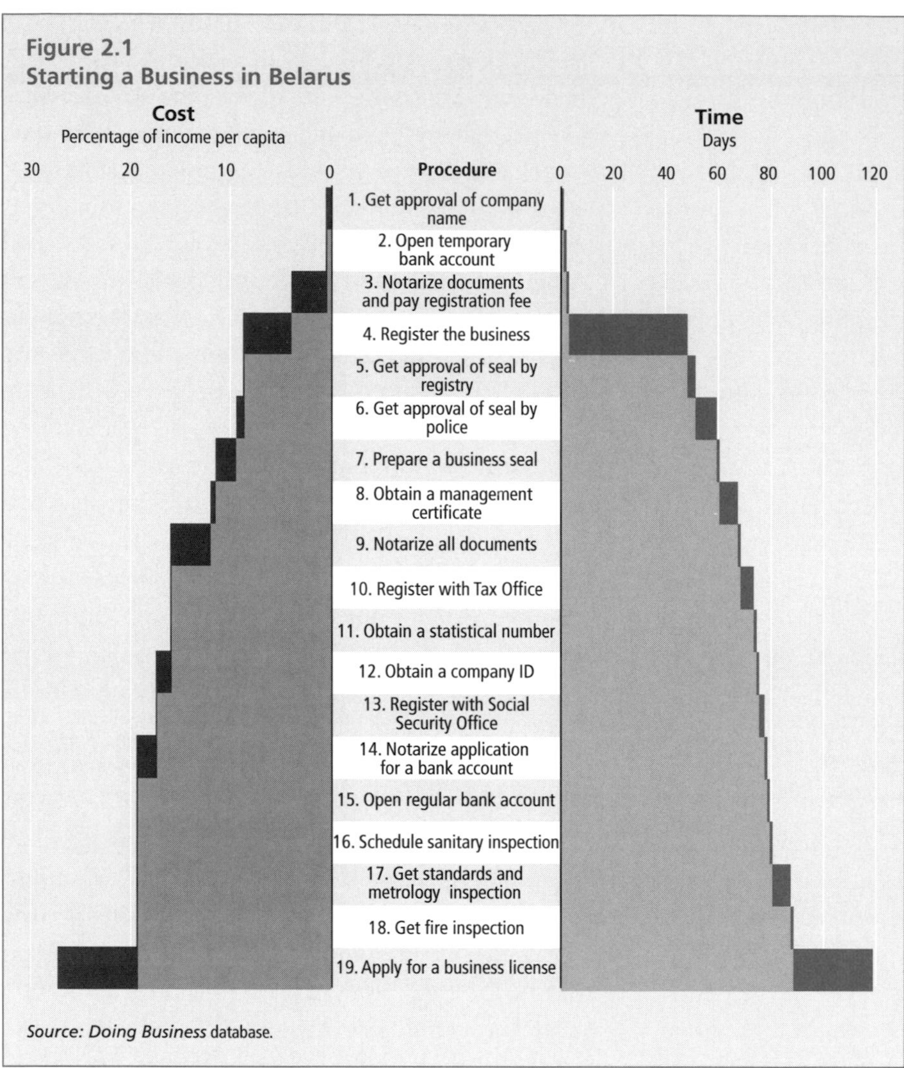

Figure 2.1
Starting a Business in Belarus

Source: Doing Business database.

Law stipulated statutory time limits for all procedures in the business entry process, but in the spring of 2001 a company in Dong Nai province obtained its registration certificate in one day, while a company in Hanoi took 123 days. At the same time, it took eight days on average to acquire a company seal in Ho Chi Minh City but 14 days in Hanoi.

No government in the world lets an entrepreneur register a new business in a single procedure, but some come close. In Canada, the entrepreneur submits the federal registration form through the online Electronic Filing Center and receives a business number within the hour. With this number, the entrepreneur applies with the Canadian Customs and Revenue Agency for tax numbers, payroll deductions, and import and export licenses. The process is identical in Australia. In Denmark, the entrepreneur also needs to register the minimum capital with a bank, while in Ireland the entrepreneur needs to order a company seal.

In other countries the process is more convoluted (figure 2.1). In Belarus, the entrepreneur needs approval of the company name from the Ministry of Justice. With this document in hand, the entrepreneur opens a temporary bank account in the name of the company to be registered. With the new bank statement, the entrepreneur needs to visit a notary public, who authorizes all the documents. The notarized documents are submitted to the state registry.

The process doesn't stop there. The entrepreneur requests an inspection of business premises from the labor ministry—and while waiting for the inspectors to come, prepares a company seal and takes a course to obtain a management certificate. The certificate is notarized and registered with the local police department. When all the documents are notarized, the entrepreneur visits the tax office and the social security office, leaves copies with them, and obtains receipts. He or she is finally ready to apply for a company identification number and a statistical number.

The end of business registration is in sight. The entrepreneur can now notarize the application of a regular bank account, and obtain the bank account—then schedule a sanitary inspection, a standards and metrology inspection, and a fire inspection. When the business passes all inspections, the entrepreneur can

Table 2.1
Frequency of Entry Procedures across Countries

Purpose of Procedure	Percent of Countries
Tax registration	93
Labor registration	87
Administrative registration	76
Bank deposit	68
Notarization	63
Health benefits	62
Notice in newspaper	36
Company seal	38
Court registration	32
Chamber of Commerce	27
Statistical Office	17
Environment	12

Source: Doing Business database.

obtain a business license. The business can start operations after 19 procedures, 118 days, and $369 in official payments.

As mentioned at the outset, two procedures—the notification of existence and the tax and social security registration—are sufficient for business registration. Other procedures, such as registering with the statistical office, obtaining environmental permits, or registering workers for health benefits (table 2.1), seem to be socially desirable. And still others, such as having local chambers of commerce approve the applicant, can limit competition.

The purpose of still other procedures is dubious, and economies with heavy regulations of entry display a bewildering variety. In Vietnam, founders needed to obtain a seal-making license from the Ministry of Public Security and have the seal made by the authorized seal makers. In the Russian Federation, founders needed (until recently) to visit the Social Pension Fund, the State Fund of Compulsory Medical Insurance, the State Committee on Statistics, and the Social Security Fund before obtaining the commercial registration certificate. They needed to revisit each of those agencies to file a copy of the certificate to get a company ID from each of them. The procedures were radically reformed in late 2002.

Which countries regulate business entry the most? When the countries are divided into groups according to their income per capita, the high-income countries have the smallest number of procedures, with a median of 7. They are followed by the upper-middle-income group, with a median of 10 procedures. The lower-middle-income countries have the highest number of procedures, around 12, while the poorest countries have a median of 11 procedures. The time to register a company is, again, the shortest in the richest countries, at less than one month. Although the registration process takes around the same amount of time—50 days—in upper-middle- and lower-middle-income countries, it is significantly higher in the poorest countries, where the median number of days is 63. In contrast, the cost of starting a business grows monotonically for companies in rich countries versus poor countries. It accounts for less than 10 percent of income per capita in the high-income group, and for an amazing 120 percent in low-income countries (figure 2.2).

Regional differences are also significant (figure 2.3). Latin American governments regulate business entry the most in terms of procedures and time; they are followed by African and Middle Eastern governments. OECD governments regulate the least. The cost of registration is extremely high in African countries—at around 190 percent of per capita income. Similarly, the minimum capital requirements—with a median of more than 700 percent of per capita income—are much higher in the Middle East and North Africa than in any other region.

Strong patterns emerge by legal origin.[14] Nordic countries have the smallest number of procedures—a median of 5—the shortest time, at 21 days, and the lowest cost, at less than 1 percent of per capita income (figure 2.4). Countries in the French civil law tradition take the longest time and have the most procedures and highest cost. But France itself is a top performer among French-origin countries. Countries in the German tradition have the largest capital requirement, more than 100 percent of income per capita, whereas the median capital requirement for English-origin countries is zero.

How do the entry indicators interrelate? Do governments choose one type of entry barrier over another? For example, very fast business registration might be more costly, so that entrepreneurs are in effect paying for better public administration. Or governments could reduce obstacles by requiring few

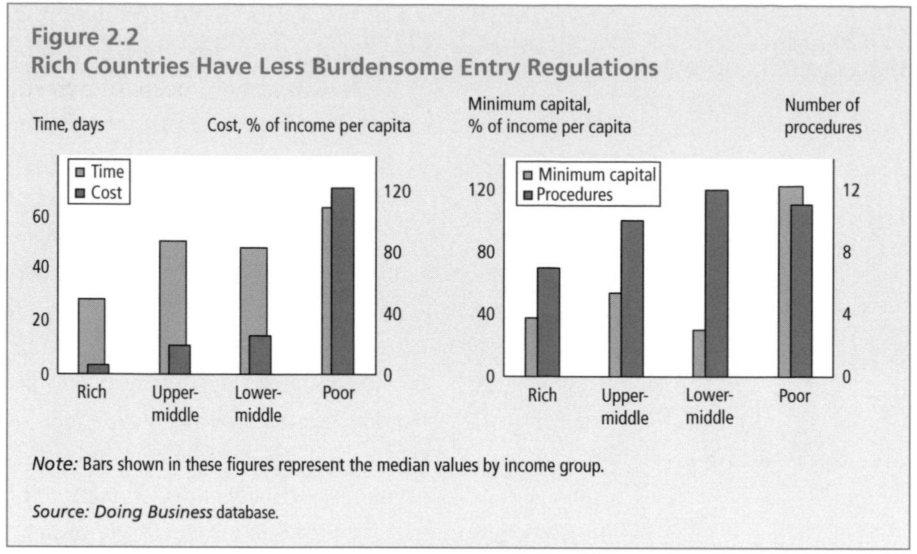

Figure 2.2
Rich Countries Have Less Burdensome Entry Regulations

Note: Bars shown in these figures represent the median values by income group.

Source: Doing Business database.

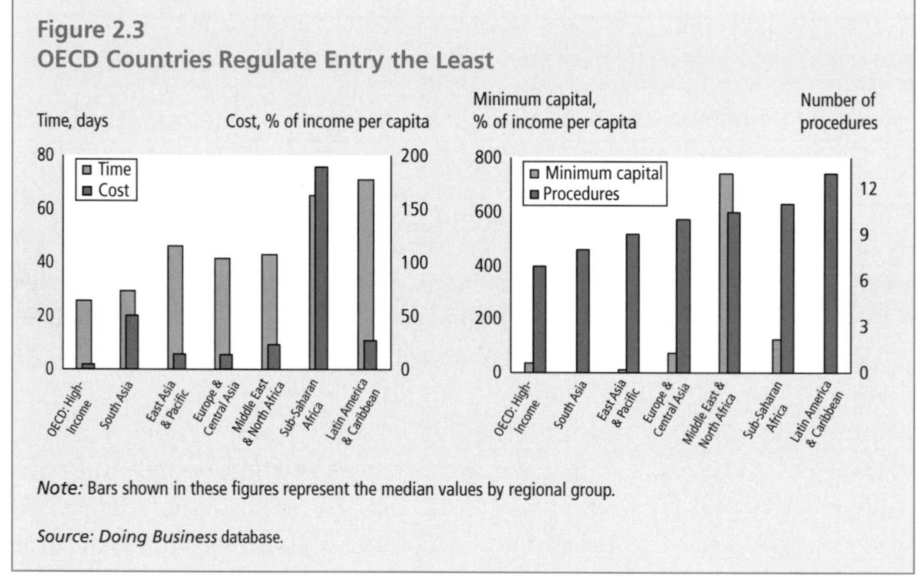

Figure 2.3
OECD Countries Regulate Entry the Least

Note: Bars shown in these figures represent the median values by regional group.

Source: Doing Business database.

registration procedures but allow only the wealthy and well-connected to register, by imposing large minimum capital requirements. This is not the case. Some governments appear to regulate starting a business in every way possible—the number of procedures, time, cost, and minimum capital requirements are highly correlated.

Are Entry Regulations Good? Some, Yes—Many, No

Do entry regulations, even when seemingly desirable, lead to better outcomes? Unregulated markets exhibit frequent failures, ranging from monopoly power to externalities. A government that pursues social efficiency might try to address these failures through regulation. The government screens new entrants to ensure that consumers buy high-quality products from desirable sellers and to reduce such externalities as pollution. By being registered, new companies acquire a type of official approval, which makes them reputable enough to engage in transactions with the general public and other businesses. If so, stricter regulation of entry should be associated with superior social outcomes.

It isn't. Compliance with international quality standards declines as the number of entry procedures rises, and pollution levels in developing countries do not fall with the introduction of environmental permits. Measures of food poisoning and job-related accidents are not lower in countries with a higher number of sanitary and health and safety regulations.[15]

Entry regulations do have real effects—mostly unwanted. Cumbersome entry procedures push entrepreneurs into the informal economy, even after controlling for income per capita (figure 2.5).[16] There, workers lack health insurance and pension benefits. Products are not subject to quality standards. It is impossible for businesses to obtain bank credit or use courts to resolve disputes. And employers cannot use state-provided training budgets for employees and school support for their children.

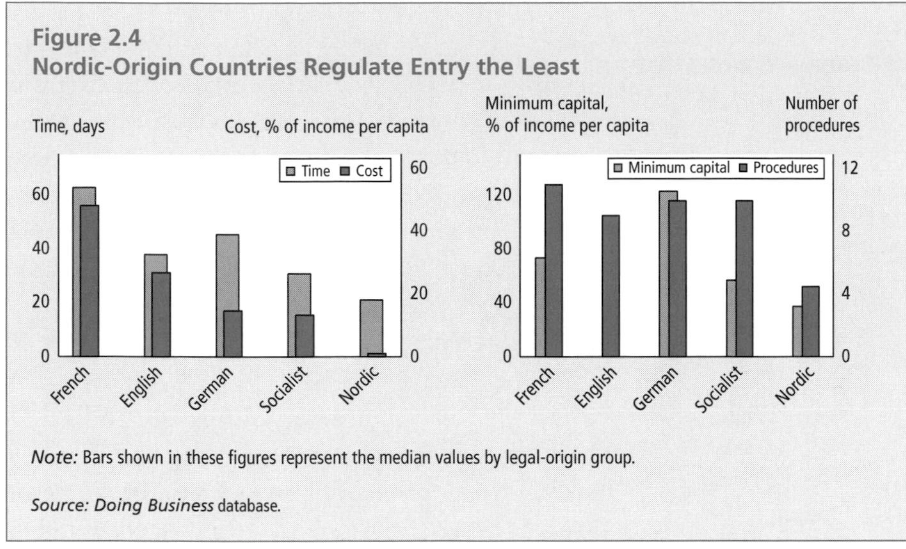

Figure 2.4
Nordic-Origin Countries Regulate Entry the Least

Note: Bars shown in these figures represent the median values by legal-origin group.

Source: Doing Business database.

implies that employment in the French retail sector would have been 10 percent higher today.[17] Another study suggests that if Italy were to adopt U.S. entry regulations, private investment as a share of manufacturing output would rise by an estimated 40 percent.[18] Although similar studies have yet to be conducted in developing countries, the effects of entry regulation are thought to be of similar magnitude—for two reasons. Distortions are larger (as the data show), so the deregulation effect is more significant. But regulatory enforcement is not as strong and therefore this effect is mitigated.

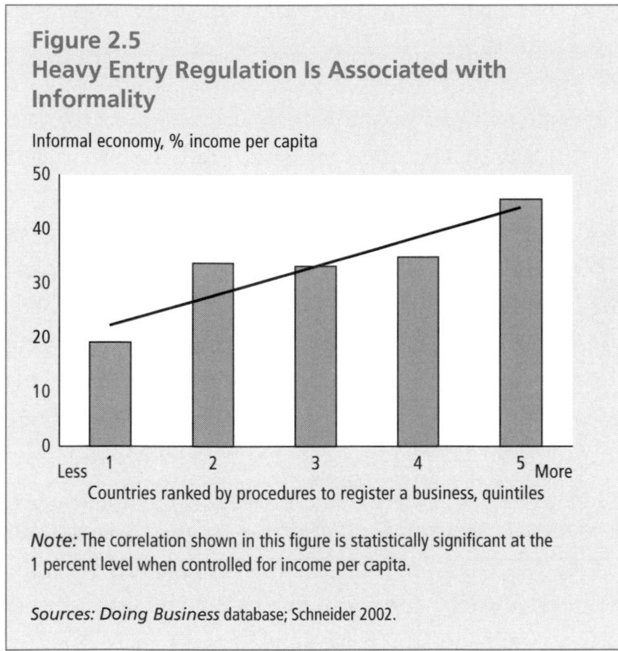

Figure 2.5
Heavy Entry Regulation Is Associated with Informality

Note: The correlation shown in this figure is statistically significant at the 1 percent level when controlled for income per capita.

Sources: Doing Business database; Schneider 2002.

Consumers face higher prices in developed and developing countries with relatively heavy entry regulations.[19] Field studies in developing countries show that foreign investors avoid investment in countries with more burdensome regulation, thereby reducing the potential welfare benefits to consumers in the country.[20]

The unwanted effects go on. Controlling for income per capita, cumbersome entry procedures are associated with higher corruption in the government offices that handle the procedures, particularly in developing countries (figure 2.6). Each procedure is a point of contact—an opportunity to extract a bribe. Llosa writes: "Such a regulatory system is not only immoral but inefficient. Within it, success does not depend on inventiveness and hard work but on the entrepreneur's ability to gain sympathy of presidents, ministers, and other public functionaries (which usually means his ability to corrupt them)."[21]

Whether entry regulation is socially desirable can also be addressed from a different perspective. If democratic countries regulated more, and if politicians in such countries responded to their constituencies, one could hypothesize that such regulations are beneficial by design. By contrast, authoritarian governments may

Some developed economies also suffer from excessive regulation and pay the price. In France, the Ministry of Industry adopted the Loi d'Orientation du Commerce et de l'Artisanat in 1974, to protect small shopkeepers and craftsmen against competition from larger retail stores. The legislation created a zoning permit requirement, at the discretion of the local municipal council. These entry requirements weakened employment growth in the formal retail sector. If the regulations had not been introduced, the analysis

Figure 2.6
Cumbersome Entry Regulation Is Associated with Corruption

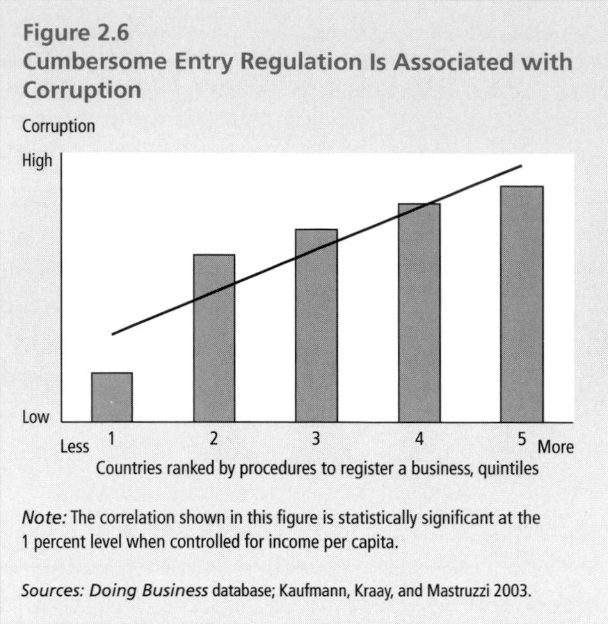

Note: The correlation shown in this figure is statistically significant at the 1 percent level when controlled for income per capita.

Sources: Doing Business database; Kaufmann, Kraay, and Mastruzzi 2003.

be more likely to be captured by incumbent firms and to have regulatory systems aimed at maximizing bribes rather than addressing market failures.[22] That is why more-representative and more-limited governments would regulate entry less. Indeed, a study on business entry regulations in 85 countries finds, holding per capita income constant, that countries with more-limited governments and greater political rights have lighter entry regulation.[23] This result is even stronger when the analysis is repeated with the *Doing Business* sample of 133 countries.

In sum, cumbersome entry regulations do not increase the quality of products, make work safer, or reduce pollution. They hold back private investment. They push more people into the informal economy. They increase consumer prices. And they fuel corruption. Governments that regulate more are less accountable to their citizens.

What to Reform?

With so many examples of successful reform in regulating business entry, there is no better time to act. Indeed, even jurisdictions with efficient business registration procedures have recently updated their regulations—Denmark in 1996, Australia in 2001, Norway in 1997, New Zealand in 1998. As have some others: Vietnam in 1999, Pakistan and the Russian Federation in 2002, and Turkey in 2003. In contrast, the laws regulating business registration in the Dominican Republic date to 1884, in Mozambique to 1888, in Angola to 1901, and in Burkina Faso to 1916. However, antiquated laws are not always to blame: in Sierra Leone, the Legal Practitioners Act of 2000 made mandatory the use of attorneys in incorporation. This requirement in effect doubled the cost of business registration.

Administrative Reform

One can start by providing prospective entrepreneurs with all necessary information, including the number and sequence of procedures, their time, and their cost. This will minimize the time lost due to not understanding procedures and will reduce the likelihood of bribes or unofficial padding of fees. In Venezuela, the relevant information is available on the Web, with a graphic presentation of the sequence (http://economia.eluniversal.com/guiadinero/micro3.shtml#). In Spain, since 1999, the ministries of economy, finance, labor, public administration, and the chamber of commerce have an informational Web site (www.ipyme.org) for entrepreneurs, showing exactly where and when to go, and what documents to bring. The site describes the documentation for completing the registration process, and the additional requirements to start a business. The associated fees, stamp duties, and notary costs are also listed. Many other countries provide such services, such as Ireland, Latvia, and Singapore. Where Internet usage is still low, the information can be provided in leaflets or posted on the wall of the registry office, as it is in Mongolia, South Africa, and Thailand.

Other countries have recently adopted new technologies to improve communication and to share information among government offices. In 2002, tax offices around Pakistan were linked electronically. While it previously took a week for an entrepreneur in Karachi to receive a tax registration number, it now takes a few hours.

Next, regulators can move to a single registration form and a single registration number. France moved to a single form in 1994, Finland in 2001.[24] Many countries—such as Croatia, Madagascar, Portugal, Serbia, and Montenegro—require several types of forms to be filed, which can be confusing and sometimes expensive. Countries also require multiple registration numbers, issued by various government agencies. In Ecuador, a new business needs five separate registration numbers: from the superintendent of companies, from the mercantile registry, from the tax office, from the social security institute, and from the ministry of labor. In contrast, South Africa moved to a single company identification number in 1998, Belgium in 2000, Italy in 2001, and Moldova in 2002.

Administrative rules can be revised to allow for statutory response times, for posting information on fee schedules, and for silent-consent rules. Fewer than a third of the *Doing Business* countries have statutory response times. But a recent study of the 2000 Enterprise Law in Vietnam shows their effectiveness. After the maximum was set at 15 days, the average response time fell from 45 days to 19 days, in about a year.[25] Posting fee schedules helped fight administrative corruption in India. Silent consent means that if entrepreneurs have not heard from the government agency within a given number of days, approval is automatic and they may continue to the next procedure. A silent-consent rule was just adopted in Bulgaria.

A more comprehensive reform is to establish a one-stop shop for company registration. In 1994, France established the *Centre de Formalite des Entreprises* (CFE), a single office where entrepreneurs can file all the declarations and documents needed to set up a new enterprise. The CFE then forwards the documents and declarations to the government departments and courts that must approve or register the new business. Similar one-stop shops have been established in Thailand (in 1997), in the Dakahlia region of Egypt (in 1999), and in El Salvador (in 2000). In establishing them, however, governments need to ensure that they dismantle other steps for business registration, to avoid creating "one-*more*-stop shops." In the Philippines, where a One Stop Action Center was established in 1987, investors continued to complain about cumbersome procedures and delays. For some administrative requirements, a double licensing procedure was in effect imposed, with the investors having to apply to both the One Stop Action Center and the licensing body.[26] One solution is to use an already-existing government agency to process the application for business registration and forward the information to other agencies. Just this kind of reform was adopted in Turkey recently. Instead of going to eight government agencies for approvals, the entrepreneur submits a single application to the Trade Registrar. The registration is issued in one day.

As Internet technology becomes widespread, registration can become electronic, through a virtual one-stop shop. Several countries already use online business or tax registration—Australia, Austria, Canada, Denmark, Hungary, Latvia, New Zealand, and Singapore. Ho Chi Minh City in Vietnam recently introduced it as well. In 2001, Italy passed a law on electronic signatures, allowing entrepreneurs to submit documents by e-mail. The trade registrar office is establishing data transmission links with companies, notaries, chambers of commerce, and trade associations. When in place, the process will radically reduce the time and cost of registering a new business. Israel, Peru, and Thailand are in the midst of similar reforms.

Online business registration has other benefits. In Korea, Hong Kong (China), Taiwan (China), and the United Kingdom, anyone may view the register of company names over the Web and confirm that the proposed company name is unique. Entrepreneurs in Bolivia can do the same through touch-tone phone, saving time in a potentially long procedure.

Legal Reform

Several legal reforms have produced good results. One is the adoption of a general-objects clause in registration, so that entrepreneurs do not need to specify the precise nature of their business activity. Company laws in all Nordic and common-law countries allow for such a clause. This not only eliminates the need for court involvement—it allows for the instantaneous, if expensive, purchase of off-the-shelf companies to create a new one.[27] The entrepreneur needs only to go to a lawyer, buy the shell company, and change its name. The general-objects clause also makes it easier for companies to reregister. Recent legislation in Belarus and Uzbekistan made it necessary for tens of thousands of companies to reregister, at great cost to their owners. The process would be automatic if the entrepreneur needed only to confirm continuing existence, under a general-objects clause.

Comprehensive reform of business entry regulations would eliminate the capital requirement. In the early days of incorporating a business, minimum capital was required for the privilege of obtaining limited liability.[28] And shareholders paid dearly. The 1855 Limited Liability Act in England mandated a minimum value of shares at £10 and a minimum number of shareholders at seven. In today's money, this would have amounted to $5,265.

Some countries still justify capital requirements—as protecting creditors, as protecting the company against insolvency, and as protecting the public from activities that could reduce social welfare. But this makes little sense. Why would a highly leveraged company that transports radioactive waste have the same capital requirement as a company that designs software? If capital requirements were commensurate with risks of creditors, shouldn't they differ across sectors? When in-kind contributions become acceptable, as they are in almost all countries, what is the actual value of minimum capital in the event of insolvency?

In about a dozen countries in the *Doing Business* sample, the capital requirement is a major obstacle to starting a business (table 2.2). In Japan, more than half of the potential business start-ups are thwarted for lack of minimum capital.[29] Not surprising, with paid-in capital at registration amounting to 71 percent of income per capita.

The desired direction of reform is to let private contracts between debtors and creditors substitute for capital rules. This is exactly what the 1982 corporate law reforms in Canada and the 1984 reforms in South Africa did.[30]

The use of notaries for the authorization of documents related to business registration can also be eliminated. Notaries are not part of the registration process in Nordic countries, and seldom are in common-law countries (only in Ethiopia, Sri Lanka, and the United Kingdom). In contrast, notaries are almost always used in Latin America, French-speaking Africa, and transition countries (figure 2.7).

Where notaries are needed to authorize documents, this is frequently the most expensive part of the company registration. In Mexico, notary costs are $875,

Table 2.2
Some Countries Have Prohibitive Capital Requirements

Country	Minimum capital requirement (US$)	Minimum capital requirement (% of income per capita)
El Salvador	11,429	549
Burkina Faso	1,435	652
Egypt, Arab Rep.	11,593	789
Niger	1,435	844
Mali	2,152	897
Yemen	8,413	1,717
Ethiopia	1,756	1,756
Cambodia	5,112	1,826
Mongolia	9,006	2,047
Jordan	42,313	2,404

Source: Doing Business database.

Figure 2.7
Notaries—An Unnecessary Burden

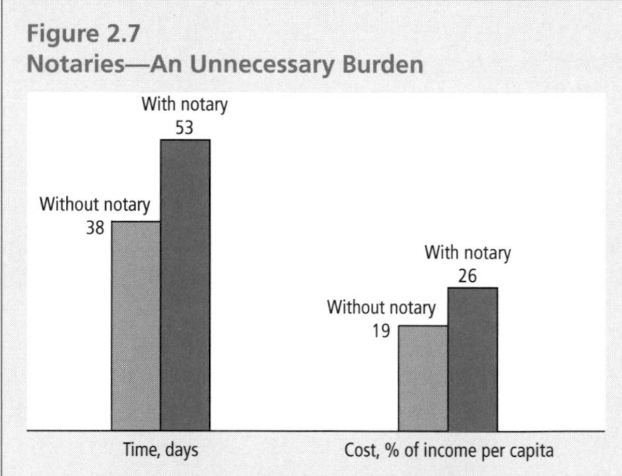

Note: Bars shown in these figures represent median values for countries with and without notary involvement in business registration. Differences in medians are statistically significant at the 1 percent level for the time measures but significant only at the 13 percent level for the cost measure.

Source: Doing Business database.

almost 80 percent of the total costs. In Turkey, notarization costs $780, 84 percent of the registration cost. In Guatemala it costs $850, 73 percent; in Slovenia, $920, 67 percent; and in Angola, $2,800, 51 percent.

Why do some countries still have notaries involved in business registration? It is hard to tell, but history is replete with examples of institutions that have outlived their usefulness. Notaries were a large part of Italian trading relations with foreign partners in the twelfth through fifteenth centuries, particularly in the Levant. The Papacy saw a good source of income in notarial services and made notaries papal administrators. Notaries quickly lost their importance in England in 1534, when Henry VIII broke from the Roman Catholic Church and made it a criminal offense to apply to the Vatican for a notarial appointment. In contrast, notaries retained their role in France, Spain, and Italy. Colonization ensured their existence in many countries around the world.

The service a notary provides—checking the identity of company founders and company officers—is routinely performed by public administrators for many other services. And clerks at the business registry are as able as notaries to confirm identity.

In many developed countries, business registration is an administrative process and the courts take no part in it. But in almost all French-speaking African countries and in most transition countries, business entry is a judicial process and court approval is necessary (figure 2.8). Judicial approval tends to be a very long procedure. In Bulgaria the business registration process takes 30 days, 21 of them spent at the court. In Slovenia the court process takes 37 days, and in the Czech Republic it takes 45 days.

Slovakia is drafting legislation to convert to an administrative process. So is Serbia and Montenegro. Reform is not expensive, because the fees from registration services cover costs. Evidence of the cost-effectiveness of such administrative registration is available from the 2001 reform in Italy and the ongoing reform in Serbia and Montenegro. The cost of setting up a system of administrative registration in Serbia and Montenegro is estimated at $1.5 million, with annual operating costs of about $1.1 million. Compare this with projected annual revenues of $1.8 million from registering businesses. So the investment would be paid off fully in less than three years.[31]

Figure 2.8
Courts Are Bottlenecks

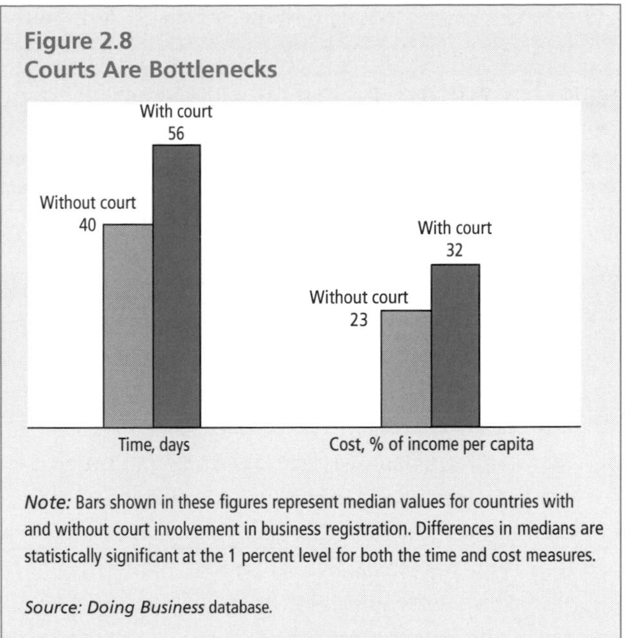

Note: Bars shown in these figures represent median values for countries with and without court involvement in business registration. Differences in medians are statistically significant at the 1 percent level for both the time and cost measures.

Source: Doing Business database.

Legal reforms involve rewriting the company law, but they would significantly speed up business entry and reduce its costs. They would also free commercial courts from the large number of registration and reregistration cases. Instead, commercial judges could focus on their primary role—resolving disputes.

The adoption of some or all of these administrative and legal reforms will generate additional entrepreneurial activity, as in Austria after the adoption of the 1999 Young Enterprise Law. The law eliminated all registration-related costs and removed some procedural burdens. The number of new registrations shot up from 19,000 a year before 1999 to about 26,000 a year after that. In Vietnam, the Enterprise Law of January 2000 spurred the creation of 50,000 new private enterprises, almost 75 percent of the total.

Notes

1. De Soto 1989.
2. The first corporations were Egyptian burial societies. Their independence was founded in primeval rights of

association and respect for the sacred. In Europe the first corporations were monastic orders, led by the order of St. Benedict (480–547). As Benedictine monasteries swept north into Europe from Italy and East from Ireland, selling wine, cheeses, brandies, and breads, they also became the first transnational corporations. The proceeds were used to establish and later incorporate universities and libraries. Soon, medieval towns around Northern Europe started adopting corporate charters. Corporations took off in 16th-century England, chartered by the Crown for the pursuit of mercantilist policies, with a designated public purpose: to establish ferries, canals, water systems, toll roads, bridges, banks, colleges, and colonial enterprises, such as the East India Trading Company of 1601 and the Massachusetts Bay Company of 1628.

3. In particular, the 1832 Reform Act gave voting rights to all men of households with annual revenue of £10.
4. Pistor and Berkowitz 2003.
5. Berle and Means 1932.
6. Adam Smith, *The Wealth of Nations*, 1776, p. 100, referring to companies incorporated by Royal Charter.
7. Diamond 1982.
8. Shannon 1931, Blumberg 1996.
9. Horvath and Woywode 2003.
10. Jacobs 2002. There are three economic arguments for limited liability: it encourages investors to take risks, it facilitates the distribution of risk among corporations and creditors, and it avoids high litigation costs in case of debt recovery.
11. FIAS 2003.
12. These numbers are reported in table 2.1 in Morisset and Neso 2002.
13. Trang and others 2001, Zhuravskaya 2003, and FIAS 2003.
14. The classification of legal origin follows La Porta and others 1999.
15. Djankov and others 2002.
16. Friedman and others 2000; Djankov and others 2002; Batra, Kaufmann, and Stone 2003.
17. Bertrand and Kramarz 2002.
18. Alesina and others 2003.
19. Hoekman, Kee, and Olarreaga 2001 find that a 10 percent increase in the number of procedures is associated with a 5.8 percent increase in prices in countries without liberal trade regimes. Also see Djankov and others 2002.
20. FIAS 2003.
21. De Soto 1989, p. XVII.
22. Olson 1991, DeLong and Shleifer 1993.
23. Djankov and others 2002.
24. European Commission 2002.
25. Trang and others 2001.
26. Sader 2002.
27. Buying an off-the-shelf company is a long-standing practice: it had become common to buy charters from moribund companies during the economic boom that followed the Revolution in 1688 in England.
28. Corporate capital was regarded as a trust fund to protect creditors. In the United States, this view was formulated in *Wood vs. Dummer* in 1824 (Wood vs. Dummer, 3 Mason 308, Fed. Case No. 17,944, 1924).
29. Japanese Association of Small Businesses 1999.
30. Jordan 1996.
31. Jacobs 2002.

3 Hiring and Firing Workers

During the Industrial Revolution in Britain, women received one-third to one-half the wage of men. One scholar of that period writes: "Employers did not offer a living wage to the female since they assumed that she was dependent upon a household headed by a male and therefore did not depend only on her wages for subsistence."[1] About one hundred years later, in 1882, a factory inspector describes employment relations in Russia: "The owner is an absolute sovereign. He is not tied by any law, and often applies and interprets existing legal regulations at his own discretion. The workers must obey him."[2] A third example comes from Zimbabwe, where, before independence in 1980, Africans did not benefit from minimum-wage legislation, were discriminated against in appointments to skilled jobs, and were barred from training programs.[3] Regulation can change that.

Employment law protects workers from arbitrary, unfair, or discriminatory actions by their employers. Regulations—from mandatory minimum wage to premiums for overtime work to grounds for dismissal to severance pay—have been introduced as a response to apparent market failures. The failures range from the exploitation of workers in one-company towns to discrimination on the basis of gender, race, or age to the suffering of the unemployed in the Great Depression and in the transition of formerly socialist economies.

More recent concern for social justice, particularly in developing countries, led the International Labor Organization to establish a set of fundamental principles and rights at work. They include the freedom of association, the right to collective bargaining, the elimination of forced labor, the abolition of child labor, and the elimination of discrimination in hiring and work practices.[4] They constitute the minimum regulation necessary for the effective functioning of labor markets.[5] Every country needs to adopt and enforce them.

However, if regulation in other aspects of the employment relation is too rigid, it lowers labor force participation, increases unemployment, and forces workers into the informal economy. Economic analysis shows that if the average Latin American country were to reduce its employment protection to the level found in the United States, estimated total employment would rise by almost six percentage points.[6] In some countries, the negative effects of rigid employment regulation are even larger. A 10 percent increase in dismissal costs in Peru is associated with an estimated increase in long-term unemployment of 11 percent,[7] and in India and Zimbabwe of about 20 percent.[8]

Disadvantaged groups are hurt the most. Evidence from Argentina, Chile, Colombia, France, the Russian Federation, Spain, and Tunisia shows that new entrants into the labor market—women and youths—suffer disproportionately the consequences of reduced employment opportunities.[9] As a result, many women and teenagers either remain unemployed or find employment in the informal economy. In Côte d'Ivoire, 73.3 percent of informal employees are women; in Uganda, 80.5 percent; in Peru, 57.5 percent.[10] Cross-country analyses suggest that if Mozambique were to reduce its labor regulations to the level found in Zambia, the share of informal employment might drop by as much as 13.5 percentage points, and the share of informal employment of women might drop by 18 percentage points.[11]

Rigid employment regulation is associated with more poverty in developing countries. A study of India suggests that, between 1958 and 1990, poverty in West Bengal, the Indian state with the highest labor protection, increased by 10 percent as employment opportunities were denied to poor people.[12] Almost 2 million urban poor people would have found jobs in West Bengal if the state government had not passed stricter regulation on dismissals and work hours. In contrast, the government of Andra Pradesh, another Indian state, made employment regulation more flexible in the 1980s: 1.8 million urban poor found jobs in manufacturing and service companies in the next decade.

Improving the flexibility of employment law while maintaining fundamental workers' rights requires several reforms. Among them, introducing part-time and fixed-term employment contracts, reducing the minimum wage for young workers, and allowing for shifting the work time between periods of slow demand and peak times have proven successful in several countries. Other possible reforms for countries with greater administrative capacity include providing unemployment benefits to workers in times of low demand (short-time compensation) and using a negative income tax in place of a minimum wage.

What Is Employment Regulation?

Employment regulation is one of four bodies of labor law:

- Employment regulation
- Social security laws
- Industrial relations
- Workplace safety.

This chapter is limited to employment regulation. Next year's report will provide analysis of social security laws, and the report in the year after will study industrial relations and workplace safety regulation.

Employment regulation governs the individual employment contract, including flexibility of hiring through part-time and fixed-term contracts; and conditions of employment, including maximum number of hours in a work week, premiums for overtime work, paid annual leave, and a minimum wage. It also governs flexibility of firing, including grounds for dismissal, notification rules for dismissal, priority rules for dismissal, and severance pay.

Social security laws govern the social response to needs and conditions that have a significant impact on workers' quality of life, such as old age, disability, death, unemployment, and maternity. Social security laws are present in developed countries but still nascent elsewhere.[13]

Industrial-relations laws regulate the bargaining, adoption, and enforcement of collective agreements; the organization of trade unions; and industrial action by workers and employers.

Workplace safety covers the working environment and training of workers for the use of machinery and equipment, as well as the regulation of production processes or materials that are hazardous to workers' health. Workplace safety regulation has beneficial effects for both workers and businesses.[14]

Employment regulation is fairly new, established after World War II in many advanced economies. In the aftermath of the 1973 oil shock, many developed countries tightened employment laws, especially in the area of collective dismissals. Since then, regulation has been continually undergoing reform—every developed country except the United States has made major revisions to its labor regulation since 1990. In May 2003, the German government announced reforms to reduce unemployment. The main proposals would make dismissals easier and reduce the time unemployed people are allowed to receive benefits. The reform will also make fixed-term contracts more attractive to small-business owners.

Flexibility of Hiring

The first area of employment regulation addressed in this chapter is hiring by means of part-time and fixed-term contracts. Part-time contracts have proven popular in recent reforms. Employees who value flexible work schedules—especially younger people continuing their education, women with children, and older people who work to supplement

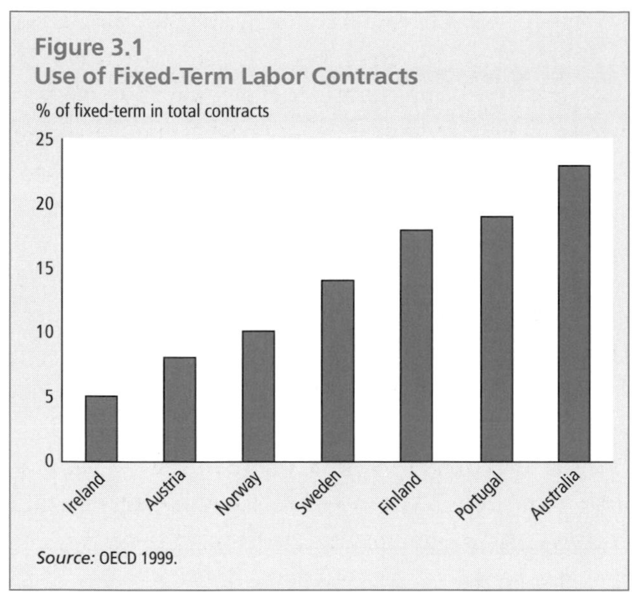

Figure 3.1
Use of Fixed-Term Labor Contracts

Source: OECD 1999.

their income—have been the main beneficiaries. By mid-2003 every country in the *Doing Business* sample allowed part-time contracts. Some countries—including France, Japan, Namibia, and Romania—exempt part-time employment from carrying the mandatory benefits of full-time workers. Part-time contracts are also easier to terminate. These two features make it attractive for businesses to hire part-time workers. In many OECD countries, where data on labor contracts are available, roughly a quarter of the workforce has part-time jobs: the Netherlands (30 percent), Australia (27 percent), Norway (27 percent), Switzerland (26 percent), New Zealand (24 percent), and the United Kingdom (23 percent).[15] In Australia, teenagers make up a large share of the part-time workforce, working under so-called casual contracts.

Fixed-term contracts ease the entry for new employees. They were established in France in 1979, to be used for the replacement of employees on leave, temporary increases in activity, and seasonal activities, as well as in contracts for disadvantaged groups such as youths and women.[16] By 1999, 10 percent of the workforce in France had such contracts. Spain adopted revisions to its labor code in the mid-1980s to allow for part-time and fixed-term contracts; by 1999, almost 30 percent of workers, primarily first-time entrants in the labor market, had such contracts. The 1996 revision of the labor code in Tunisia introduced fixed-term contracts, and by 2001 about 15 percent of the labor force had them.[17] Other countries with high rates of fixed-term contracts are Australia, Finland, Portugal, and Sweden (figure 3.1).

Many countries allow fixed-term contracts only for specific tasks. This is true for most Latin American countries—such as Argentina, Bolivia, Brazil, Guatemala, Mexico, Panama, Uruguay, and Venezuela—and for southern Europe—France, Greece, Italy, Portugal, and Spain. Chile, Japan, Mexico, and Sweden limit the duration of fixed-term contracts to one year, but many former socialist countries allow fixed-term contracts of up to five years. Poland does not regulate the duration of fixed-term contracts. Nor do Australia, New Zealand, South Africa, the United Kingdom, or Zambia.

Conditions of Employment

The legal provisions for conditions of employment cover flexibility in working time requirements, mandatory payment for non–working days (including paid annual leave and holidays), and minimum-wage legislation. In countries with a common-law tradition, substantial aspects of the employment relation are left to the individual agreement between the worker and manager.

Kenya, Oman, Singapore, Slovenia, Thailand, and the United States, among others, impose no regulation regarding daily rest. However, it is legislated at a minimum of 14 hours in Chile, Colombia, Ethiopia, Panama, and Syria. With the exception of New Zealand, all countries regulate the number of work hours. Botswana, Chile, Costa Rica, Ireland, Malaysia, Morocco, and Vietnam all allow a 48-hour workweek. France has the shortest workweek, at 35 hours, followed by Denmark, with 37 hours. Night work is generally allowed in most countries, except Albania, Belarus, Mozambique, Norway, Turkey, and Uruguay. Work on holidays is not subject to any regulation in Denmark, Hong Kong (China), Latvia, Malaysia, Singapore, and Tunisia. However, holiday work is strictly regulated in countries with a German legal tradition, including Austria, Germany, and Switzerland.

Figure 3.2
Premiums for Overtime Work—from Nothing to Double

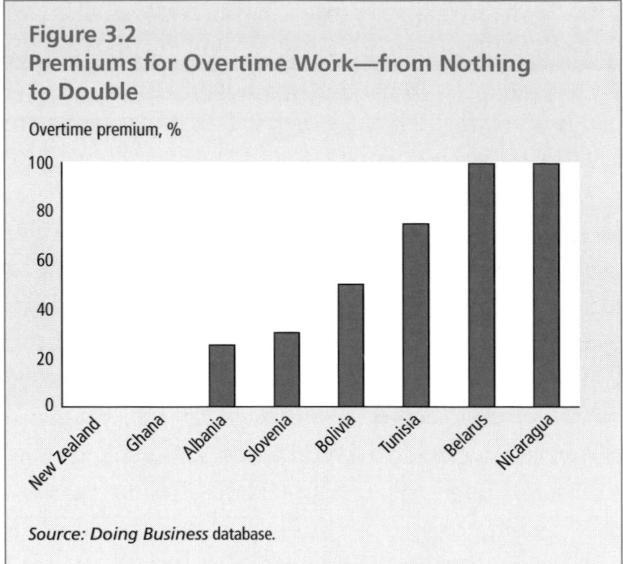

Source: Doing Business database.

Figure 3.3
How High Is the Minimum Wage

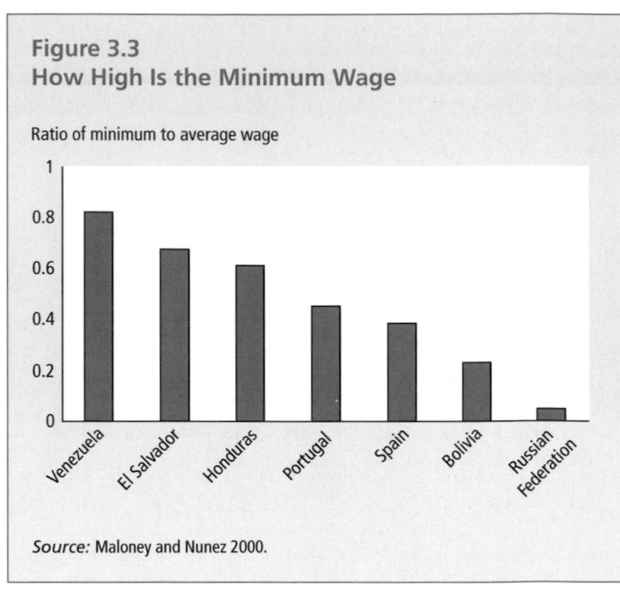

Source: Maloney and Nunez 2000.

In cyclical or seasonal industries, overtime work is often used. Burkina Faso, Cameroon, Jamaica, Hong Kong (China), New Zealand, Spain, and the United Kingdom do not have regulations on a premium for overtime work. Chad, Italy, and Mali require a 10 percent premium over wages paid for work in normal hours. Bangladesh, Belarus, India, Mexico, Nicaragua, Pakistan, Uruguay, and Uzbekistan mandate up to double pay for overtime work (figure 3.2). Some Central European countries recently revised employment regulation to allow managers to shift work time from periods of slow demand to peak periods. In Poland, such shifts must balance out within six months; in Hungary, within a year. Such reforms eliminate the uncertainty of spending longer hours at work for employees, while reducing the costs of unpredictable or cyclical demand—and overtime pay—for businesses.

The United States leaves it to individual or collective worker contracts to agree on the number of days of paid annual leave. In all other countries, the duration of annual leave is subject to regulation. The most generous annual leave is mandated in Sierra Leone (39 days), followed by Congo Republic (35 days); Ethiopia (33 days); Chad, Côte d'Ivoire, and Niger (32 days); and Burkina Faso, Egypt, Finland, Nicaragua, and Yemen (30 days).

Several OECD countries—Austria, Denmark, Finland, Italy, Norway, Sweden, and Switzerland—have no mandatory minimum wage. In some countries, like Austria, this is the result of a long social partnership among labor unions, business associations, and the government. Regulation would be redundant. In countries that regulate minimum wages, the ratio of minimum to average wages varies from 0.82 in Venezuela to 0.05 in the Russian Federation (figure 3.3). If this ratio is high, businesses are unwilling to hire less experienced workers, discriminating against youths or mothers returning after maternity leave, who have been out of the workforce for some time. One promising reform is to enforce a lower minimum wage for younger workers, as established in Chile in 1989, generating a significant increase in job opportunities for recent graduates. The introduction of similar apprentice wage laws was a common reform in other Latin American countries in the 1990s.[18]

Flexibility of Firing

Flexibility of firing encompasses grounds for dismissal, procedures for dismissal, notice periods, and severance payments. The rules on grounds for dismissal vary from "contract at will"—as in Ghana, Israel, and the United Kingdom, where the employment relation may be terminated by either party at any time—to allowing termination of contracts under a narrow list of "fair" causes such as redundancy—as in France—to not considering redundancy as a fair cause for dismissal—as in Bolivia, the Republic of Korea (before 1998), and Portugal. The 1998 reforms of employment regulation in the Republic of Korea allowed for dismissal on the grounds of redundancy or economic restructuring.

The procedures for dismissal often require notification or even approval by unions, workers councils, the public employment service, a labor inspector, or a judge. Some countries also mandate retraining and reassignment to another job in the enterprise—and establish priority rules for dismissal or re-employment of redundant workers. In Tunisia, companies must notify the labor inspector of planned dismissals in writing one month ahead, indicating the reasons and the workers affected. The inspector may propose alternatives to layoffs. If these proposals are not accepted by the employer, the case goes to the regional tripartite committee comprised of the labor inspector, the employer organization, and the labor union. The committee decides by a majority vote (if the inspector and union reject the proposal, no dismissal is possible). It may also suggest retraining, reduced hours, or early retirement. Only 14 percent of dismissals end up being accepted. As a result, annual layoffs are less than 1 percent of the workforce, compared with more than 10 percent in the average OECD country.

Even if employers are permitted to dismiss workers, regulations may impose notice periods and severance payments (figure 3.4). In Croatia, employers need to give workers three months' advance notice and pay six monthly salaries in severance for employees with tenure in the business of more than 20 years.

Figure 3.4
Severance Payments—from Nothing to 20 Months' Pay

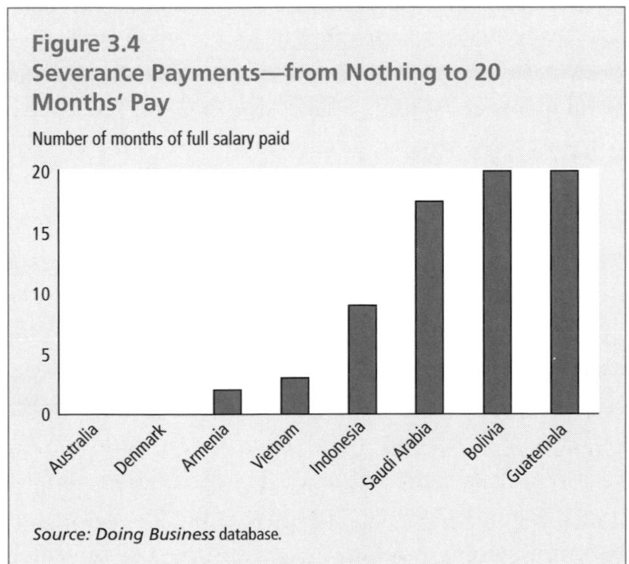

Source: Doing Business database.

Often workers feel unjustly dismissed or may not receive sufficient compensation. One recourse is to file a lawsuit against their employers. In many countries, such disputes are handled by specialized labor courts or tribunals. Until recently, representatives of employee and employer organizations frequently sat on the jury, alongside professional judges. This practice has often led to protracted judicial procedures and difficulties in reaching compromise. In Brazil, prior to 1999, an equal number of representatives from labor unions and business associations served as judges on the Labor Conciliation Board, the regional labor courts, and the supreme labor court. The average labor dispute took almost three years.[19] In 1999 this practice was abolished and only professional lawyers could become labor judges. Also, judges were given lifelong tenure, which reduced their susceptibility to political pressure. By 2001, the time needed to resolve disputes was halved.

Large Divergences in Practice

Employment regulation is an area of great divergence among developed countries. A comparison of New Zealand and Portugal, two OECD members with similar income per capita, illustrates the differences in regulatory scope.

- Fixed-term contracts may be entered into in New Zealand for any reason, and no maximum duration is prescribed by law. In Portugal, such contracts are allowed only for specific tasks, such as substitution for another worker or seasonal activity, and are temporary.
- Working times and leave times in New Zealand are regulated by collective bargaining and individual employment contracts. In Portugal, the constitution regulates work and leave times, remuneration, and working conditions.
- New Zealand mandates no premium for overtime work. There are no restrictions on night work, and paid annual leave is 15 days. In Portugal, the

premium for overtime work ranges from 50 percent to 75 percent. There are restrictions on night work, and paid annual leave is 24 days.

- New Zealand allows "contracts at will," which can be terminated with notice by either party. Portugal has a public policy list of fair grounds for dismissal, mandatory early notice, and priority rules for re-employment of redundant workers.
- In New Zealand "reasonable notice," usually one week, is required to dismiss a worker, and there is no regulation on the amount of severance pay. In Portugal, the standard dismissal notice is 60 days, and the severance pay for workers with 20 years of tenure is mandated by the law to be 20 months of wages.

The transplantation of employment laws during colonial days created stark differences between developing countries as well.

- In Ghana, a common-law jurisdiction, fixed-term contracts are allowed for any reason, and there is no maximum duration. In Mozambique, a former Portuguese colony with a similar income per capita, fixed-term contracts are allowed only for seasonal activities.
- In Ghana, leave and remuneration are negotiated in individual employment contracts. In Mozambique, the constitution regulates them, with minimum annual leave of 22 days.
- Ghana allows contracts at will. Mozambique's labor code lists fair grounds for dismissal and imposes stringent procedural limitations, such as mandatory notification of the government and priority rules for re-employment of redundant workers.
- Labor regulation in Ghana imposes no severance payment for dismissal. In Mozambique, the length of notice is regulated at 12 weeks, and the severance payment for a worker with 20 years of experience is 30 months of wages.

To document the systematic differences in employment regulation across countries, three indices were constructed by studying the letter of the law and conducting surveys of labor lawyers in each country (table 3.1). The methodology is simple: if the regulation restricts the ability of managers and workers to negotiate the employment contract, a value of 100 is entered, zero otherwise. For example, fixed-term contracts are allowed in Venezuela only for temporary tasks, while in Vietnam they are allowed for any task. On this component of the hiring index, Venezuela gets a 100, Vietnam a 0. Similarly, managers have to give fair cause for dismissal in Cameroon, but not in Jamaica. On this component of the flexibility-of-firing index, Cameroon gets a 100, Jamaica a 0. The scores are averaged across the components of each index to get the value of the index itself.[20] Table 3.1 details the components of each index (flexibility of hiring, conditions of employment, and flexibility of firing). Averaging across these three indices yields the index of employment regulation, where higher values represent more rigidity in employment regulation.

Employment regulation is more flexible in developed countries. Austria, Denmark, Hong Kong (China), New Zealand, Singapore, the United Kingdom, the United States, and New Zealand are among the 10 countries with the most flexible employment protection (table 3.2). Malaysia, Papua New Guinea, and Zimbabwe are also in this group. The countries with the most rigid employment regulation include six Latin American countries (Brazil, Mexico, Panama, Paraguay, Peru, and Venezuela), and Angola, Belarus, Mozambique, and Portugal.

Some countries have very strong protection in one of the indices but not in others. Greece, Taiwan (China), and Tunisia restrict the use of fixed-term contracts. Hungary and Poland are among the countries with the most regulation on conditions of employment. Belarus, Mexico, and Peru have strict regulations on firing. In general, however, the indicators of labor regulation tend to move together: restrictions on hiring go with restrictions on firing (figure 3.5), as well as with more rigid conditions of employment.

Which groups of countries have the most flexible regulation? Rich countries have the lowest average scores on all indices (figure 3.5). Nordic-origin countries regulate employment relations the least in conditions of employment but less so in dismissals, in which

Table 3.1
Rigidities in Employment Regulation

Hiring	Part-time contracts	Is part-time employment prohibited? Are part-time workers exempt from mandatory benefits of full-time workers? Is it easier or less costly to terminate part-time workers than full-time workers?
	Fixed-term contracts	Are fixed-term contracts allowed only for fixed-term tasks? What is the maximum duration of fixed-term contracts (in months)?
Conditions of employment	Hours of work	What is the mandatory minimum daily rest? What is the maximum number of hours in a workweek? What is the premium for overtime work? Are there restrictions on night work? Are there restrictions on weekly holiday work?
	Leaves	What is the number of legally mandated days of annual leave with pay in manufacturing? Is paid time off for holidays mandatory?
	Minimum wage	Is there a mandatory minimum wage? Are conditions of employment specified in the constitution?
Firing	Grounds for firing	Is it unfair to terminate the employment contract without cause? Does the law establish a public policy list of "fair" grounds for dismissal? Is redundancy considered a fair ground for dismissal?
	Firing procedures	Must the employer notify a third party before dismissing one redundant employee? Does the employer need the approval of a third party to dismiss one redundant worker? Must the employer notify a third party before a collective dismissal? Does the employer need the approval of a third party before a collective dismissal? Does the law mandate retraining or replacement prior to dismissal? Are there priority rules applying to dismissal or layoffs? Are there priority rules applying to re-employment?
	Notice and severance payment	What is the legally mandated notice period (in weeks) after 20 years? What is the severance pay as a number of months for which full wages are payable after covered employment of 20 years?
	Job security	Is the right to job security specified in the constitution?

Source: Doing Business database.

English-origin countries have the lightest regulation. Across regions, East Asian economies regulate the least and Latin American countries the most, even after the significant labor reforms in the 1990s. This result is consistent with previous studies. For example, in 1999 the cost of firing a full-time worker was equivalent to 93 days of wages in Latin America, twice the 45 days in the OECD.[21] Countries in the socialist legal tradition have the second-strictest labor regulation.

What Are the Effects of Employment Regulation?

The fact that employment regulation arose in response to market failures does not mean that today's regulations are optimal. Their design may have been poor to begin with. And what was appropriate in, say, 1933, when Portugal adopted its constitutional protections of workers, may not be appropriate today, because circumstances, technology, and business organization have changed.

Indeed, although employment regulation generally increases the tenure and wages of incumbent workers, strict regulatory intervention has many undesirable side effects. The first is to limit job creation. Quarterly job creation in Portugal, the most heavily regulated labor market in the sample, is 59 percent of that in the United States, one of the ten least regulated labor markets. With fewer new jobs available, Portuguese workers stay in jobs they do not like.[22] Conversely, the relaxation of labor regulation in the United States

Table 3.2
Indexes on Employment Regulation

Flexibility of hiring		Conditions of employment		Flexibility of firing		Employment laws	
Most-flexible regulation							
China	17	Hong Kong (China)	22	Hong Kong (China)	1	Singapore	20
Czech Republic	17	Zimbabwe	22	Singapore	1	United States	22
Namibia	17	Denmark	25	Uruguay	3	Malaysia	25
Nigeria	17	Malaysia	26	Papua New Guinea	4	Denmark	25
Papua New Guinea	17	Singapore	26	United States	5	Papua New Guinea	26
Australia	33	United States	29	Japan	9	Hong Kong (China)	27
Canada	33	South Africa	36	United Kingdom	9	Zimbabwe	27
Denmark	33	Sweden	39	Australia	13	United Kingdom	28
Poland	33	Norway	39	Austria	14	Austria	30
Uganda	33	Kuwait	40	Malaysia	15	New Zealand	32
Least-flexible regulation							
Brazil	78	Nicaragua	90	Brazil	68	Paraguay	73
Chad	78	Mongolia	90	Panama	68	Peru	73
Greece	78	Paraguay	90	Peru	69	Mozambique	74
Guinea	78	Turkey	91	Ukraine	69	Venezuela, RB	75
Thailand	78	Poland	92	Mexico	70	Belarus	77
Venezuela, RB	78	Hungary	92	Belarus	71	Mexico	77
El Salvador	81	Ukraine	93	Russian Federation	71	Angola	78
Mexico	81	Chad	93	Paraguay	71	Brazil	78
Panama	81	Rwanda	94	Portugal	73	Portugal	79
Taiwan (China)	81	Bolivia	95	Angola	74	Panama	79

Note: Indexes range from 0 to 100, with higher values indicating more-rigid regulation. The employment-laws index is the average of the flexibility-of-hiring, conditions-of-employment, and flexibility-of-firing indexes.

Source: Doing Business database.

since the 1950s has helped increase new employment opportunities by as much as 150 percent.[23] When a Portuguese business decreases employment, it is 40 percent less likely to increase it when the economy picks up than a U.S. company is. This result is corroborated by other studies that show jobless recoveries in economies with heavily regulated labor markets.[24] It means that some workers remain in perennial unemployment.

A second effect is to reduce the flexibility of the workforce: workers who have endured long unemployment spells tend to have obsolete skills. Unemployment duration is three times higher in Portugal than it is in the United States, and more than twice as high in Brazil and Spain, two other heavily regulated markets.

Third, flexible labor regulation is associated with higher R&D investment in technologies. In particular, businesses in low-employment-protection countries in the OECD have almost 30 percent higher investment in R&D than businesses in OECD economies with rigid employment laws.[25] Why? Because organized labor frequently resists attempts to acquire new technology, particularly if it is perceived to displace workers. In addition, stringent regulations on firing may push managers into reorganizing the production process in ways that provide employment for displaced workers, which in turn reduces incentives to buy the latest technology.

Fourth, restrictions on hiring and firing have been shown to result in smaller firm size, and to leave economies of scale unexploited in manufacturing and some services (the evidence is primarily from OECD economies).[26]

All of these effects—less job creation, longer unemployment spells and the related skill obsolescence of

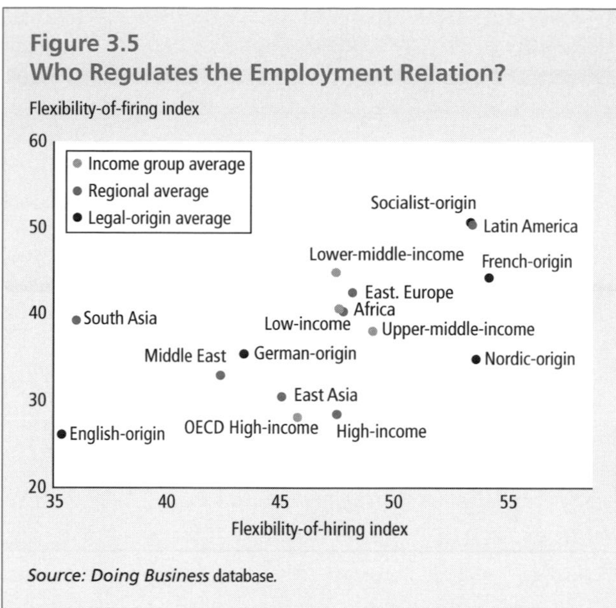

Figure 3.5
Who Regulates the Employment Relation?

Source: Doing Business database.

workers, less R&D investment, and smaller company size—may serve to reduce productivity growth.[27]

Surveys of managers also show that employment regulation is a burden on businesses in many developing countries. One survey asks managers to rank eight areas with regard to the burdens regulations impose on the operation and growth of their businesses: business licensing, customs and foreign trade restrictions, foreign currency and exchange regulations, employment regulations, environmental regulations, fire and safety regulations, tax regulations and their administration, and high effective tax rates.[28] Employment regulations were seen to be the major obstacle to improving productivity in Bangladesh, Brazil, Chile, Colombia, India, Panama, Portugal, Thailand, Tunisia, and Venezuela. And they were rated the second-most-important obstacle to productivity growth in Argentina, Bolivia, Ecuador, Mexico, and Uruguay.[29]

More worrying, employment regulation limits the opportunities of disadvantaged groups to come out of poverty. Excessive regulation is associated with higher unemployment,[30] especially for youths and women (figure 3.6). Cross-country analyses suggest that if France were to make its labor regulations as flexible as those in the United States, the employment rate might increase by up to 1.6 percentage points. The effect is even larger for Spain, at 2.3 percentage points. Women would benefit the most, with more than 70 percent of the new jobs. Using the employment regulation index in this chapter, a reduction in the value by a third would be associated with a 10-percentage-point fall in the unemployment rate of young women.

Without job opportunities in the formal economy, many people join the unofficial sector (figure 3.7). There, workers have no social protection whatsoever. Cross-country studies show that a reduction of the employment regulation index by a third is associated with a 14-percentage-point decline in informal employment and a 6.7-percentage-point fall in output produced in the informal economy.

What to Reform?

Reform is taking place, but it is often hotly contested by labor unions and frequently falls short or is reversed.[31] In the early 1980s, Spain introduced more-flexible legislation on fixed-term contracts, only to roll it back in the latter part of the decade. In 1996, the Peruvian government tried to reduce severance payments by 50 percent. The ensuring

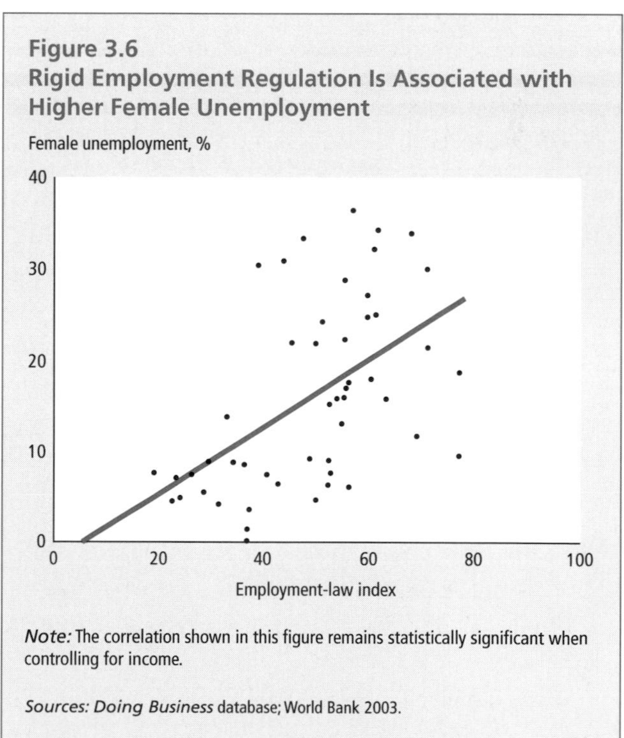

Figure 3.6
Rigid Employment Regulation Is Associated with Higher Female Unemployment

Note: The correlation shown in this figure remains statistically significant when controlling for income.

Sources: Doing Business database; World Bank 2003.

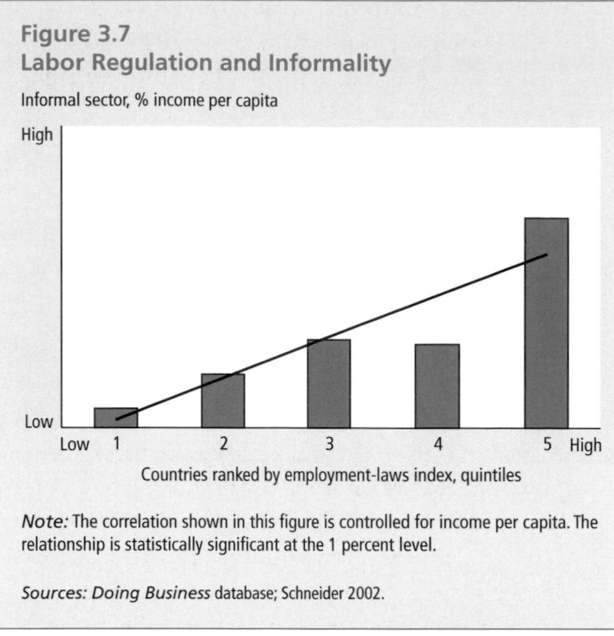

Figure 3.7
Labor Regulation and Informality

Note: The correlation shown in this figure is controlled for income per capita. The relationship is statistically significant at the 1 percent level.

Sources: Doing Business database; Schneider 2002.

uproar caused the government to instead increase severance payments. In 1998 Argentina revoked temporary employment contracts—which had been introduced in 1995 as the main component of the labor reform.

A general reform oriented toward less regulation of labor markets has yielded positive results in Latin America and in some transition economies. Six Latin American countries have reformed their employment legislation in the last decade: Argentina, Brazil, Colombia, Guatemala, Panama, and Peru. With the exception of Panama, all of those countries introduced temporary contracts. The contracts have lower dismissal costs, and employers usually pay lower payroll taxes. Among transition economies, the Czech Republic, Estonia, Hungary, Poland, the Russian Federation, and the Slovak Republic made the hiring of workers more flexible and reduced regulation on conditions of employment. Estonia's reform is the most far-reaching: it reduced regulation on both the hiring and the firing of workers. There has been little reform of employment regulation in Africa and South Asia.

Several other types of reforms of employment regulation have eased the burden on businesses and provided better job opportunities for poor people:

- Many OECD countries have introduced flexible part-time and fixed-term contracts. These contracts bring groups less likely to find jobs, such as women and youths, into the labor market. Germany increased the duration of fixed-term contracts to eight years, whereas Poland has eliminated the time limit.
- Several countries have either introduced apprentice wages (Colombia) or lowered the minimum wage for new entrants (Chile).
- Some countries (Hungary, Poland) have made it possible for employers to shift work time between periods of slow demand and peak periods without the need for overtime payment.
- Other countries have eased regulations on firing. Serbia and Montenegro reduced the severance payment for a worker with 20 years of tenure from 36 months to 4.

As countries adopt better technology for monitoring the labor market and build up their administrative capacity, they will be able to undertake more-sophisticated reforms. Several OECD countries have implemented legislation on short-term compensation, which provides employees with job security at times of low demand. If an employer cannot gainfully occupy a worker during slack times, a government fund covers the payment for such periods. Belgium, Italy, and Sweden have good experience managing such funds.[32] This innovation reduces employee turnover and shifts the burden away from the business. Further reform may include a negative income tax in place of a mandatory minimum wage. Such a tax would give people an incentive to join the workforce in entry-level jobs while alleviating the burden on unemployment insurance.[33]

Notes

1. Valenze 1985, p. 36.
2. Turin 1934, p. 34.
3. Fallon and Lucas 1991.
4. ILO 1998, 2000, 2001, 2002, 2003.
5. Becker 1971. Economic studies show that the presence of such fundamental rights improves productivity.
6. Heckman and Pages 2000.
7. Saavedra and Torero 2000.

8. Fallon and Lucas 1991. The negative effects of rigid labor regulations are not limited to developing countries. The introduction of high severance payments in France after World War II is estimated to have increased long-term unemployment by 4.4 percentage points (Lazear 1990).
9. Hopenhayn 2001 for Argentina; Montenegro and Pages 2003 for Chile; Kugler 2000 for Colombia; Dolado and others 1997 for Spain; World Bank 2002a for the Russian Federation; Abowd and others 1999 for France.
10. Betcherman 2002.
11. Botero and others 2003.
12. Besley and Burgess 2003.
13. Mulligan and Sala-i-Martin 2000; Botero and others 2003.
14. Fishback 1998.
15. OECD 1999. All figures are for 1999.
16. Blanchard and Landier 2000.
17. World Bank 2003.
18. Gill, Montenegro, and Domeland 2002.
19. World Bank 2002b.
20. The rankings generated by this index are consistent with the rankings on regulation on dismissals in Latin American countries (Heckman and Pages 2000) and the rankings on employment protection in the OECD (OECD 1999). The correlations are highly significant, at 0.63 and 0.73, respectively. These provide support for the robustness of the methodology.
21. Heckman and Pages 2000.
22. Blanchard and Portugal 2001.
23. Holmes 1998.
24. Betcherman and others 2001.
25. Nicoletti and others 2001.
26. Nicoletti and Scarpetta 2003.
27. Scarpetta and Verdier 2002; Montenegro and Pages 2003.
28. Batra and others 2003.
29. World Bank 2003. Two other surveys, by the chamber of commerce in 2000 and 2001, also show that 52 percent of managers of private manufacturing enterprises perceived labor regulations to be excessive, especially in the textiles, chemicals, and construction industries.
30. Scarpetta 1996 for the OECD; Fields and Wan 1989 for South Asia; Aidt and Tzannatos 2003 for a recent survey of country studies.
31. Gill, Montenegro, and Domeland 2002.
32. Van Audenrode 1994.
33. Blanchard 2002.

4 Enforcing Contracts

Imagine that a new client comes to a textile company and orders shirts. The client and the company manager sign a contract for payment on delivery. But at delivery, the client refuses to pay in full. What happens next? In New Zealand, the company manager will show the client the contract and ask for payment. The client is likely to pay. In Poland, the company manager will show the contract to the client and ask for payment. The client is likely to refuse to pay. In Côte d'Ivoire, the company manager would probably not deal with the new client unless the client could provide references from other textile companies or from companies that operated in the same region.[1] In Vietnam, the client might not bother going to the company without having at least half of the money available for an advance payment.

Why the differences? The answer lies in the efficiency of courts—the main institution enforcing contracts. New Zealand has a very efficient court system. Polish courts take a long time to resolve disputes. Courts in Vietnam and Côte d'Ivoire are considered inefficient. In the words of a Vietnamese enterprise manager interviewed in 1999: "The court is weak, and no entrepreneurs use it."[2] Weaknesses in the legal system span countries and centuries. Going back 400 years, Shakespeare's *Hamlet* lists court delays among the calamities of life: "The oppressor's wrong, the proud man's contumely, the pangs of despised love, the law's delay."[3] In the absence of efficient courts, fewer transactions take place, and those transactions involve only a small group of people linked through kinship, ethnic origin, and previous dealings.

Courts have four important functions. They encourage new business relationships, because partners do not fear being cheated. They generate confidence in more complex business transactions by clarifying threat points in the contract and enforcing such threats in the event of default. They enable more sophisticated goods and services to be rendered by encouraging asset-specific investments in their production. And they serve a social objective by limiting injustice and securing social peace. Without courts, commercial disputes often end up in feuds, to the detriment of everyone involved.

Companies that have little or no access to courts must rely on other mechanisms, both formal and informal—such as trade associations, social networks, credit bureaus, and private information channels—to decide with whom to do business. Companies may also adopt conservative business practices and deal only with repeat customers. Transactions are then structured to forestall disputes. Whatever alternative is chosen, economic and social value may be lost.[4]

Four types of reform of contract enforcement have proven successful:

- Establishing information systems on caseload and judicial statistics has delivered a large payoff. Judiciaries with such systems (for example, in the Slovak Republic) can identify their primary users and the biggest bottlenecks.
- Taking out of the courts transactions that are not disputes—such as the registering of new business entities—can free up resources for commercial litigation. Because such reform may require new laws, governments can in the meantime reorganize the workflow in the courts so that

clerks, not judges, are responsible for company registration.

- Simplifying the procedures is often warranted for commercial disputes, especially in developing countries. For example, summary debt collection proceedings of the type recently established in Mexico alleviate court congestion by reducing procedural formalism. When default judgments—automatic if the defendant does not appear in court—are introduced as well, delays are cut significantly.
- Modifying the structure of the judiciary may allow for small-claims courts and specialized commercial courts. Several countries with small-claims courts—such as Japan, the United Kingdom, and New Zealand—recently increased the maximum claim eligible for hearing at the court. In other countries, such as Botswana and India, local courts deal with small cases and pass disputes concerning larger amounts to the higher courts. Where the judiciary is least developed, as in Angola, Mozambique, and Nepal, specialized courts are premature. Instead, reformers are allowed to introduce summary proceedings within general-jurisdiction courts or have specialized judges in the general court, with a focus on the execution of judgments.

Ease of Contract Enforcement

Using a hypothetical business transaction, lawyers in 133 countries were asked to describe how a company would go through the courts to recover its overdue payment. The survey covers the procedure-by-procedure evolution of a commercial case before courts in the country's most populous city. Respondents were given the amount of the claim (half of income per capita), the location and main characteristics of the litigants, the presence of city regulations, the nature of the remedy requested by the plaintiff, the merit of the plaintiff's and defendant's claims, and the social implications of the judicial outcome.[5] These standardized details enabled the respondent law firms to describe the procedures explicitly—and to determine the duration and cost of each procedure.

On the basis of their responses, three indicators of the efficiency of contract enforcement were constructed:

- The number of procedures, mandated by law or court rules, that demand interaction between the parties to the dispute or between them and the judge or court officer.
- The cost, as a share of income per capita, incurred during dispute resolution—comprising court fees, attorney fees, and payments to other professionals.
- The estimated time to resolve a dispute, measured as the number of days from the moment the plaintiff files the lawsuit in court until the moment of settlement or actual payment. Separate estimates are made for the average time until the completion of process, trial, and enforcement. Comparisons with studies of actual court practices in several Latin American countries show remarkable consistency in the length of time between filing and settlement (see figure 1.3).[6]

Three examples illustrate the striking differences in the efficiency of contract enforcement across countries. In Slovenia, the creditor must complete 22 procedures and spend 1,003 days to get paid (figure 4.1). It will cost more than $360, or 7.2 percent of the claim amount (3.6 percent of income per capita), in attorney and court fees. In Tunisia, it takes only 14 procedures and 7 days to take a debt recovery case from filing to enforcement of judgment. There are no requirements to appoint a lawyer or initiate a protest procedure before a public notary. The creditor files a claim in court, and the court issues a summons to the debtor. The cost is 8 percent of the claim (4 percent of income per capita). In Guatemala, it takes 19 procedures and 1,460 days to enforce the contract, with 40 percent of the claim amount going to attorney and court fees.

What procedures are common in resolving commercial disputes? In 61 percent of the *Doing Business* sample, the case is handled by a general-jurisdiction court (table 4.1). In some other countries in the sample—including Canada, Denmark, France, Italy, Japan, the Netherlands, Norway, and Singapore—it is handled by a specialized court. Almost all countries use professional judges, but in a few countries, including

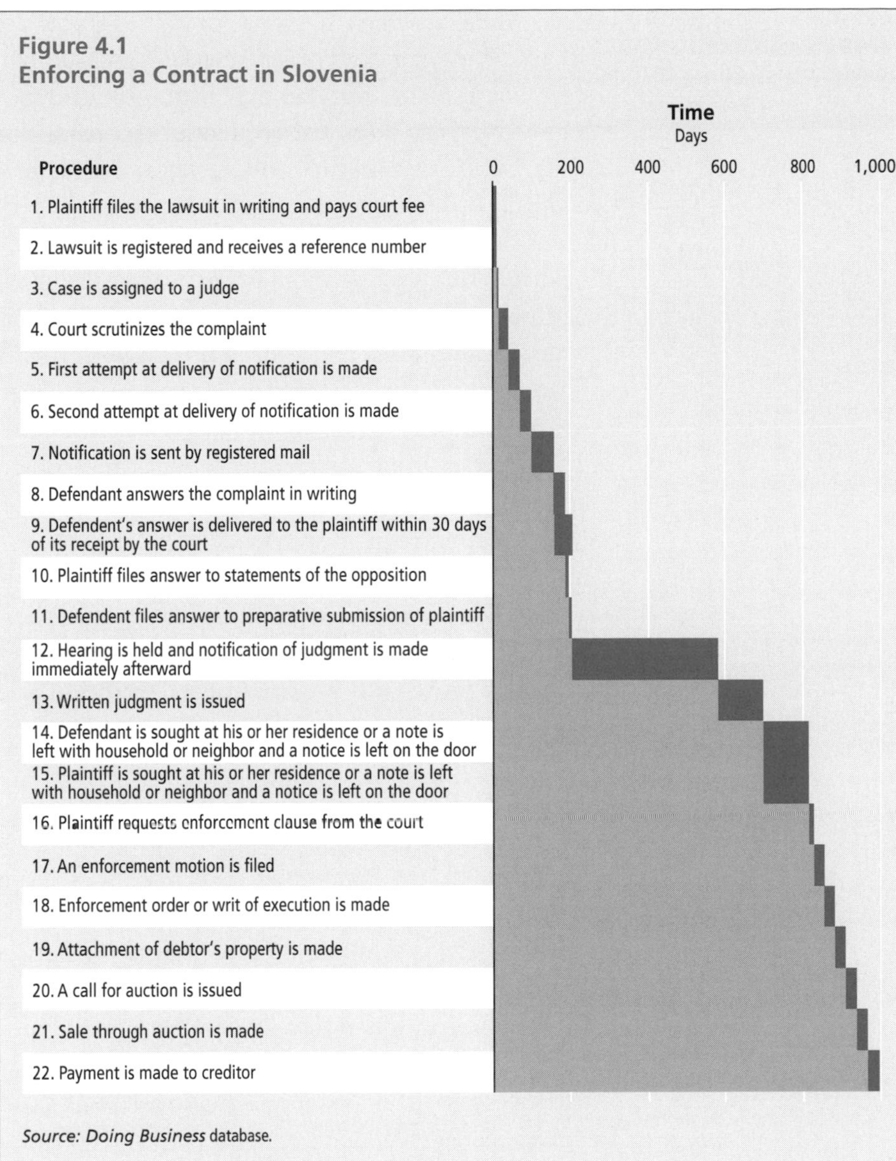

Figure 4.1
Enforcing a Contract in Slovenia

Source: Doing Business database.

for resolving commercial disputes, as in Brazil, Zambia, and Zimbabwe. In Uganda it takes only 16 procedures and about three months to resolve a dispute. Several other countries—Belgium, Hong Kong (China), New Zealand, and the United States—have small-claims courts, but the maximum claim amount is smaller than the one specified in the hypothetical experiment here. For example, in New York the jurisdictional limit for small-claims courts is $3,000. In New Zealand, the limit of the disputes tribunal is NZ$7,500, or $3,125. If the amount is less than NZ$12,000 and both parties agree, the disputes tribunal still determines the claim. Similar arrangements are available in the small-claims courts of most Australian states.

Countries also differ in the use of written arguments in court proceedings. Almost every one has written requirements for the filing, process, judgment, and enforcement-of-judgment stages. But only about a third require that all the evidence be written and that final arguments be submitted in written form. Only half require written notification of judgment. Most Latin American countries (such as Colombia, Ecuador, Honduras, and Venezuela) and some Middle Eastern countries (such as Morocco) require written documentation for every procedure.

In just over half the sample, the complaint must be justified by citing relevant parts of the law instead of presenting the complaint on equity grounds and letting a judge determine its admissibility in court.

Australia, Turkey, and Zambia, lay judges preside over the case. Few countries make the use of lawyers for legal representation mandatory, but many plaintiffs hire lawyers anyhow. Mandatory representation is the norm in Argentina, Bolivia, Italy, Morocco, Spain, and some other Latin American and Middle Eastern countries. More generally, mandatory legal representation is a feature of French-civil-law countries. Rich countries tend to have specialized courts, relying less on professional judges and legal representation.

Small-claims courts, which tend to follow simpler procedures than general courts, are sometimes used

Table 4.1
Frequency of Procedures in Contract Enforcement

Procedure	Frequency
Use of professionals	
General-jurisdiction court	61%
Professional judge tries the case	94%
Use of an attorney is mandatory	20%
Written arguments are required for—	
Filing	92%
Service of process	94%
Opposition	72%
Evidence	40%
Final arguments	30%
Judgment	86%
Notification of judgment	55%
Enforcement of judgment	98%
Legal justification	
Complaint must be legally justified	55%
Judgment must be legally justified	87%
Judgment must be on law (not on equity)	63%
Statutory regulation of evidence	
Judge cannot introduce evidence	39%
Judge cannot reject irrelevant evidence	11%
Out-of-court statements are inadmissible	69%
Mandatory prequalification of questions	23%
Oral interrogation only by judge	15%
Only original documents and certified copies are admissible	50%
Authenticity and weight of evidence defined by law	30%
Mandatory recording of evidence	69%
Control of superior review	
Enforcement is suspended until resolution of appeal	56%
Comprehensive review in appeal	85%
Interlocutory appeals are allowed	80%
Other statutory interventions	
Mandatory pretrial conciliation	13%
Service of process by judicial officer required	51%
Notification of judgment by judicial officer required	28%

Source: Doing Business database.

This presents another hurdle for businesses, because it forces them to seek legal advice. Only a fifth of common-law countries require legal justification of the complaint, four-fifths of civil-law countries, including Austria, France, Germany, and Spain. Denmark is the only civil-law country that does not require reference to a specific law at any stage of the proceedings. Canada, Ghana, Malaysia, New Zealand, and Singapore are other countries that do not require legal justification.

The requirements for who may introduce evidence and how they may do so are frequently responsible for causing delays in contract enforcement. More than two-thirds of the sample countries have statutory regulations on out-of-court statements and the recording of evidence, but fewer than a fifth have regulations on the admissibility of irrelevant evidence, the prequalification of questions, and oral interrogations. Those are primarily Latin American countries (such as Guatemala and Honduras). But Portugal, Mozambique, and the Nordic countries (Denmark, Finland, Norway, and Sweden) have few statutory regulations of evidence. Italy imposes no regulations whatsoever.

In 56 percent of the countries, enforcement is suspended if an appeal is filed, and the suspension lasts until the appeal is resolved. Nearly all countries, particularly those with a German legal tradition and Middle Eastern countries other than Egypt and Jordan, allow for comprehensive review in appeal and for appeals during trial. Pretrial mediation is mandatory in Albania, Bolivia, Cameroon, Madagascar, Malawi, Mali, Nicaragua, the Philippines, and Uzbekistan. Latin American countries and former French and Portuguese colonies in Africa have the most stringent mediation requirements.

Richer countries tend to have fewer procedures to resolve disputes—especially common-law countries (Australia and the United Kingdom), but also Denmark, Norway, and Switzerland (table 4.2). Several poorer countries, like Jamaica, Tanzania, Tunisia, and Zimbabwe, also have few procedures. African countries impose the greatest number of procedures, with Angola, Burundi, Cameroon, Chad, the Democratic Republic of Congo, and Sierra Leone among the 10 countries with the largest number. Those are joined by three Latin American countries—Mexico, Paraguay, and Puerto Rico—and the Kyrgyz Republic and Oman.

Complex legal procedures frequently cause long delays. One example is described in the autobiography of Goethe, Germany's poet.[7] On taking his law degree in 1771, the young Goethe began practicing before the Reichskammer Court in Wetzlar. Here is a description of what he found at the courthouse: "A monstrous

Table 4.2
Number of Procedures to Enforce a Contract

The fewest ...		... and the most	
Australia	11	Angola	46
Norway	12	Paraguay	46
United Kingdom	12	Cameroon	46
Zimbabwe	13	Mexico	47
Denmark	14	Sierra Leone	48
Jamaica	14	Chad	50
Switzerland	14	Oman	54
Tanzania	14	Puerto Rico	55
Tunisia	14	Congo, Dem. Rep.	55
Taiwan, China	15	Burundi	62

Source: Doing Business database.

Table 4.3
Days to Enforce a Contract

The fastest ...		... and the slowest	
Tunisia	7	Bosnia and Herzegovina	630
Netherlands	39	Italy	645
New Zealand	50	Lebanon	721
Singapore	50	Nigeria	730
Botswana	56	Angola	865
Japan	60	Ethiopia	895
Armenia	65	Poland	1000
Nicaragua	65	Slovenia	1003
Lithuania	74	Serbia and Montenegro	1028
Korea, Republic of	75	Guatemala	1460

Source: Doing Business database.

chaos of papers lay swelled up and increased every year. Twenty thousand cases had been heaped up, and double that number was brought forward." It was not unusual for a case to remain on the docket for more than 100 years. One case, filed in 1459, was still awaiting a decision in 1734.

Conditions have greatly improved since Goethe's time. Countries with very different characteristics have managed to achieve quick dispute resolution (table 4.3). Among them are common-law countries such as Botswana, Singapore, and New Zealand, which take less than two months. Japan, the Republic of Korea, and Lithuania, all in the German legal tradition, also have expeditious procedures, as do the Netherlands, Nicaragua, and Tunisia, three French-legal-origin countries.

Tunisia, the world's leader in speedy resolution of commercial disputes, is perhaps the most surprising. Its current procedures were put in place only in 1996. The courts employ a special proceeding, called injunction to pay, for recovering a debt claim. The process lasts one week and includes 14 procedures, from the moment of filing the claim with the tribunal cantonal in Tunis until the moment the creditor receives payment. On the plaintiff's application, the judge will order payment if the debt claim is well justified.

Botswana uses an expedited court proceeding that does not require a trial. Such summary procedures are available mostly in common-law countries, although some civil-law countries also have them. In Botswana, the creditor would apply for a summary procedure in cases where the defendant is unable to raise any credible opposition to the plaintiff's complaint. The debtor may request leave to defend, but the request will be denied by the court. The whole process requires 22 procedures and lasts 56 days.

Four transition economies (Bosnia and Herzegovina, Poland, Serbia and Montenegro, and Slovenia) join three African countries (Angola, Ethiopia, and Nigeria), as well as Guatemala, Italy, and Lebanon, as countries with the longest delays. For Italy, the explanation lies in the country's lax appeals process, which allows disruption of the proceedings at any point during the trial. Guatemala takes the longest time to enforce a simple commercial contract—four years, on average.

The greatest differences across countries are in the costs of proceedings (table 4.4). Several economies, both developed (Austria, the Netherlands, the United Kingdom, the United States, and Taiwan [China]) and developing (Brazil, Jordan, Mongolia, Uzbekistan, and the Republic of Yemen) impose negligible costs. But in several countries—the Democratic Republic of Congo, Côte d'Ivoire, India, and the Philippines—the costs are almost equal to income per capital or double the claim amount. In Burkina Faso, the Dominican Republic, Indonesia, the Kyrgyz Republic, and Malawi, the costs are two or more times income per capita. Why, then, would businesses take disputes to court?

Another important factor in deciding whether to use the courts is the predictability of resolving a dispute. Lawyers were asked the minimum and maximum

Table 4.4
Cost of Dispute Resolution

The cheapest ...	US$	% of income per capita
Jordan	5	0.3
United States	120	0.4
Yemen, Rep. of	2	0.5
Netherlands	120	0.5
United Kingdom	120	0.5
Taiwan, China	68	0.5
Austria	240	1.0
Mongolia	7	1.8
Uzbekistan	13	2.1
Brazil	83	2.4
... and the most expensive		
Côte d'Ivoire	572	83.3
Congo, Dem. Rep.	800	92.3
India	444	95.0
Philippines	1086	103.7
Madagascar	304	120.2
Burkina Faso	375	172.8
Kyrgyz Republic	730	254.7
Indonesia	1754	269.0
Dominican Republic	9250	440.5
Malawi	920	520.6

Source: Doing Business database.

Figure 4.2
Uncertainty in Contract Enforcement

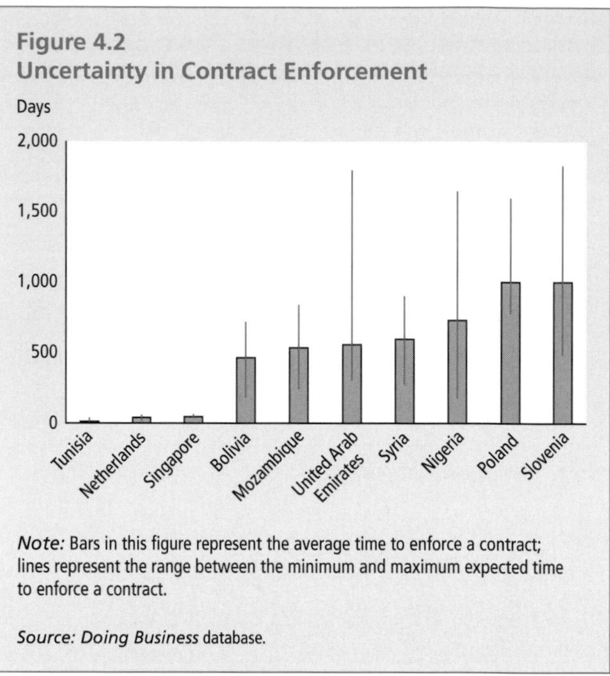

Note: Bars in this figure represent the average time to enforce a contract; lines represent the range between the minimum and maximum expected time to enforce a contract.

Source: Doing Business database.

expected number of days for enforcing a contract in the hypothetical case. The range can be large. In Nigeria, for example, it takes 730 days on average to resolve a dispute, but it can take as much as 1,643 days; in Slovenia the range is 480 to 1,825 days, and in the United Arab Emirates, 300 to 1,800 (figure 4.2). Uncertainty is positively correlated with the average time to resolve disputes: countries with inefficient courts are also likely to have uncertain outcomes. Analysis shows that the effect of uncertainty is only about a third as significant as the effect of average time in explaining the use of courts by businesses. In other words, focusing reform on reducing the length of judicial process has a high value.

Which Courts Are Socially Desirable?

Courts should be fast, fair, and affordable. Legal experts argue that the three attributes are difficult to balance. The main reason to regulate procedures in commercial dispute resolution is that informal justice is vulnerable to subversion by the rich and powerful. If one of the disputants is more economically or politically powerful than the other, he can encourage the judge to favor him, using either bribes or threats. In practice, fewer procedures are associated with both reduced time and cost, and with perceptions of improved fairness. Analysis of data from a World Bank survey of more than 10,000 enterprises in 82 countries establishes that a lower number of procedures is associated with more fairness and impartiality in the legal system (figure 4.3). It is also associated with more honesty, more consistency, and more public confidence in courts.

History supports these findings. In 17th-century England, debt disputes were decided by lay courts, presided over by the local mayor and a clerk.[8] Procedures were simple—the plaintiff wishing to initiate a lawsuit needed only to go to the town hall on a court day and enter a complaint with the clerk. Proceedings were oral, and rulings were not subject to appeal. Courts were accessible to everyone—rich and poor.

Another example of the attractiveness of fast and affordable resolution of commercial disputes comes from the Spanish *consulados* of the Middle Ages.

Figure 4.3
A Small Number of Procedures Is Associated with Perceived Fairness

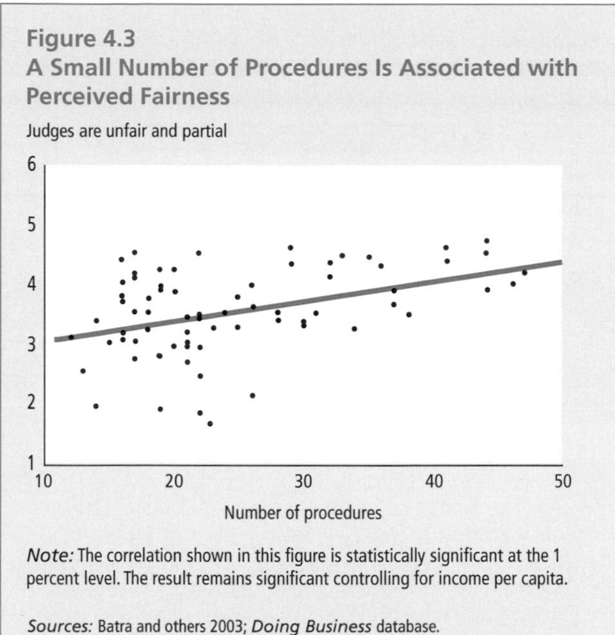

Note: The correlation shown in this figure is statistically significant at the 1 percent level. The result remains significant controlling for income per capita.

Sources: Batra and others 2003; *Doing Business* database.

Figure 4.4
Procedural Complexity Is Associated with Greater Corruption

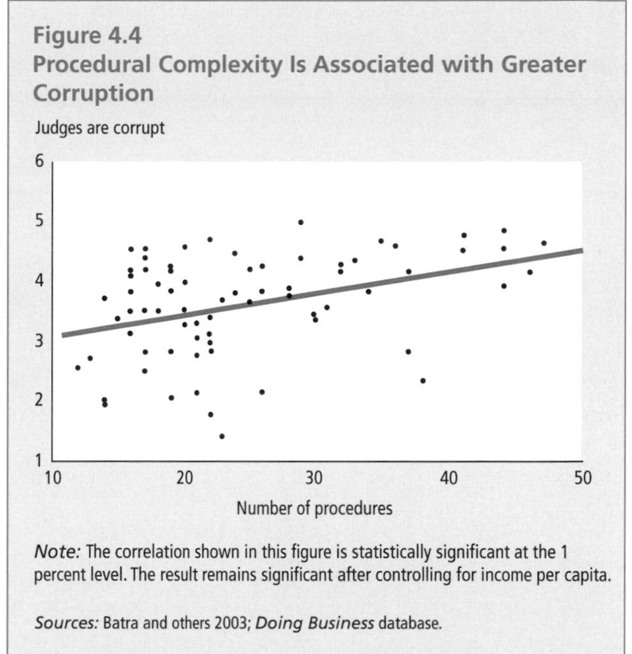

Note: The correlation shown in this figure is statistically significant at the 1 percent level. The result remains significant after controlling for income per capita.

Sources: Batra and others 2003; *Doing Business* database.

Originally used as maritime courts, these private commercial courts spread in Spain in the late thirteenth and fourteenth centuries, starting with Valencia in 1283. In 1592, a consulado was established in Mexico City to serve the needs of Spanish merchants in the New World. In the request for a royal charter, the Mexico City merchants guild asserted that "every day there arise many lawsuits and disputes, disagreements and differences over company accounts, consignments, freights and insurance, risks, damages, leakage and spillage, losses, failures, and defalcations. The settlement of such matters in ordinary courts proves costly and time-consuming."[9] Staffed by merchants, the consulados used oral proceedings to resolve disputes.

In addition to leading to other poor outcomes, legal complexity facilitates corruption. In surveys conducted in the Slovak Republic in 2000, more than 80 percent of entrepreneurs indicated the slowness of the courts to be among the three main obstacles to doing business—and that giving "something special" to a court clerk or judge was necessary to speed the process along.[10] Between a third and a half of the respondents found Slovak commercial judges to be corrupt. More generally, a higher number of procedures is associated with more opportunities in the judicial system for extracting bribes (figure 4.4).

If the efficiency and fairness of dispute resolution in court are questionable, companies use other ways to structure transactions so that disputes do not occur. A 1996 enterprise survey of six African countries—Burundi, Cameroon, Côte d'Ivoire, Kenya, Zambia, and Zimbabwe—studied perceptions of court inefficiency. In Burundi only 15.4 percent of respondents said courts were effective for dispute resolution. In Cameroon, Côte d'Ivoire, Kenya, and Zambia, only about 20 percent of respondents thought that courts could be used for recovering unpaid debt. More than 70 percent of supplies were procured from a single supplier. The average supplier/customer relationship was 10 years, and infrequent orders accounted for less than a fifth of total orders.[11] In these and some other countries, information from private credit bureaus or public credit registries is increasingly being used by lenders to compensate for poor enforcement systems.

Another survey, of small entrepreneurs in Vietnam in 1999, found that only 9 percent of respondents would consider using courts to resolve disputes.[12] One entrepreneur said, "They normally create problems. In Vietnam no one believes we have a good legal system." Instead, entrepreneurs rely on social networks for information about new customers.[13] In 40 percent of transactions, payment is made in advance.

What Explains Differences in Court Efficiency?

Richer countries have more-developed judicial systems—and more resources to establish specialized courts, to train judges and support staff, and to bring the latest technology to the courtroom. Comparing countries by income quartiles, the richest jurisdictions have the lowest median cost, at 6.6 percent of income per capita (figure 4.5); the shortest median time, at 210 days; and the lowest number of procedures, 18. Upper-middle-income countries have the longest time, with a median of 270 days, followed by the poorest countries at 248 days and lower-middle-income countries at 225 days. The poorest countries have the highest costs, at 31 percent of income per capita, and have the largest number of procedures, 30.

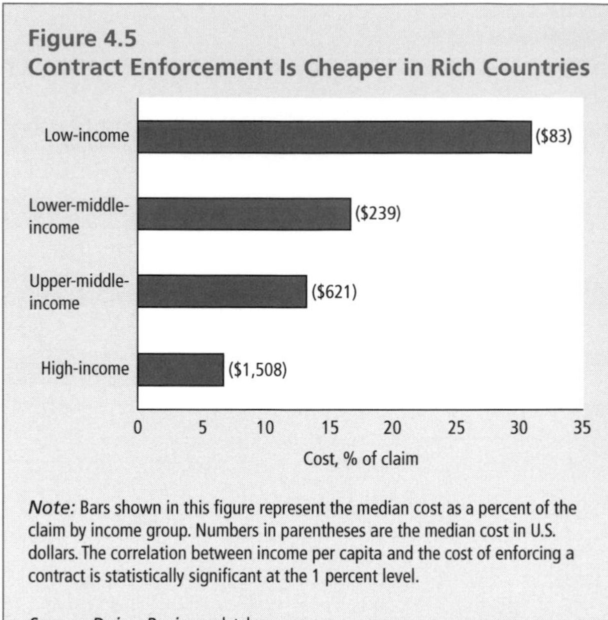

Figure 4.5
Contract Enforcement Is Cheaper in Rich Countries

Note: Bars shown in this figure represent the median cost as a percent of the claim by income group. Numbers in parentheses are the median cost in U.S. dollars. The correlation between income per capita and the cost of enforcing a contract is statistically significant at the 1 percent level.

Source: Doing Business database.

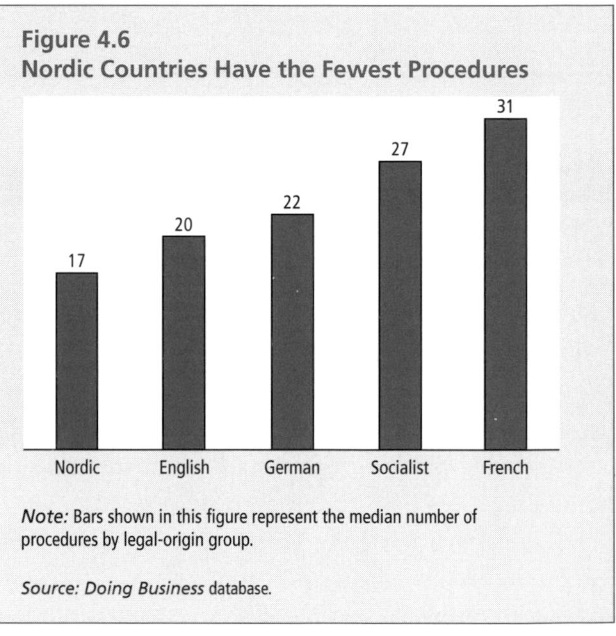

Figure 4.6
Nordic Countries Have the Fewest Procedures

Note: Bars shown in this figure represent the median number of procedures by legal-origin group.

Source: Doing Business database.

Legal tradition is also associated with the efficiency of contract enforcement. Nordic countries have the fewest procedures (17), the shortest time (139 days), and the second-lowest (after Germanic countries) cost, at 9 percent. Countries in the French legal tradition have the most procedures (31), and the second-longest time and cost (300 days and 13.7 percent). Germanic countries have low costs but a long duration (a median of 348 days). They have the third-fewest number of procedures, 22 (figure 4.6). Legal tradition is not destiny, however. Tunisia—a relatively poor Middle Eastern jurisdiction in the French legal tradition—is a premier example of efficiency.

Among civil-law countries, Latin American jurisdictions have the most onerous contract enforcement, in the number of procedures and time. It takes a median of one year, 30 procedures, and 17 percent of income per capita to resolve a dispute. Only Sub-Saharan Africa has higher median costs—at 46 percent. OECD (high income) countries take the shortest time (median of 200 days), have the lowest cost (6.2 percent of income per capita) and the fewest procedures (18).

The complexity of judicial processes is the main channel for the income and legal tradition of countries to affect the efficiency of contract enforcement. Common-law countries, mainly wealthier ones, have the lowest procedural complexity. Seven of them—Australia, Canada, Ghana, Jamaica, New Zealand, the United Kingdom, and Zambia—make the top-10 list. Japan, Taiwan (China), and Turkey are the other three. In contrast, Latin American countries have the highest procedural complexity, with Argentina, Costa Rica, El Salvador, Guatemala, Nicaragua, Panama, Peru, and Venezuela among the 10 countries with the most complex procedures. They are joined by France and Spain. The complexity of procedures is associated with higher cost and longer duration (figure 4.7).

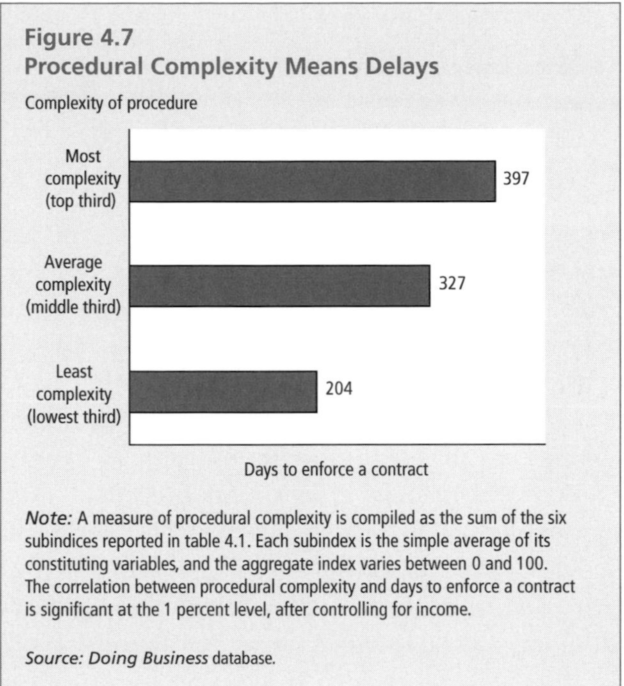

Figure 4.7
Procedural Complexity Means Delays

Note: A measure of procedural complexity is compiled as the sum of the six subindices reported in table 4.1. Each subindex is the simple average of its constituting variables, and the aggregate index varies between 0 and 100. The correlation between procedural complexity and days to enforce a contract is significant at the 1 percent level, after controlling for income.

Source: Doing Business database.

These results are indicative beyond the debt recovery case amounting to half of income per capita. The procedures described here would be similar in other types of commercial disputes—delivery of faulty goods, breach of confidentiality agreements, illegal use of intellectual property, use of shared public resources, failure to deliver on time, and so forth. Also, the background study[14] tested for robustness by varying the contract amount to 5 percent of income per capita and the nature of the commercial dispute by using a landlord/tenant dispute. The findings remain essentially the same.

What to Reform?

Four types of reform have proven successful in improving the efficiency of contract enforcement: establishing information systems and judicial statistics, taking nondispute cases out of courts, simplifying judicial procedure, and creating specialized courts and specialized procedures within courts.

Establishing Information Systems and Judicial Statistics

Lack of reliable information systems and workflow statistics limits the ability of judicial reformers to identify bottlenecks. It also hinders the monitoring of reform progress and the accountability of court administrators. Judicial statistics should include data on the number of petitions and cases at various stages of the judicial process—as well as court users' characteristics, the nature of disputed cases, the amount of the dispute, and the use of appeals, all fundamental for court management. Such data could also be made public, to increase transparency and accountability.

Such countries as Brazil, the Dominican Republic, and Mexico have recently piloted management information systems in the courts. The results are encouraging, especially when bolstered by other reforms. For example, the pilot in Mexico revealed that more than 60 percent of the cases do not go beyond the initial filing of claim. This gave the government a more accurate picture about the actual caseload of judges—who, like judges everywhere else, claimed an ever-increasing workload. In the Dominican Republic, the pilot established that almost a third of the cases in civil courts are not disputes but are company registrations and consensual divorce filings. These cases can be handled by court clerks, which will increase the productivity of judges. The pilot in Brazil documented the need for simplification of debt collection procedures—even with summary judgments, payments took years to collect. As a result, default judgments have been suggested.

New technology and information systems can also have a very direct impact on court efficiency. When a system of automatic case assignment was implemented in the Slovak Republic, the time between filing and the first hearing was reduced from 73 to 27 days, and the number of procedures between filing and first hearing went from 23 to 5.[15] What was previously done by sending a paper file from one office to another is now done electronically, with several court officers able to view the file simultaneously, thereby further reducing delay. Automatic case assignment is also an anticorruption device, eliminating the possibility of litigants "paying" to have certain judges assigned.

Taking Nondispute Cases Out of Courts

In many countries, particularly those with a civil-law tradition, courts are tied up with cases such as the creation or voluntary dissolution of a company, where there is no dispute. Since such cases are usually numerous, they demand considerable court resources. At the Sofia District Court, Bulgaria's largest, 23 judges handle all types of cases, with eight judges dealing exclusively with business registration and re-registration. Removing court approvals from the business registration process and adopting an equivalent administrative procedure would increase judges' time by half.

Another example comes from the courts in neighboring Serbia and Montenegro. In Belgrade, 18 of the 95 commercial judges work exclusively on registration and re-registration cases. If these cases were handed to an administrative agency, as is currently proposed in the draft law, the judiciary would have about 25 percent more time to spend on disputes. Company registration has recently been taken out of the courts in Honduras, resulting in reducing the cases that courts need to deal with every year by 8,000.

Taking nondispute cases out of the judiciary often requires new legislation, which may take time. In the meantime, the judicial process can be reorganized to give more responsibility to court clerks in handling such cases.

Simplifying Judicial Procedures

Several areas of reform to simplify procedures have been explored: introducing oral procedures, simplifying the notification process, limiting the number and timing of appeals, reducing or eliminating the need for legal justification, and simplifying the regulation on evidence. Such simplification is associated with less time and cost to resolve disputes (figure 4.8).

In countries where written elements dominate, judges do not have direct contact with witnesses and other sources of evidence. This absence of direct contact, together with piecemeal rather than continuous trials, causes delays. Reforms targeted at introducing oral procedures in dispute resolution increased court efficiency in 18th-century Prussia, and more recently in Italy, Paraguay, and Uruguay.[16] In pilot reforms in Argentina, new oral procedures reduced the average time of cases from three years to less than six months.[17]

In some countries, the defendant is notified directly by the plaintiff or the plaintiff's attorney, or simply by letter. In others, the defendant cannot be held accountable unless an appointed court officer serves the claim. In Bulgaria, notification of defendants was identified as the major factor in causing long delays in commercial cases. The court was obliged to notify the defendant in person before the case could commence. With a creative defendant, this process could—and did—take years. So the code of judicial procedure was revised in 2000. Now, after the first notification fails, it is enough to post a second notification on the court's announcement board and in the official gazette. Mexico reformed its notification procedure even further, allowing for default judgment if the defendant does not appear on the first hearing.

In most countries, the enforcement of judgment is automatically suspended until resolution of the appeal; this suspended judgment substantially reduces the value of the first-instance judgment. In others, the suspension of enforcement is either not automatic or even not allowed, which is associated with less time to resolve disputes (figure 4.8). One solution in the former case is to charge interest on delayed judgment to allow the winning party to recoup the cost of delay.

In Tanzania, one of the main procedural changes with the establishment of the specialized commercial section of the high court was to bar appeals during

Figure 4.8
Simple Rules Are Associated with Less Time and Lower Cost

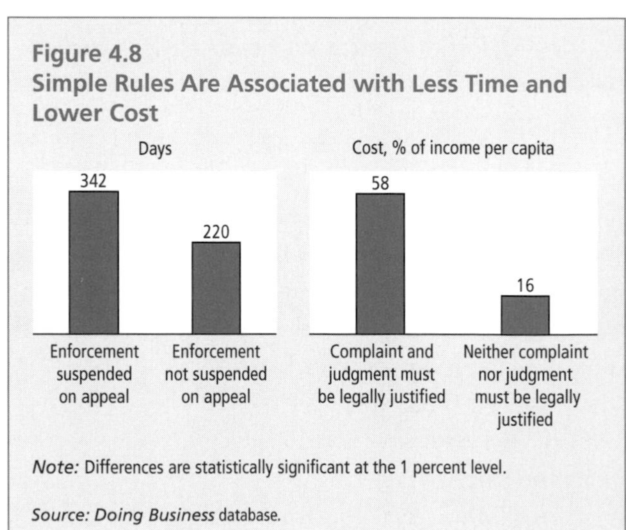

Note: Differences are statistically significant at the 1 percent level.

Source: Doing Business database.

trial. Parties must wait until the court reaches a final judgment before making an appeal. This prohibition has eliminated an average of nine months in appeal time.[18] However, the problem arises again when a final appeal is launched. Even countries with specialized courts typically do not have a separate appeals court. Appeals on judgments by specialized courts are pulled together with all other cases, as in Tanzania, and take a long time to resolve; consequently, much of the value of specialized courts is lost.

Establishing Specialized Courts

Specialized courts generally entail procedural simplification aimed at "mass production" in small-claims courts, commercial courts, or administrative tribunals. Creating small-claims courts or expanding their jurisdiction has been popular in the last two decades. Small-claims courts have substantially reduced time to disposition and are typically much cheaper than regular courts.

Specialized commercial courts are associated with faster and cheaper contract enforcement in wealthy countries such as Germany, Japan, and the Netherlands—but also in developing countries such as Ecuador, India, and Tanzania. One reason for the greater efficiency is that judges become expert in handling commercial disputes. Perhaps more important, commercial courts often have less formal procedures: the use of oral arguments is often permitted even in countries where the general courts require written procedures. Countries with specialized courts or specialized commercial sections in the general courts are about 50 percent faster in resolving commercial disputes—218 days versus 349 days, on average.

Specialized commercial courts are much less prevalent in civil-law countries, although this was not the case historically.[19] In many civil-law countries, specialized commercial courts were established and later abandoned. In Spain, the commercial courts were abolished in 1868, after the revolution. In Portugal, commercial courts were abolished in 1932, with the procedure unified under the Code of Civil Procedure in 1939. Commercial courts suffered the same fate in Brazil, again with a new Code of Civil Procedure.

In contrast, commercial courts retained their importance in France. Born out of the merchant courts in medieval fairs, they were established as permanent courts in 1563, during the reign of Charles IX. The enactment of the Commercial Code in 1806 enlarged their jurisdiction considerably by broadening the legal definition of commercial transactions. Later judicial reforms, such as those in 1958, did not diminish their importance.

Some Latin American countries have reintroduced specialized courts. In 1996, for example, specialized debt collection was established in the four major cities of Colombia: Bogota, Medellin, Cali, and Barranquilla. The judge is responsible for the seizure, attachment, appraisal, and auctioning of property to repay debt. By the year 2000, 75 percent of cases were being resolved within a year, and the number of pending commercial cases fell by 5,000. Also, the number of filed cases increased, from about 4,000 a year to 11,000 a year.

Several other countries are following suit. In India, the first Debt Recovery Tribunals were established in 1994. By 2003, 30 such tribunals had started operating in various cities around the country. Only financial institutions may file cases for claims greater than one million rupees. A recent evaluation finds, "Debt Recovery Tribunals are seen as a vast improvement over traditional courts as regards time and appropriate procedure."[20] In 2001, Ghana instituted a fast-track court, a specialized section in the high court. It has better technology, including a computerized system for case management, automated transcripts, and random assignment of cases. Judicial process is also eased. The court's success has prompted plans to establish fast-track sections in all regional capital courts.

Some countries that already have small-claims courts—such as Japan, the Netherlands, and the United Kingdom—have recently increased the permissible amount that may be tried in such courts, a reform vastly popular among litigants.[21] Other countries—such as Australia and New Zealand—have allowed litigants to agree on raising the disputed claim's limit if they consent to try it in the small-claims court. Such reforms are likely to result in further efficiency gains.

In 1999 a commercial court was established in Dar-es-Salaam, Tanzania, as a specialized division of the high court. It has jurisdiction over larger commercial disputes, with claims amounting to more than 10 million Tanzanian shillings, about $12,500.[22] It also has a higher fee, about 4 percent of the claim, but it handles cases in three to four months on average, much shorter than the two to three years needed for an average commercial dispute in the high court. Assessing its fairness and efficiency, one local lawyer says: "The commercial court has proven its efficiency. It has built much-needed public confidence in the commercial community, so much that many companies now insist on a dispute clause in their contracts. This clause provides that any disputes not resolved amicably by the parties will be referred to the commercial court."[23]

By mid-2003 the specialized court in Dar-es-Salaam was having growing pains. Countries planning to introduce specialized commercial courts can learn much from its experience. After two years of success, it became inundated with cases. Plaintiffs were willing to pay the 4 percent fee to see their case resolved within six months or less. The greater demand was not met by putting in more judges, clerks, and stenographers. A special court of appeals was also needed.

In countries where the judiciary is still developing, specialized courts are likely to be premature. Instead, having specialized judges or establishing specialized commercial procedures within the general courts and focusing them on execution, as was recently proposed in Mozambique, is likely to pay higher dividends. A specialized commercial court may quickly become a victim of its own success, as in Tanzania. Or it may stretch judicial resources beyond the capacity of some poorer countries. The lesson: enacting new procedures, training judges in the subject matter, educating lawyers about the new court, and giving it wide publicity yield sustainable results only if those steps are matched by committing more resources as the demand expands. In Tanzania, a second commercial court is being planned in Arusha, which will take some of the burden off the one in Dar-es-Salaam.

Out-of-court resolution mechanisms are a better short-term solution in countries with only a rudimentary judiciary; they are also important in other developing countries. One example illustrates their benefits. The Ministry of Justice in Argentina, troubled by the long delays in commercial litigation, implemented a pilot project in 20 civil courts, requiring them to send commercial cases to mediation. Both the ministry and Fundación Libra, an NGO, trained mediation staff.[24] The results exceeded expectations. Of more than 32,000 cases that went through mediation between April 1996 and April 1997, only a third were returned to the courts. Mediation took only two months on average—a far cry from the three to four years that it generally takes to resolve commercial disputes. Voluntary mediation has enjoyed tremendous success in several other countries—among them Albania, Bangladesh, Bolivia, Ecuador, and the Russian Federation.[25]

Legal reform experts have come forward with further lists of recommended changes. Their lists include hiring more judges, improving the selection of judges, managing judicial careers, improving judicial administration and entrance into the legal profession, ensuring the independence of the judiciary, and meeting resource needs.[26] Some reforms work admirably, such as the training of judges in Malawi, Sri Lanka, and Uganda. But training can be expensive. In many rich countries it requires substantial resources—France spends more than $23 million a year on judicial training; the Netherlands, $20 million; and the United States, $17 million.[27]

Furthermore, not everything that looks good as a reform succeeds. Increasing the number of judges to deal with case overload is often recommended, even though the evidence shows that it does not increase efficiency.[28] Why? Because it treats the manifestation (overworked judges), not the cause (procedural complexity), of judicial inefficiency.

One needs only to look at history. In the sixteenth century, France and England had economies of roughly equal size, probably generating similar numbers of commercial disputes. But the extensive complexity of the judicial process in France required more judges. One legal historian writes, "The total number of royal judges in France must certainly have exceeded 5,000. In contrast, from 1300 to 1800 the judges of the English central court of common law and Chancery rarely exceeded 15."[29] The staggering difference in the number of state-employed judges is due to the fact that English courts depended on the local administrators (municipal councils) to serve as jurors.

Beyond enhancing—or at least not preventing—out-of-court resolution mechanisms, there is little governments in poor countries can do in the short term. Private parties will find ways to do transactions—by writing contracts that are easier to monitor and enforce (such contracts might specify the use of leasing agreements or give title to assets in exchange for loans) and by relying on reputation mechanisms for enforcing contracts. Business takes place in societies with courts as dysfunctional as those in Angola or Congo. Though far from efficient, such contracts may be the best way of doing business given the circumstances. In such countries, other institutions that facilitate contract enforcement, such as credit information registries, take on great importance.

Notes

1. Bigsten and others 2000.
2. McMillan and Woodruff 1999.
3. Shakespeare, *Hamlet: Prince of Denmark*, act III, scene 1.
4. Informal substitutes for courts are usually expensive to maintain. One study investigates the contractual arrangements in the Indian software industry and finds that the lack of court enforcement results in 15–20 percent higher transaction costs (Banerjee and Duflo 2000). Similar costs of inefficient contract enforcement are reported in transactions among Romanian firms (Murrell 2003). The true cost is higher and is reflected in the foregone opportunities in new transactions.
5. For further description, see the data notes in the *Doing Business* Indicators tables.
6. Studies of actual court files provide the best method of accurately documenting the duration of judicial procedure. In addition, such studies show who litigates, what the main types of disputes are, what amounts are claimed, how litigation ends, and how often judgments are appealed. See Hammergren (2003) for a survey of existing studies.
7. Goethe 1969.
8. Muldrew 1993.
9. Cedula Reales 448, Archivo Municipal de Mexico, 1590.
10. World Bank 2001a.
11. Bigsten and others 2000.
12. McMillan and Woodruff 1999.
13. Informal networks that use collective action as a means to enforce contracts have been known for centuries; they include the Maghribi traders (Greif 1993), the merchant courts in the Champagne fairs (Milgrom, North, and Weingast 1990), and the German Hansa (Dollinger 1970). But informal mechanisms break down when the opportunity cost of deviation is reduced, as occurs with increased competition (Woodruff 1998) or costly substitution (Clay 1997).
14. Djankov and others 2003.
15. World Bank 2001b.
16. Botero and others 2003.
17. World Bank 2001b.
18. Finnegan 2001.
19. Zweigert 1983.
20. PriceWaterhouseCoopers 2001, p. 5.
21. Botero and others 2003.
22. Finnegan 2001.
23. Sinare 2000, p. 4.
24. World Bank 2001b.
25. World Bank 2002a.
26. Hammergren 2000.
27. World Bank 2002b.
28. See Dakolias 1999. For evidence to the contrary, see Djankov and others 2003.
29. Dawson 1960, p. 71.

5 Getting Credit

An entrepreneur with a promising business idea can obtain credit as easily in Maputo or Jakarta as in London or New York. In 1996, a real estate developer in Maputo decided to build a luxury homes complex. He invited the head of a local large bank, a fellow golf club member, for lunch, and described his idea. For $10 million, 50 homes could be built to house the middle class. Other than the land, the entrepreneur could offer no collateral or monetary contribution. No matter. Within a week, he received $4 million to start the work.

The same year, a young entrepreneur in Jakarta proposed an even grander idea to the second-largest Indonesian bank. With a population nearing 200 million, the country needed its own national car—and he could produce it. He would need $800 million to secure the participation of a foreign partner who would bring in the technology. The entrepreneur could not offer collateral. But the banker needed none. The name of the entrepreneur: Hutomo Mandala Purta, son of (then) President Suharto.

Credit is as easily obtainable in Maputo or Jakarta as in London or New York. By the right people. For everyone else, obtaining credit in most developing countries involves a lot of frustration and likely rejection. Few bother.

In most countries, banks will not extend credit without assurances that borrowers are creditworthy and that it will be possible to recover the debt if there is a default. As a consequence, entrepreneurs with promising business opportunities cannot obtain loans if the bank does not have enough information on the value of the property and the credit history of the borrower—and if the legal system does not protect creditors.

Two types of institutions expand access to credit and improve its allocation: credit information registries or bureaus, and creditor rights in the country's secured-transactions and bankruptcy laws.[1] They operate best together—information sharing allows creditors to distinguish good from bad clients, while legal rights to enforce claims help in the event of default. Sometimes, information-sharing mechanisms remedy poor legal protection. Public credit registries can also help remedy the lack of private credit bureaus in poor countries. What is often termed "credit culture" is in fact an outcome of the underlying institutions.

From the excommunication of usurers in the medieval church to the homestead protections in the United States, regulations protect borrowers from unscrupulous creditors. But well-intended shielding of borrowers is often misguided—in the words of one distinguished lawyer, "in its zeal to protect debtors [the law] precludes them from becoming borrowers."[2] Good credit institutions define property rights for both creditors and debtors, making everyone better off. Collateral and insolvency regulations define the rights of creditors to recover their loans. In addition, collateral regulation helps debtors by extending the right of property title to the right to use property as security for finance. Information-sharing institutions enable debtors to build reputational collateral.

Countries have chosen different paths to expand access to credit. Poor countries are as likely to have public credit registries and strong creditor rights as developed countries, although their enforcement of regulation is weaker than that of developed countries. But private credit bureaus are much more prevalent in developed financial markets. Countries in the common-law tradition rely more on creditor protections in the law. Civil-law countries, especially in the French tradition, use public credit registries more frequently.

What can governments do to help creditors believe they will be repaid? Establishing appropriate regulations for the operation of private credit bureaus is a critical start. Removing legal restrictions to exchanging credit information, unambiguous endorsement of credit bureaus by central banks, and well-designed consumer protection and privacy laws will create incentives for the sharing and proper use of good-quality credit information. In some cases—especially in poor countries where commercial incentives for private bureaus are low—establishing public credit registries has helped remedy the lack of private information sharing, or complemented private bureaus by focusing on banking supervision. The design of the registries influences their impact: broader coverage of the credit market and regulations on collection, distribution, and quality of information are associated with larger credit markets.

Legal creditor protections can be improved by reforming collateral law: introducing summary enforcement proceedings, eliminating restrictions on which assets may be used as security for loans, and improving the clarity of creditors' liens through collateral registries and clear laws on who has priority in a disputed claim to collateral. More-efficient courts are crucial for the legal protections to take effect. Reforms of insolvency laws are sometimes necessary—as discussed in the next chapter.

Sharing Credit Information

Every lender gathers information on the creditworthiness of potential borrowers. A debtor's history with a bank is also an important way to build a good track record.[3] Credit registries make borrowers' reputations accessible to other creditors. By facilitating information exchanges among lenders, registries help creditors sort good borrowers from bad, price loans correctly, and reduce the costs of screening. When borrowers know that their reputation will be shared among lenders, they have additional incentives to repay. And because credit histories are available, borrowers benefit from lower interest rates, as banks compete for good clients.[4]

Informal reputation mechanisms have helped lenders allocate credit for centuries.[5] But they are appropriate only for small-scale business activities or among a close-knit group of merchants and lenders. As formal financial intermediaries developed, so have the institutions to help them allocate credit. Formal institutions for credit information sharing emerged in the 17th century in Paris, where notaries exchanged data on debtors' creditworthiness—and in Amsterdam, where the municipality initiated a precursor to the modern public credit information registry.[6] In the 18th century, private credit reporting businesses emerged in the United States, evolving into today's Dun & Bradstreet (D&B). Back then, D&B delivered its reference books to subscribers under lock and key. In the 19th century, mutual-protection societies developed in Germany.[7]

The credit information industry has grown at an astonishing pace, facilitated by rapid technological advances and financial deepening.[8] Today D&B transmits credit information on more than 60 million businesses worldwide. Yet credit-information-sharing organizations differ greatly. Some concentrate on business or trade credit. Those are typically "inquiry driven," and rely mainly on information available through public sources, direct investigations, and trade creditors. Others focus on consumer credit and facilitate direct exchange among financial institutions.[9] Although many such registries (also known as bureaus) operate nationally, there is growing international consolidation. The largest, Experian, has more than 40,000 clients in 50 countries, with annual sales in excess of $1.7 billion.

Institutions sharing credit information also differ in ownership structure. The first publicly owned credit information registry was established in Germany in 1934 after the banking crises of the Great Depression. Since then, many governments have followed suit, with distinct waves in Latin America after the macroeconomic instability of the 1980s—and more recently in transition countries. Some public credit registries, such as those in Germany and Turkey, were started to monitor systemic risk and began distributing information to lenders only later. Others—including those in Bangladesh, Bulgaria, France, Mozambique, and Taiwan (China)—were established to help lenders allocate credit effectively.

How prevalent are credit information registries? Surveys conducted for this report show that private bureaus that facilitate exchange of information among financial institutions operate in 57 countries and in every developed country but France.[10] Public credit registries operate in 68 countries, and are being established in Albania, Armenia, and Panama (figure 5.1).[11]

What is the coverage of institutions sharing credit information? On average, private bureaus cover 321 borrowers per 1,000 people, ranging from more than 800 borrowers per 1,000 people in Canada, New Zealand, Norway, and the United States to less than 1 in newly established registries in Ghana and Pakistan (table 5.1).

Public credit registries cover much less information. The average registry contains records on 40 borrowers per 1,000 inhabitants and 44 percent of the value of credits to gross national income (GNI). But there is significant variation, from the extensive scale of Portugal's, with 496 borrowers per 1,000 inhabitants and 130 percent of credit to GNI, to Nigeria's and Serbia and Montenegro's, with less than one borrower per 1,000 people and credit to GNI below 1 percent.

Rules and Regulations on Public Credit-Information Sharing

Public credit registries vary greatly in the extent to which their design supports lending transactions. The first difference is in the rules on collection of information. More than two-thirds of registries record only loans above a minimum size. Minimum loan cutoffs average $87,000 but can be more than $1 million, as in Germany and Saudi Arabia, indicating a focus on monitoring systemic risk. Other regulations on collection mandate whether nonbank lenders may submit data, as in Belgium, Bolivia, France, Taiwan (China), and Vietnam, and whether defaults must be erased when loans are repaid. The duration of historical data collected also varies: for Venezuela it is two years; for Honduras, three years; and for Mozambique and Tunisia, 10 years.

Second is the scope of information distribution. Some public registries distribute data only on the total indebtedness of the borrower, as in Austria, Germany, Saudi Arabia, and the United Arab Emirates. Others provide demographic data, court judgments, loan repayment patterns, utility payments, credit inquiries,

Table 5.1
How Much Credit Information Is Available?
Number of borrowers (firms/individuals) per 1,000 people

Private Bureaus				Public Credit Registries			
Top 10		Bottom 10		Top 10		Bottom 10	
Norway	945	Spain	48	Portugal	496	Niger	0.6
New Zealand	818	Israel	47	Spain	305	Mozambique	0.6
United States	810	Belgium	42	Chile	209	Central African Republic	0.5
Canada	806	Guatemala	35	Argentina	149	Rwanda	0.4
Japan	777	Portugal	24	El Salvador	130	Cameroon	0.4
Ireland	730	Philippines	22	Malaysia	105	Saudi Arabia	0.3
Australia	722	Hungary	15	Venezuela	97	Nigeria	0.2
Germany	693	Sri Lanka	9	Peru	92	Congo, Rep. of	0.2
United Kingdom	652	Pakistan	0.5	Ecuador	82	Chad	0.2
Poland	543	Ghana	0.2	Belgium	68	Serbia and Montenegro	0.1

Source: Doing Business database.

Figure 5.1
Which Countries Have Credit Registries?

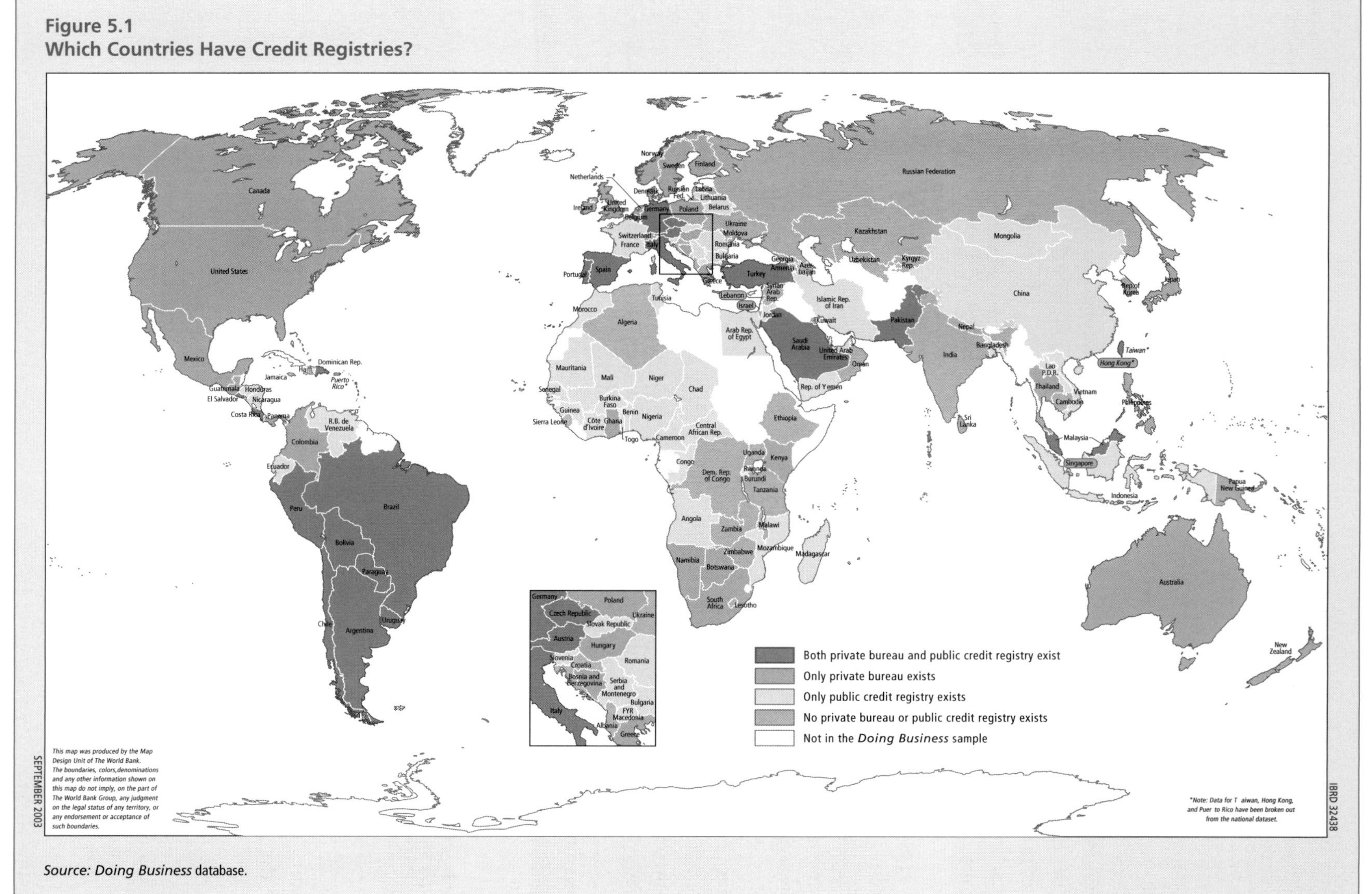

Source: Doing Business database.

and ratings—for example, Belgium, Ecuador, Romania, Taiwan (China), Venezuela, and Vietnam distribute at least two such types of detailed information. Another distinction is whether registries distribute positive or negative information, or both. Positive data include total loans outstanding, assets, and personal information, which helps in identifying total indebtedness and assessing capacity to repay a loan. Negative data reveal past defaults and arrears, and provide more information on willingness to repay commitments. About 70 percent of public registries distribute both negative and positive data, 25 percent only positive, and the remaining 5 percent only negative information (Belgium [before mid-2003], the Dominican Republic, and Turkey).

Third is the access to information, including who may use the registry and whether data are available for distribution within a day, electronically, and free of charge. In 39 percent of countries, only bank supervisors and institutions that submit data have access. In 41 percent, information is available only on the creditor's own customers. Lenders in countries of the West African Monetary Union wait almost three months to receive paper copies of the data. Public registries charge no fees, except in Belgium, Brazil, Bulgaria, Ecuador, Egypt, Italy, Mozambique, Pakistan, Romania, Taiwan (China), and Vietnam.[12]

Fourth is the quality of information, including how current the data are, and the safeguards in place to ensure that information is accurate. Two-thirds of countries impose legal penalties for reporting inaccurate data and conduct statistical checks for errors. One-third are required by law to respond to borrower complaints.

To gauge the ability of registries to support credit transactions, the *Doing Business* team constructed an index of the rules and regulations concerning the collection, distribution, accessibility, and quality of public registries (table 5.2), measured on a scale of 0 to 100.

The results quantify the variation in design and indicate whether a public registry is oriented more to serving lenders or to supervising banks. The Taiwan (China) registry includes an extensive range of information on borrowers and their loans regardless of the loan size, and was designed with the main purpose of serving lenders. It also has the highest score, 70. The recently established public credit registries in Mongolia and Vietnam are designed to provide an extensive range of information and score 68 and 67, respectively. In contrast, the registries in most West African countries—Benin, Burkina Faso, Côte d'Ivoire, Mali, Niger, and Senegal—provide little information to facilitate credit markets.

Table 5.2
Public Registries Differ in Design

Top 10		Bottom 10	
Country	Score	Country	Score
Taiwan (China)	70	Yemen, Rep. of	38
Mongolia	68	Morocco	33
Vietnam	67	Serbia and Montenegro	33
Austria	66	Niger	22
Spain	64	Mali	22
Lithuania	63	Benin	22
Belgium	63	Côte d'Ivoire	22
Argentina	61	Burkina Faso	22
Italy	61	Senegal	22
Portugal	61	Togo	22

Note: Scores range from 0 to 100, with higher values indicating that the structure of the public registry is designed to serve lenders.

Source: Doing Business database.

Regulations on Private Credit Information Sharing

Private bureaus are generally better designed for facilitating credit transactions than public registries are (figure 5.2). They are also far more likely to report that lenders are their primary clients. In addition, they tend to be specialized in listing either firms or individuals, while almost all public credit registries cover both. Private bureaus collect information from a more extensive range of sources, such as trade creditors, retailers, courts, and other public records. They distribute longer historical data and more types of data, have fewer restrictions on access, and provide such other services as credit scoring, monitoring of borrowers, fraud detection, and sometimes even debt collection. Because fewer private bureaus have minimum loan size requirements, they may be better placed to cover consumers, entrepreneurs, and small businesses.

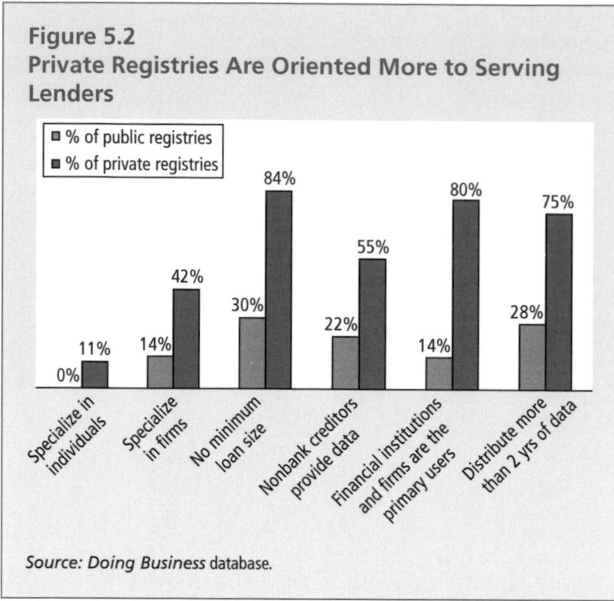

Figure 5.2
Private Registries Are Oriented More to Serving Lenders

Source: Doing Business database.

Private bureaus are formed in response to commercial opportunities and market conditions,[13] but government regulation also plays a part.[14] Bank secrecy, privacy, and data protection regulations mandate what information may be shared, while regulations on information disclosure and access affect the availability of data.

Countries differ significantly in their approach to regulating credit information.[15] Some governments help create a market for credit registries by requiring that lenders obtain credit reports before granting loans, as in Bangladesh, Belgium, Colombia, Ecuador, Malaysia, Nicaragua, and Pakistan. The absence of interest-rate restrictions on commercial lending (as in 70 percent of surveyed countries) also helps create demand. Countries that require unique identifiers—such as national ID or social security numbers—greatly facilitate bureau activity by allowing payment information from different sources to be attributed to one borrower. At the other extreme, some governments either do not permit private registries to operate, such as those in Azerbaijan, the Islamic Republic of Iran, and Mongolia—or they limit licenses for credit registries, as in Finland and (after recent changes) in Thailand. Some countries have few relevant laws or none at all, thereby restricting registries because there is no clear legal basis for operating. In Uzbekistan, a credit registry has obtained its business registration but cannot operate because it does not have the legal foundation to collect and distribute data.

Three other areas of regulation influence credit information sharing: the content of credit information that may be shared legally, the rules on access to information sources and disclosure, and the rights of the borrower to obtain credit information.

The extent of credit information that financial institutions may share with private bureaus is typically proscribed by secrecy provisions in banking laws and by data protection or privacy laws. In some countries, sharing positive data is restricted, as in Australia, Nicaragua, and Portugal. In others, sharing is forbidden except in cases of criminal prosecution, tax evasion, or money laundering. Consent clauses in lending contracts can circumvent bans on sharing in some countries. But interviews with bureaus and banks indicate that such circumvention is insufficient, because lenders usually want official government endorsement before sharing. In the majority of countries, information may be shared only if the borrower authorizes doing so or if there is "permissible purpose." Many countries regulate the amount of historical information that may be shared, with almost half requiring that information on defaults be eliminated after the default is repaid, thus preventing a banker from acquiring a full picture of the loan applicant's history (table 5.3).

Table 5.3
Regulating Private Information Sharing—Country Examples

Negative Information Only	Remedied Defaults Must Be Erased	Limits on Retaining Historical Data
Australia	Czech Rep.	Brazil
Chile	Chile	Germany
Finland	Hong Kong (China)	Italy
Hong Kong (China)	Portugal	Nicaragua
New Zealand	Switzerland	Panama
Nicaragua	South Africa	Peru
Portugal	Thailand	South Africa

Sources: Doing Business database, Jentzsch 2003a.

Open access to public information sources—such as databases, court judgments, notary records, trade registries, and financial statements—facilitates the creation and operation of private bureaus. In some countries the sources exist but various regulations constrain access to them, as in China, the Dominican Republic, Egypt, Jordan, Nigeria, the Russian Federation, Syria, Uzbekistan, and the Republic of Yemen. The degree of centralization of the information sources also matters. In Argentina and Morocco, court judgments are publicly available but can be seen only at the local level, greatly increasing the costs of gathering information. Regulations requiring firms to disclose information publicly and accounting standards ensuring standardization of information can also enhance information sharing. Although all countries require that publicly traded companies disclose financial statements, only around half require private companies to do so. Such regulations help compensate for other restrictions. For example, although laws in Finland restrict information sharing among banks, public records are open—so the private bureau gathers comprehensive data on borrower defaults from court records.

Countries also apply a range of measures to protect borrowers' rights to information on their creditworthiness. The measures affect businesses as well as consumers, because credit scoring for small businesses is based largely on personal profiles of their owners. Regulatory protections include the right of borrowers to see their own information, to correct errors, to be notified in the event of an adverse action, to stop its disclosure in case of dispute, and to know to whom it was disclosed. Those measures ensure the proper use of information and enhance the quality of data, because they establish incentives for credit registries to maintain accurate information. Such provisions are being adopted at an increasing rate in North America and Europe.[16]

Legal Rights of Creditors

In deciding whether to extend credit and at what interest rate, lenders need to know what share of debt they can recover if a borrower defaults. Since secured-transaction laws were first codified in ancient Rome, one of the main ways for creditors to recover bad debt has been with collateral.[17] Collateral laws enable firms to use their assets as security to generate capital—from the farmer in Bolivia pledging his cows as collateral for a tractor loan to the securitization of loan portfolios that drives mortgage finance markets in the United States.[18] Collateral strengthens the incentives of debtors to repay their loans. By providing creditors with the right to an asset on default, collateral also reduces a lender's costs of screening loan applicants. And well-designed collateral agreements can facilitate the efficient sale or liquidation of bankrupt firms.

For those reasons, collateral is a major determinant in lending decisions across countries. Patterns in the use of collateral show it to be especially important for small firms in obtaining loans.[19] It is also important in poorer countries. In some developing countries, overcollateralization indicates poor enforcement—collateral is necessary, but less valuable than in rich countries, because the prospects of recovering it are dim. For example, banks in Malawi, Moldova, and Mozambique typically secure more than 150 percent of a loan's value. Interviews with lenders indicate that they will always first attempt to negotiate repayment on default. But collateral provides insurance for recovering bad loans when negotiation fails.

The value of collateral depends largely on the ease of creating and enforcing security agreements, which are far from equal across countries. In the Dominican Republic and Peru, stamp duties and taxes to create a security agreement can add up to 4 percent of the total debt; in Nigeria, 2 percent; and in Tunisia, 1.9 percent. In contrast, costs are negligible in France, Japan, the United Kingdom, and the United States.

And what happens if a borrower defaults? Lawyers in more than 130 countries were surveyed for a hypothetical case of collecting on a bad loan secured by business equipment. It takes a week for a creditor to seize and sell collateral in Germany, Ireland, Tunisia, and the United States. But it can take five years in Bosnia and Herzegovina, Brazil, and Chile. In Albania, recent reforms allow creditors to seize and

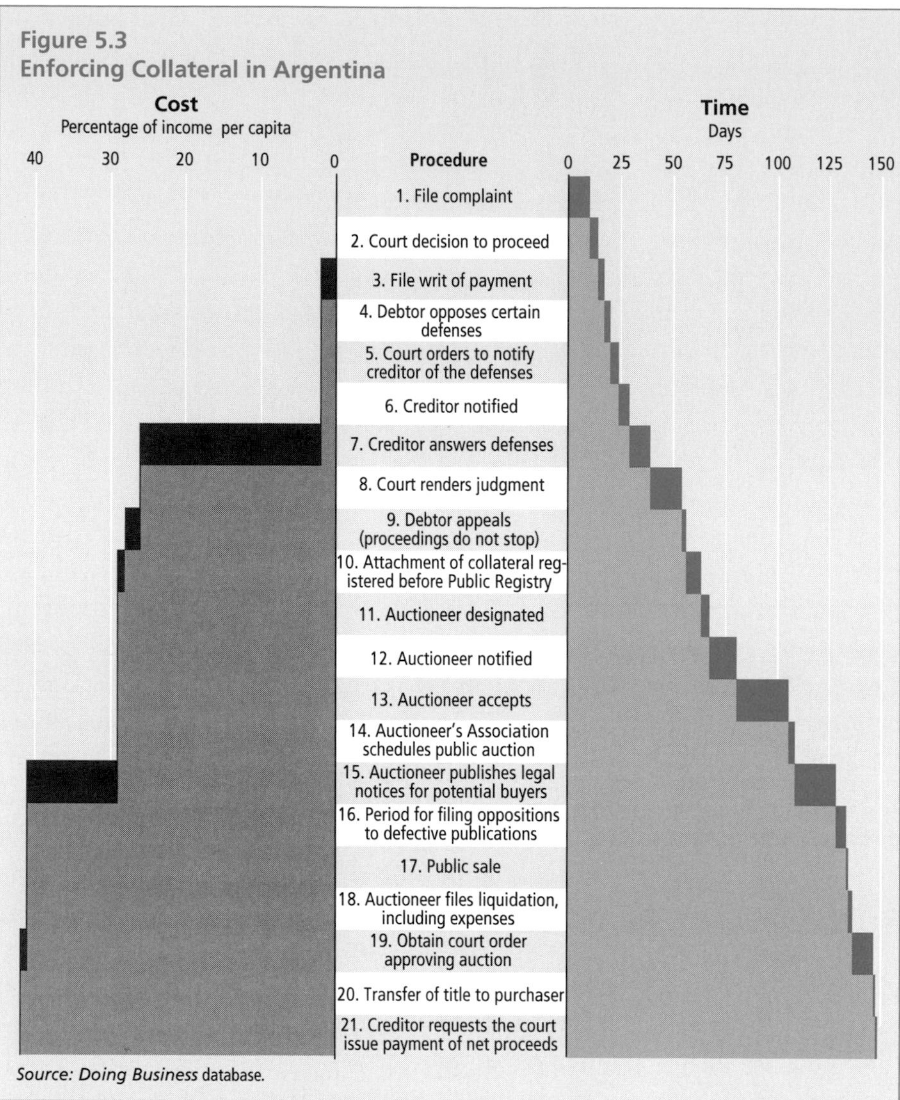

Figure 5.3
Enforcing Collateral in Argentina

Source: Doing Business database.

sell collateral without court involvement. The process takes a week. Such "private" mechanisms exist in a quarter of the sample countries. In Chile, the creditor files a claim with the court, and the court must declare default and order a bailiff to seize assets, before there is public auction. The debtor may appeal the process at every stage. In Argentina, enforcing collateral in the hypothetical good-case scenario takes 148 days and costs 42 percent of income per capita (figure 5.3).

The value of collateral also depends on the efficiency of the insolvency regime; creditors are concerned about recovering collateral if a debtor firm goes bankrupt.[20] Bankruptcy laws define who controls the insolvency process, who has rights to the property of a bankrupt firm and with what priority, and the efficiency of realizing the rights. Without legal protections along each of those dimensions, creditors will either increase the price of loans to adjust for the additional risk—possibly beyond the reach of some entrepreneurs—or not make loans at all. The overall effect is to reduce access to credit.

Four powers of secured creditors in reorganization and liquidation procedures have been shown to enhance credit:[21]

- Whether there are restrictions, such as creditor consent, when a debtor files for reorganization, as opposed to cases where debtors can seek unilateral protection from creditors' claims by filing for rehabilitation (as in the United States in Chapter 11 of the bankruptcy code).
- Whether secured creditors can seize their collateral after the decision for reorganization is approved—in other words, whether there is no "automatic stay" or "asset freeze" imposed by the court.
- Whether secured creditors are paid first out of the proceeds from liquidating a bankrupt firm.
- Whether creditors or an administrator are responsible for managing the business during the resolution of reorganization, rather than having a bankrupt debtor continue to run the business.

Of the four, priority payment for secured creditors in liquidation is the most widespread—in 62 percent of countries (figure 5.4). Countries that do not rank

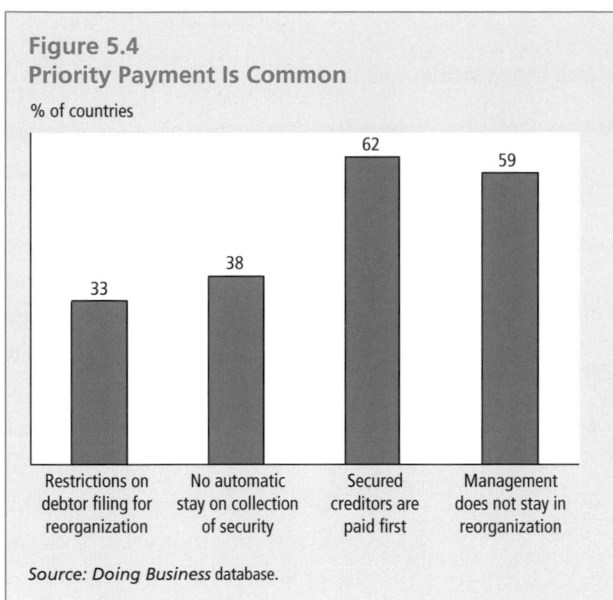

Figure 5.4
Priority Payment Is Common

Source: Doing Business database.

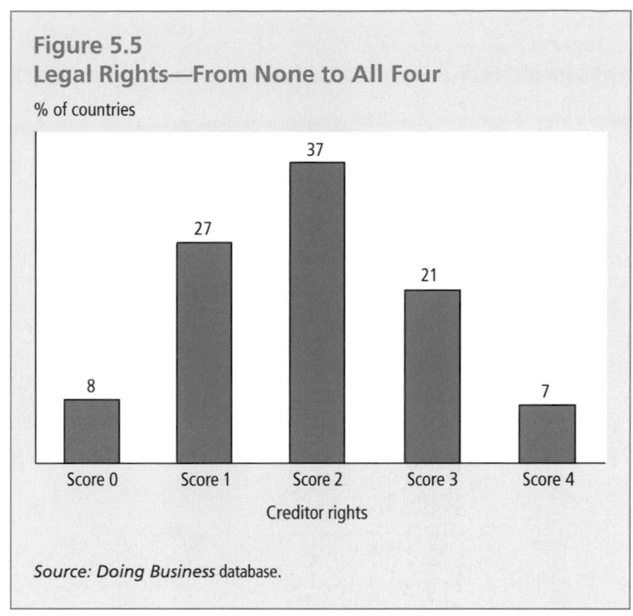

Figure 5.5
Legal Rights—From None to All Four

Source: Doing Business database.

secured creditors first usually favor employee and tax claims. In Turkey, government claims have priority over secured creditors. In France, Hungary, Poland, Peru, and the Russian Federation, labor claims get higher priority, usually including wages, benefits, and severance payments, as well as other labor claims accumulated during the period of insolvency. Brazil, Greece, India, Mexico, Romania, and West African countries give priority to both labor and government claims over secured creditors.

Around 60 percent of countries require that an administrator run the business during the reorganization. In the remaining jurisdictions, the bankrupt debtor retains the main responsibility for decisions on ordinary business, as in Argentina, Austria, Brazil, Chile, China, Greece, Italy, the Philippines, Sweden, Turkey, and the United States.

Restrictions on entering into reorganization and on a creditor's seizing and selling of collateral after a reorganization petition has been approved are less common; they exist in around every third country. In nine countries, laws do not provide for any reorganization procedure at all (Bosnia and Herzegovina, Egypt, Jordan, Kenya, Nepal, Panama, Syria, Uganda, and Zambia). In others, filing for reorganization provides automatic safe harbor from creditors' claims by means of an automatic stay. The type of automatic stay also varies significantly. In Indonesia, creditors must wait 90 days before they may enforce their security, and in Canada, 30 days. But in Benin, creditors may not enforce security until after the insolvency process, which takes 40 months on average—a duration that increases the cost and uncertainty of enforcement.

Eight percent of countries, including poor countries such as Colombia, Tunisia, and Yemen, but also France, provide none of the legal protections measured in the index. Only nine jurisdictions provide all four legal rights to creditors: Hong Kong (China), Kenya, Lebanon, New Zealand, Nicaragua, Nigeria, Panama, the United Kingdom, and Zimbabwe (figure 5.5). In the United Kingdom, a secured lender has the power to immediately appoint an administrator to take over the management of a bankrupt company and enforce security, thereby effectively blocking the possibility of a debtor's entering into a reorganization proceeding without creditor consent. The administrator is given wide powers, thereby providing the secured lender with complete control of the process and a first priority of payment.

Other aspects of secured lending regulations facilitate credit. Broadening the scope of security—the type of assets, debt, borrowers, and lenders that may be part of a security agreement—is one example.

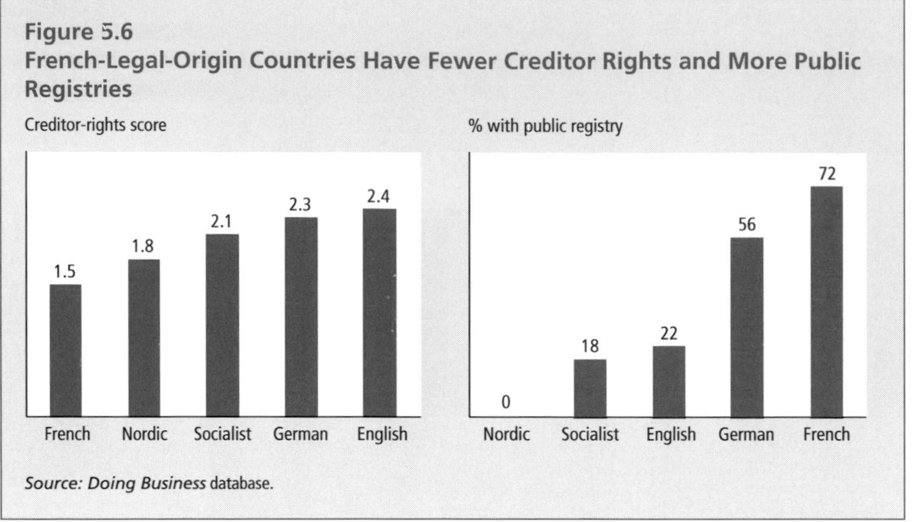

Figure 5.6
French-Legal-Origin Countries Have Fewer Creditor Rights and More Public Registries

Source: Doing Business database.

The clarity of property rights—through property registries and laws that provide creditors with priority access to the collateral in disputes outside insolvency—is also critical for giving lenders more certainty of what can be recovered on default. And the levels of intervention by courts and other public officials in creating and enforcing security may present significant obstacles to borrowers and lenders alike. (Quantitative measures of those and other aspects of laws on secured lending will be presented in *Doing Business in 2005*.)

Explaining Patterns in Creditor Protections

Do governments in rich countries "buy" good institutions? Not always. Surprisingly, poor countries are as likely as rich countries to have laws protecting creditor rights. Poor countries are also as likely to establish public credit registries.[22] Those in developed countries collect more credit information and have broader regulations on the quality of information. But they do not provide more access or distribute more types of information than the ones in poorer countries. For this reason, public credit registries in developing countries are more likely to report financial institutions as primary users. But the cost and time needed to create and enforce security is higher in developing countries. And they have weaker regulatory environments for information sharing as well as weaker enforcement of laws.

Do countries "inherit" good institutions? Yes. Legal tradition is the key determinant of creditor protections. Creditor-rights scores average 2.4 (out of a maximum of 4) in common-law countries, but only 1.5 for countries with French legal heritage (figure 5.6). Lenders also face more delays and higher costs of enforcing collateral in French-origin countries. But perhaps as a remedy for poor creditor rights, French-legal-origin countries are more likely to establish public credit registries. Three-quarters of them have public registries, compared with a quarter of common-law countries and no Nordic countries.

Do countries copy their neighbors' institutions? Looking across regions, Latin American countries are more likely to have public credit registries—71 percent, compared with only a third of OECD economies. They also have the fewest creditor rights—1.7 on average. Transition countries have an average score of 2.3. There are no other important differences across regions.

Public agencies are sometimes built to compensate for the lack of private institutions. In almost 80 percent of countries, there was no private bureau when the public credit registry was established. Countries without private bureaus are a third more likely to have public registries than countries with private bureaus (59 percent, compared with 39 percent), and those registries are more likely to report that they serve lenders rather than banking supervisors.

The presence of private bureaus is strongly associated with country wealth, although the regulatory framework for information sharing is also important. Highly concentrated lending markets—in which lenders have less incentive to share information because they could lose the rents they extract from knowing their customers—reduce the likelihood of a private registry.

Governments establish public registries as a remedy for poor protection of legal creditor rights. Countries with a public registry have significant lower

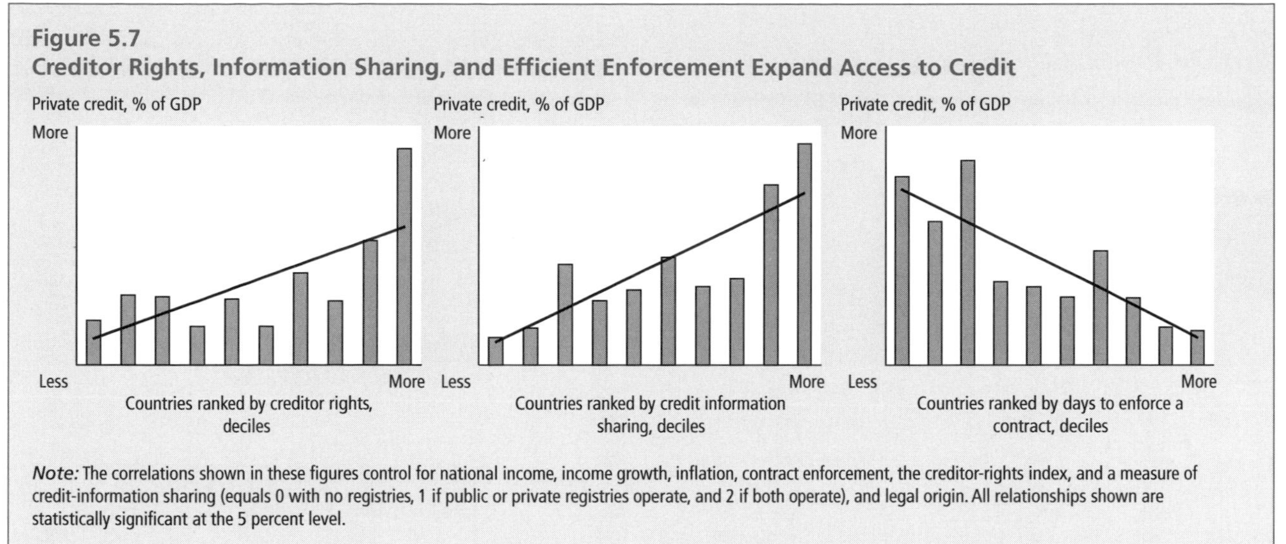

Figure 5.7
Creditor Rights, Information Sharing, and Efficient Enforcement Expand Access to Credit

Note: The correlations shown in these figures control for national income, income growth, inflation, contract enforcement, the creditor-rights index, and a measure of credit-information sharing (equals 0 with no registries, 1 if public or private registries operate, and 2 if both operate), and legal origin. All relationships shown are statistically significant at the 5 percent level.

Source: Doing Business database.

creditor-rights scores—1.6—compared with other countries—at 2.3. By providing more information for creditors to use in assessing risk, governments may compensate for creditors' weaker rights to enforce security on default. The result is consistent with banks' use of creditor protection. In Latin America—the region with the weakest legal protections—banks report that they give twice as much weight to information from credit registries as on collateral when making loan decisions.[23]

A similar substitution is evident between the ease of contract enforcement and the presence of public registries. Countries that score lower on rule of law and have more bureaucratic contract enforcement are much more likely to have a registry, a relationship suggesting that registries help remedy poor enforcement through courts. Because the reputational effect provides an incentive for borrower repayment, creditors can rely on registries as a form of contract enforcement before the fact rather than go through the courts on default. That reliance is also important where lenders face social pressure not to enforce claims.[24]

What Is the Impact on Credit Markets?

Institutions sharing credit information (public registries and private bureaus), stronger creditor rights, and better enforcement systems are associated with deeper credit markets across countries (figure 5.7).[25] The fact that the impact of information sharing is greater when controlling for creditor rights suggests that information sharing compensates for poor legal protection. The effect of creditor rights is much stronger when controlling for measures of enforcement, a finding suggesting that without enforcement, laws on the books are not enough to protect creditors.

What protections work best in which circumstances? Although both information sharing and creditor rights are good for credit market depth, the relative importance of the various creditor protections depends on country wealth. For the poorer half of the sample, information sharing has greater impact than creditor rights. But in the richer countries, the effect of credit information sharing is less significant than that of creditor rights. Legal protections—important everywhere—have more impact in rich countries.

Country wealth is an important factor for understanding whether information sharing is better organized publicly or privately. When the impact of public and private registries is analyzed separately, the effect of private bureaus on credit depth is positive and significant. The effect of having a public registry, though positive, is statistically insignificant. But this result masks important differences by income group. In the poorer half of the sample, both private bureaus and public registries are associated with more private credit, although the effect of private bureaus is larger. In poor countries, public registries help compensate for weak creditor rights, poor enforcement, and the lack of private registries. The effect of a public registry on credit-market depth in poor countries is even greater when it has achieved high coverage and high scores on the public registry rules index—that is, with broader rules on collection, distribution, access, and quality.[26]

In developed countries, public credit registries have a positive but insignificant association with private credit. This analysis does not capture their indirect impact. Registries perform supervisory functions as well as serving lenders (especially in wealthier countries), and such functions may have benefits not analyzed here.

Who benefits the most? Well-connected and large firms may find it easy to get loans without credit histories, especially in rich countries. Smaller firms in poor countries, for which information is scarce or of poor quality, gain the most. The relationship between the presence of information-sharing registries and a firm's access to formal sources of finance is significant and positive, more so in poor countries and the most for small firms in poor countries (figure 5.8).

Figure 5.8
Information Sharing Is Associated with a Firm's Access to Formal Finance

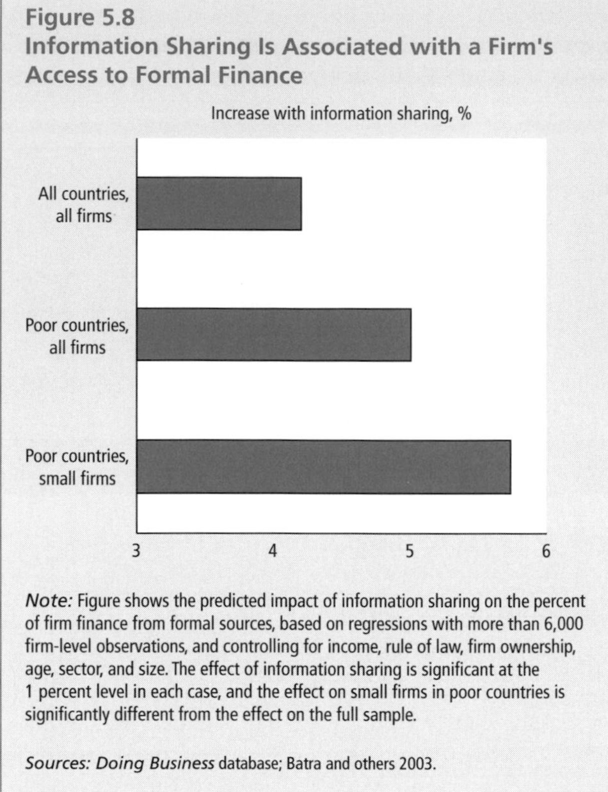

Note: Figure shows the predicted impact of information sharing on the percent of firm finance from formal sources, based on regressions with more than 6,000 firm-level observations, and controlling for income, rule of law, firm ownership, age, sector, and size. The effect of information sharing is significant at the 1 percent level in each case, and the effect on small firms in poor countries is significantly different from the effect on the full sample.

Sources: Doing Business database; Batra and others 2003.

The presence of private bureaus and public registries is also associated with a lower spread between lending and deposit rates.[27] Other studies have shown that stronger creditor rights and more information sharing are associated with lower default rates.[28] Firms in countries with information sharing are less likely to report obstacles to obtaining financing and show evidence of credit constraints.[29] Countries with stronger legal creditor protections have larger debt markets, and higher rates of capital investment and productivity growth.[30] The overall link between the development of financial markets and growth is well established.[31]

Country case studies show that introducing information sharing improves credit markets. In Chile, the establishment of a credit registry increased lending.[32] Studies of the U.S. market show that more credit information provides more power to predict defaults.[33] Simulations in European countries show that moving from no information sharing, to sharing negative data only, to sharing both positive and negative information, reduces bad loans dramatically. Lenders agree. In a survey of banks in 34 countries, more than half reported that sharing credit information reduces default rates and loan processing time and costs by 25 percent or more (figure 5.9).

What to Reform?

Facilitating Information Sharing

Establishing regulations to facilitate the sharing of credit information through private bureaus is the critical first step for poor countries and rich countries. Other steps include permitting and providing incentives for the sharing of both positive and negative information (as Hong Kong and Belgium did in mid-2003) and keeping past defaults on record. Separately, the scope of disclosure laws on financial statements can be expanded in many countries. Eliminating restrictions on access to public records can be accelerated by better technology and storage of information. Ensuring strong borrower

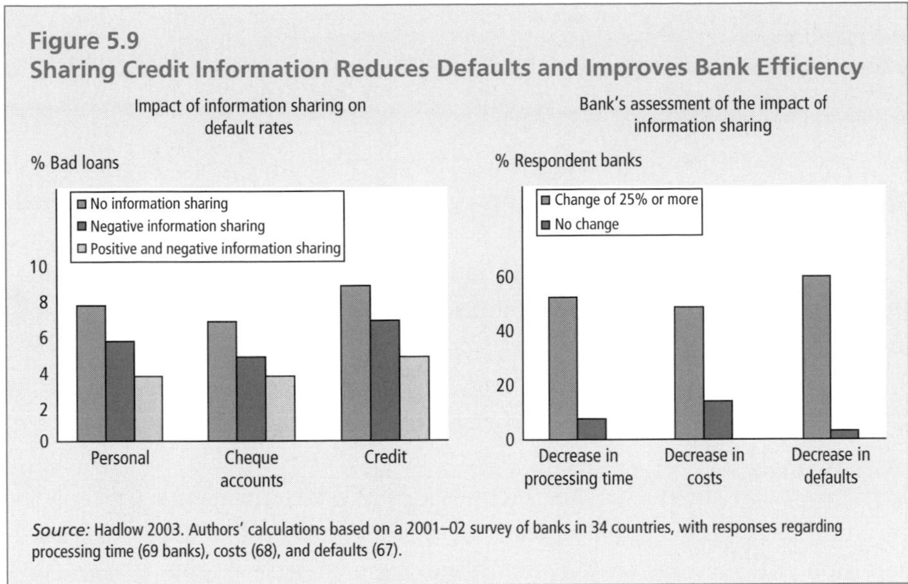

Figure 5.9
Sharing Credit Information Reduces Defaults and Improves Bank Efficiency

Source: Hadlow 2003. Authors' calculations based on a 2001–02 survey of banks in 34 countries, with responses regarding processing time (69 banks), costs (68), and defaults (67).

rights to access and correct information protects borrowers and improves the quality of information.

The impact of restrictions on information sharing—or of ambiguity in its regulation—can be severe. In Thailand, two credit bureaus have operated for several years. But in 2003 a new law imposed large fines and criminal liabilities on participating financial institutions for minor violations in sharing information, even though there are no procedures to ensure that data are shared according to the law. Both credit bureaus shut down their operations when the law was passed and reopened only five months later, when clarifying regulations were issued.

In the past few years, almost every country in the former Soviet Union has tried to set up a private credit bureau. A major impediment is the secrecy provision in the banking or data protection laws. In Armenia, Georgia, Kazakhstan, and the Russian Federation, there is a debate about whether requiring the borrower's authorization would be sufficient. Some believe so, but others think the potential liability for credit bureaus is too high. No private bureaus operate in those countries.

Laws on credit reporting help overcome lenders' unwillingness to share information—from fear of losing good borrowers to competitors, unfamiliarity, or concern over liability related to privacy or bank secrecy. But laws are rigid and must be designed in sufficiently general terms to reflect a rapidly changing industry. Laws also take time to be approved. Alternative government and central bank support has proven effective in many countries. Such support has taken the form of central bank directives, standards, penalties for noncompliance (as in Mexico, where the central bank imposes 100 percent provisioning requirements when data are not submitted to the bureau), or even letters of endorsement to banks (as in India and the Dominican Republic). Bureau codes of conduct, as in Singapore and under development in Saudi Arabia, are another more flexible way to set standards and build consensus among lenders, government, and borrowers. The extensive consultative process to develop a code of conduct not only facilitates lender compliance but also improves data quality by allowing parties to reach agreement on feasible standards and formats.

Entry of one of the major international credit reporting firms can accelerate the process of establishing a private credit registry. In the Czech Republic, Guatemala, India, and Mexico, private bureaus are being formed in joint ventures with foreign firms, which provide technical assistance and expertise. Countries need to ensure that there are no legal obstacles to such foreign investment.

Especially in poor countries and those with highly concentrated lending markets, such measures may be insufficient to attract private investment. Establishment of a public registry may offer the advantage of rapid setup because it uses central bank regulation rather than new laws. Direct enforcement by bank supervisors can counter lenders' unwillingness to comply. And establishing a public registry is cheap. The one in Mozambique cost only a few hundred thousand dollars to establish.

The design of registries is important for expanding access to credit in poor countries. Broader rules on the collection, distribution, and quality of information can expand coverage, with help from new technology. In Malaysia, a new online real-time system was introduced in the public registry in 2001. It provides coverage of all loans, instant responses to requests, and more frequent updates of credit information among financial institutions. Almost half of the registries surveyed reported the intention to upgrade their technology.

In establishing public registries, care must be taken not to stifle private information sharing. As the credit market matures, or as private initiatives materialize, public registries can be restructured to complement them by focusing on overall supervision and sharing data with the private registries, as happened recently in Mexico. The registries in Argentina, the Dominican Republic, and Peru share data with private bureaus (Bolivia will soon follow suit). Some successful strategies entail more extensive private-public partnership. Sri Lanka's credit registry was set up in 1990, with 51 percent of the capital held by the central bank, the rest shared among commercial financial institutions. The government's shareholding declines as more institutions join the registry. In Singapore and Thailand, the government initiated the establishment of private bureaus.

Legal Rights To Create and Enforce Security

Efficient courts are essential for enforcing the rights of creditors, especially unsecured lenders. For secured lending, reforms beyond the courts are necessary, simplifying the steps and reducing the costs of creating and enforcing security. Eliminating the stamp duties and taxes for creating collateral agreements—as well as the requirements to notarize documents—can substantially reduce costs. Introducing out-of-court enforcement enhances the powers for secured creditors to recover debt, as in Albania, Germany, Thailand, and the United States.

Summary enforcement proceedings through the courts are another effective reform. Moldova's reforms in 2001 introduced a fast-track enforcement procedure. Ten days after notifying the debtor, the other creditors, and the collateral registry of a default, creditors may file an enforcement order with the court. All that is required is evidence of the notifications, the default, and the security agreement. Within three days, the judge reviews the documentation and issues an enforcement judgment. There is no judicial analysis of the cause of dispute.

Such measures increase the importance of well-written security contracts, to avoid problems in the judicial review of valid documentation. And they radically change the debtor's incentives to appeal and delay the enforcement process. In Moldova, appeals are possible but must be undertaken in a separate trial. If the debtor loses, he will bear all costs. So appeals are likely only if the debtor has a genuine dispute or grievance. Since the reform, the time to seize and sell security has fallen from more than three years to around 70 days.

More-comprehensive reforms also address the scope and clarity of rights in security agreements. To begin with, countries must allow the debtor to retain possession and use collateral. Doing so is still impossible in Serbia and Montenegro, where the lender must take possession of assets to have a valid charge—hardly a practical solution for borrowers who pledge business equipment. Also important, especially for small firms, is introducing instruments that allow security for a changing pool of assets such as inventory, receivables, property that will be realized in the future (e.g., crops), or a whole enterprise especially for small firms.

Clear rules that anticipate and resolve priority conflicts are essential in defining the property rights of secured creditors. Registries of collateral agreements, where lenders can check for existing liens, also support the clarity of property rights. In the United States, a lender can check by searching in an electronic registry of almost all collateral agreements. Not so in three-quarters of the world, where registries are limited to certain types of property, such as land (including many rich countries such as Germany and France). New technology makes such registries inexpensive. In a few countries, such as New Zealand, the collateral registry interface is operated by the credit bureau. That is a win-win reform. The bureau benefits from additional information on borrowers, and the government benefits by having a sophisticated electronic registry administered by experts in information technology systems.

Finally, effective reforms of secured lending require attention to insolvency as well as collateral laws. The powers allocated to secured creditors in the insolvency process are a crucial determinant of access to credit. Good collateral instruments facilitate other goals of bankruptcy also—by providing the right incentives for liquidating unviable companies and rescuing viable firms. For example, liquidation is more efficient when collateral is concentrated in the hands of one main creditor, and sale as a going concern is more likely if the whole enterprise is pledged as security for a loan.[34]

Notes

1. Other related factors, such as macroeconomic conditions, banking supervision, and ownership, are not discussed here.
2. Garro 1998.
3. Campbell and Kracaw 1980; Diamond 1984, 1991; Stiglitz and Weiss 1988.
4. Jappelli and Pagano 1993; Miller 2003.
5. Greif, Milgrom, and Weingast 1994; Besley 1995.
6. Hoffman, Poste-Vinay, and Rosenthal 1998.
7. Olegario 2003.
8. Miller 2003.
9. The distinction between business and consumer reporting is not as clear as it may seem. Some registries cover both consumers and firms. Also, business loans are often made on the basis of personal credit histories, especially for small firms and entrepreneurs. This chapter covers both.
10. Applying a broader definition of private credit registries that do not directly exchange information among financial institutions (which includes investigative-style credit reporting firms), private registries operate in approximately 70 percent of sample countries.
11. The survey of public and private credit registries was developed with the World Bank's Credit Reporting Systems Project (http://econ.worldbank.org/programs/2245).
12. Although fees might restrict access, they may also indicate orientation toward serving lenders. This variable is therefore not included in the index of public credit registries.
13. Jappelli and Pagano (1993) build a theoretical model showing how information sharing depends on market conditions such as competition and heterogeneity and mobility of borrowers.
14. Jappelli and Pagano 2000; Jentzsch 2003a.
15. Surveys were conducted of the legal departments of private and public credit registries, as well as banking supervisors. *Doing Business* project and Jentzsch 2003a.
16. Jentzsch 2003b.
17. Keinan 2001.
18. In the strict sense; in a secured transaction, the ownership title of the asset remains with the debtor. But in many countries there are common functional equivalents known as "title finance"—for example, leasing and conditional sales—whereby the creditor takes title of the asset.
19. Mann 1997; Hill 2002.
20. Survey estimates from secured transactions lawyers indicate that (on average) around 30 percent of collateral enforcement is *inside* insolvency proceedings. However, in many countries, including Nigeria, Albania, Bolivia, and the Russian Federation, experts estimate that the rate is under 10 percent.
21. This four-point measure of creditor rights was developed by La Porta, Lopez-de-Silanes, Shleifer, and Vishny (1998) and originally covered 49 countries as of 1995.
22. Except that those in the richest quartile are less likely than those in the upper-middle quartile to have a public credit registry.
23. Miller 2003.
24. See, for example, Besley 1995.
25. Banks and nondepository financial institutions (IFS line 22d).
26. Djankov, McLiesh, and Shleifer 2003.
27. The analysis excludes three outliers with extremely high spreads (Brazil, Uruguay, and Bolivia).
28. Jappelli and Pagano 2002.
29. Love and Mylenko 2003; Galindo and Miller 2001.
30. La Porta and others 1998; Levine 1998.
31. King and Levine 1993; Levine 1997; Rajan and Zingales 1998; Demirguc-Kunt and Maksimovic 1998.
32. Fuentes and Maquieira 2001.
33. Barron and Staten 2003.
34. Bolton and Scharfstein 1996; Hill 2002.

6 Closing a Business

The penalty for declaring bankruptcy in Ancient Rome was slavery or being cut to pieces. The choice was left to the creditor. By the Middle Ages, the treatment of insolvent debtors had softened considerably. In Northern Italy, bankrupt debtors hit their naked backside against a rock three times before a jeering crowd and cried out, "I declare bankruptcy."[1] In French medieval cities, bankrupts were required to wear a green cap at all times, and anyone could throw stones at them. In England, bankrupt debtors were thrown into prison, were often pilloried, and occasionally had one ear cut off.[2]

The English bankruptcy law of 1732 was the first modern bankruptcy law. The United States introduced its first bankruptcy law in 1800, copying the English law.[3] France, Germany, and Spain adopted their first bankruptcy laws in the early nineteenth century. Imprisonment still featured as a common punishment, and bankruptcy was seen a means to liquidate financially distressed companies and distribute their remaining assets among creditors. A rudimentary rehabilitation procedure—designed to reorganize the debt of a bankrupt firm so that it could continue operating—was developed in Austria in 1914 but was rarely used. Similar procedures were introduced in Spain in 1922, in South Africa in 1926, and in Belgium, France, Germany, Netherlands, and the United States in the 1930s.[4]

A modern reorganization procedure did not appear until 1978, when Chapter 11 was adopted in the United States. In the next 25 years a wave of bankruptcy reforms brought reorganization procedures to Italy in 1979, France in 1985, the United Kingdom in 1986, New Zealand in 1989, Australia and Canada in 1992, Germany in 1994 and 1999, Sweden in 1996, and Japan and Mexico in 2000, to name a few. By mid-2003 every country in the *Doing Business* sample, with the exception of Cambodia, had bankruptcy laws.

Today's bankruptcy regimes differ enormously in their efficiency and use. Canada, Ireland, Japan, Norway, and Singapore take less than a year to resolve bankruptcy. Brazil, Chad, and India take more than a decade. In Norway and Singapore it costs about 1 percent of the value of the estate to resolve insolvency. In the Czech Republic, the Philippines, Thailand, Uganda, and Venezuela, it may cost as much as half the estate to go through formal bankruptcy. Angola, Bangladesh, Burundi, Mozambique, and Togo have bankruptcy laws on the books, but they are almost never used. In Belarus and Uzbekistan, bankruptcy is used mostly to liquidate dormant enterprises.

Bankruptcy is still in its infancy in many countries, and reform continues even in the best-performing jurisdictions. The average age of the bankruptcy law in the 10 best-practice countries—Belgium, Canada, Finland, Ireland, Japan, the Republic of Korea, Latvia, the Netherlands, Norway, and Singapore—is six years. Some countries, such as Egypt and Pakistan, are in the process of revising their bankruptcy laws. Brazil and Spain just did.

Three areas of reform hold the most promise. One is choosing the appropriate way of dealing with insolvency given a country's income and institutional capacity. Poor countries are generally better off with effective debt enforcement outside of insolvency than with complicated bankruptcy laws and specialized courts. The second is increasing the involvement of stakeholders in the insolvency process rather than relying on a court to make business decisions. The third is training judges and bankruptcy administrators in insolvency law and practice.

Some countries, such as Latvia and Mexico, have recently taken steps in all those areas, with significant improvements in the efficiency of bankruptcy procedures. Others, such as Germany and Japan, have reformed their bankruptcy law to make it more attractive to debtors. A third group of countries, including Argentina and Moldova, focuses on the training of judges. Still other countries, such as Tanzania and Thailand, have reformed their judicial structure to allow for specialized courts or specialized sections within courts.

Countries with ill-functioning judiciaries are better off without sophisticated bankruptcy systems. There is a general misperception that bankruptcy laws are needed to enforce creditors' rights. In practice, the laws usually exacerbate legal uncertainty and delays in developing countries. Private negotiation of debt restructuring under contract law—and as discussed in the previous chapter, the efficient enforcement of secured-debt contracts outside insolvency under collateral law—will succeed better.

What Are the Goals of Bankruptcy?

The goals of bankruptcy are universal. The first goal is to maximize the total value of proceeds received by creditors, shareholders, employees, and other stakeholders. Businesses should be rehabilitated, sold as a going concern, or liquidated—whichever generates the greatest total value. The second goal is to rehabilitate viable businesses and liquidate unviable ones. In other words, bankruptcy law should be neither hard on good businesses nor soft on bad ones. The third goal is a smooth, predictable transition in claims priority between good and bad financial states of the company—to reduce investors' risk. That goal is achieved by maintaining the absolute priority of claims in bankruptcy. Good bankruptcy laws generally achieve the three goals. Bad ones do not. As a result, they make everyone worse off—both debtors and creditors.

Goal #1: Maximizing Value

The value of a bankrupt business is maximized when less of it is dissipated in the direct and indirect costs of bankruptcy—and the debtor is liquidated, sold, or rehabilitated quickly. If bankruptcy is expensive and drawn out, both the distressed companies and their creditors will avoid it. Even if resolution is successful, large costs are likely to drain the resources of already-distressed company. Similarly, if the bankruptcy process lasts too long, the focus of management will be on immediate, process-related tasks rather than on strategic issues. And suppliers and customers will be likely to find a way to cease dealing with the bankrupt business.[5] For example, the bankruptcy process in Brazil takes 10 years and as a result is seldom used.

A survey of bankruptcy lawyers and judges, conducted in cooperation with the International Bar Association, estimated the time it takes to complete the insolvency procedure and its cost, as a share of total estate value.[6] The estimates refer to the insolvency of a domestic company running a hotel in the downtown area of a country's most populous city. The main features of hypothetical case are as follows.[7] The company's only significant asset is the real estate on which the hotel operates. The hotel is mortgaged to a domestic commercial bank, its main creditor. The company has 201 employees and 50 suppliers, and its revenues are fixed as a multiple (1,000 times) of the per capita income of each country. That amounts to $34 million in annual revenues in the United States, or $240,000 in Madagascar. The revenues were calibrated to match the business volume of a medium-size hotel business.

The company is controlled by a majority shareholder and is not publicly traded. It defaults on its bank loan after a difficult financial year but continues to operate and make payments to unsecured creditors. The bank prefers to liquidate its security in the fastest and cheapest way, while management and the main owner try to keep the company in operation. The unsecured creditors—holding 99 percent of the claims by number but only 26 percent by value—support the rescue effort. The company is assumed to be worth more as a going concern than it would be in a piecemeal liquidation.

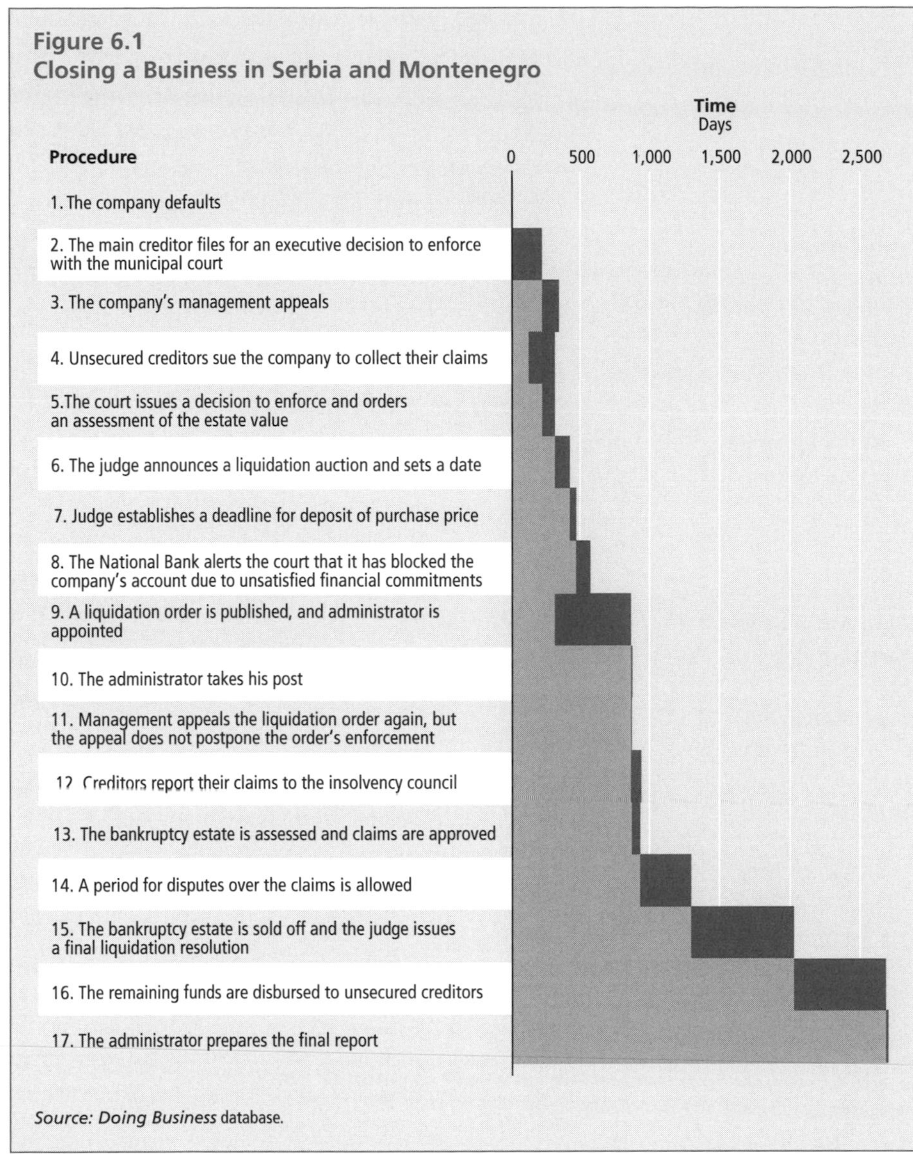

Figure 6.1
Closing a Business in Serbia and Montenegro

Source: Doing Business database.

On the basis of this hypothetical case, lawyers and judges in all countries completed a survey on the sequence of procedures and their timing in the insolvency process. Time is measured in days, as the respondent attorneys answer questions about the duration of each bankruptcy procedure. Cost is defined as the cost of the entire bankruptcy process, including court costs, insolvency practitioners' costs, and the costs of independent assessors, lawyers, and accountants. For a hypothetical financially distressed company in Argentina, from the moment it files for bankruptcy to the actual resolution, the insolvency process lasts two years and nine months and costs 18 percent of the value of the bankruptcy estate. The outcome is inefficient, because the assets are sold piecemeal, even though the business would be worth more as a going concern. In Serbia and Montenegro, the liquidation procedure takes more than seven years and costs about 38 percent of the value of the bankruptcy estate (figure 6.1). In practice, creditors can and do appeal the final distribution of proceeds, so the process could take another six months to one year.

Canada, Finland, Japan, Norway, and Singapore are among the top 10 countries in both shortest time and lowest cost (table 6.1). The Czech Republic, the Philippines, and Serbia and Montenegro are among the bottom 10. Developed countries have more-efficient bankruptcy procedures, especially in their duration. All 10 of the jurisdictions with the fastest procedures are high-income countries. Nine of the 10 countries with the cheapest bankruptcy procedures are high-income countries.

Across regions, South Asian jurisdictions have the most time-consuming bankruptcy procedures. Their average duration is more than five years, and the average cost is 9 percent of the bankruptcy estate (figure 6.2). The most expensive bankruptcy procedures are in East Asian countries, averaging 20 percent of the estate. Other regions where bankruptcy is costly are Africa, Europe and Central Asia, and Latin America. Countries in those regions also have fairly long procedures, 3 to 4 years. In contrast, OECD

Table 6.1
Time and Cost of Bankruptcy Procedures

Fastest	Slowest
Ireland	India
Japan	Chad
Singapore	Brazil
Canada	Czech Republic
Taiwan (China)	Mauritania
Belgium	Serbia and Montenegro
Finland	Panama
Norway	Indonesia
Australia	Chile
Hong Kong (China)	Philippines
Cheapest	**Most Expensive**
Singapore	Macedonia, FYR
Finland	Israel
Norway	Venezuela, RB
Netherlands	United Arab Emirates
Colombia	Uganda
Georgia	Chad
Kuwait	Czech Republic
Japan	Serbia and Montenegro
Canada	Panama
New Zealand	Philippines

Source: Doing Business database.

countries have procedures that are cheap (less than 8 percent of the estate value) and short (less than two years on average). With the exception of developed countries, insolvency is a long and expensive process all around the world. In developing countries it takes three or more years and costs 15 percent of the estate value.

Some other findings on time and cost:

- Nordic proceedings are the fastest, at around two years on average, and also the cheapest, at 4.5 percent of the estate value. Finland and Norway are among the world's top countries on time and cost.
- English-legal-origin countries are the second-fastest legal-origin group in resolving insolvency, at 2.7 years.
- In French-civil-law countries, insolvency lasts on average 3.7 years, and it costs 15 percent of the estate value.
- In transition countries, the process lasts around three years and costs 7 percent of the estate value.
- Some of the poorest countries seem more efficient than many of the middle-income countries. This finding is in part a product of the low use of the judicial system in poor countries. For example, interviews in Bangladesh reveal that bankruptcy is almost never used as a mechanism for resolving distress. The only cases that go through the bankruptcy system deal with state-owned enterprises; those cases typically involve the write-off of debt.

In the United States, there are more than 55,000 corporate bankruptcy cases each year, 20 per 100,000 population.[8] In the United Kingdom, there are some 40,000 a year, about 75 per 100,000 population. In contrast, about 500 bankruptcy cases were started in Spain, about 1 per 100,000 population. In some developing countries, bankruptcy is often used: more than 1,000 bankruptcy petitions were filed in Malaysia in 2002, about 17 per 100,000 people. In Belarus, Egypt, and Uzbekistan, nearly 1,000 cases are filed each year. But most of them are requests for the liquidation of dormant enterprises (as in Belarus) or serve merely as a threat point for private negotiations of debt restructuring (as in Egypt, where fewer than 5 percent of filings result in true bankruptcy proceedings). In Albania, the Democratic Republic of Congo, Burundi, Ghana, Haiti, Honduras, Laos,

Figure 6.2
Time and Cost to Resolve Bankruptcy

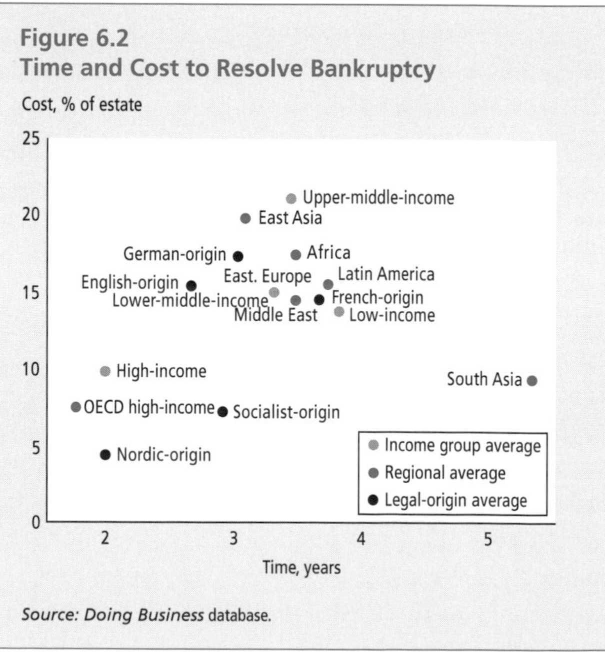

Source: Doing Business database.

Madagascar, and Vietnam, bankruptcy is almost never used by private companies or banks. (In such countries, the cost documented in the survey refers to bankruptcy of state-owned enterprises or subsidiaries of foreign firms.)

Goal #2: Rescuing a Viable Business

Bankruptcy law is often oriented to closing down unviable companies. But sometimes the bias toward discontinuing business leads to the premature liquidation of companies in temporary distress—and thus a loss of value to society.

The hypothetical case in the survey allows an investigation of whether viable companies will be rescued. It is a rescue case, since the company is more valuable as a going concern than closed down and sold off piecemeal.[9] The efficient outcome is defined as any bankruptcy procedure (rehabilitation, foreclosure, or liquidation) that results in a going-concern sale without an interruption in operations, or a successful rehabilitation with management dismissed. The sale of the hotel to a real-estate developer who will convert it into office space is less efficient than a sale to another hotel operator would be, and value will be lost even if the firm is sold to a hotel operator if operations are interrupted. The company may also remain in the hotel business with its present owner and management. That would not be efficient, though, because the management was in part responsible for the poor financial situation of the company.

Several countries liquidate the indebted company immediately and auction it as a going concern. This task may be accomplished by means of liquidation procedures, as in Austria, the Netherlands, and Poland—or by foreclosure of the secured debt outside of insolvency, as in El Salvador, Jamaica, New Zealand, and Singapore (table 6.2). Nordic countries, such as Denmark and Sweden, have bankruptcy regimes oriented toward quick liquidation. Finland allows the failing company to propose a rehabilitation plan—but the process hinges on creditors' approval, and swift liquidation typically follows. Several other jurisdictions—Belgium, Canada, Colombia, the Islamic Republic of Iran, the Republic of Korea, Peru, Portugal, Thailand, and the United States—allow rehabilitation proposals, but they are typically followed by liquidation as a going-concern sale. Japan, Spain, Taiwan (China), and Vietnam successfully adopt a rehabilitation plan wherein management is replaced. (Here it should be noted that some countries achieve the efficient outcome but do not reach other goals of insolvency—for example, in Poland a firm would be sold as a going concern in liquidation but at a high

Table 6.2
Jurisdictions with Efficient Bankruptcy Outcomes

Liquidation (going-concern sale)	Foreclosure (going-concern sale)	Unsuccessful Rehabilitation Followed by Liquidation (going-concern sale)	Successful Rehabilitation with Management Replaced
Austria	Australia	Belgium	Albania
Botswana	El Salvador	Canada	Ireland
Denmark	Ethiopia	Colombia	Japan
Netherlands	Haiti	Finland	Kazakhstan
Poland	Hong Kong (China)	Iran, Islamic Rep. of	Latvia
Slovak Republic	Israel	Korea, Rep. of	Norway
Sweden	Jamaica	Mexico	Senegal
Uganda	Kuwait	Peru	Spain
Venezuela	New Zealand	Portugal	Tanzania
	Serbia and Montenegro	Thailand	Taiwan (China)
	Singapore	United States	Vietnam

Note: In Ireland the efficient outcome is achieved through successful adoption of a plan whereby the firm is sold as a going concern.

Source: Doing Business database.

Figure 6.3
Nordic-Origin and High-Income Countries Have Efficient Bankruptcy Outcomes

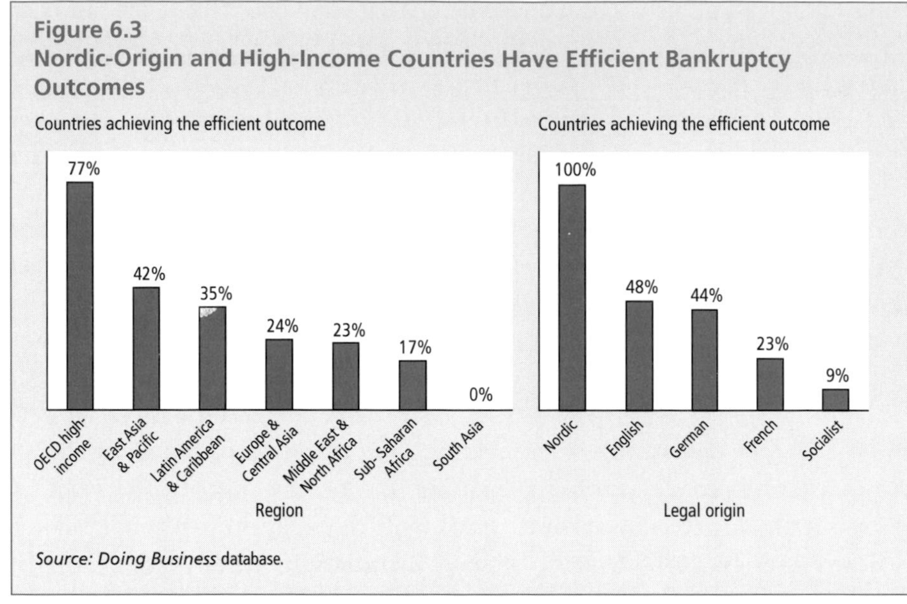

Source: Doing Business database.

cost—18 percent of the estate—and without priority payment for the secured lender.)

Almost all high-income countries achieve the efficient outcome. It is always achieved in Nordic countries, and in about half of common-law and German-legal-tradition jurisdictions (figure 6.3). Empirical studies of Swedish bankruptcy show that liquidation leads to successful sale as a going concern in more than three-quarters of the cases.[10] The probability of achieving an efficient outcome is 42 percent in East Asia, 23 percent in French-legal-tradition jurisdictions, and only 9 percent in socialist-legal-tradition jurisdictions. South Asia is the least efficient, with no country achieving the efficient outcome. One in three countries in Latin America achieves the efficient insolvency outcome, compared with one in four in Eastern Europe and Central Asia and one in five in Africa.

Goal #3: Keeping the Order of Claims Stable

The bankruptcy system ensures the stability of creditors' claims between normal times and times of financial distress. Senior claims need to be paid off before any others. Stability of priority is important for two reasons. First, senior creditors will be reluctant to lend if they do not have a predictable priority to their claim after a company is in bankruptcy. Second, having different priorities inside and outside of bankruptcy can result in perverse incentives, with some creditors wasting resources to induce management either to forestall or to precipitate bankruptcy.

Bankruptcy laws favor secured creditors, employees' claims, or taxes.[11] In the hypothetical case, two-fifths of the countries, including Armenia, Botswana, Panama, and Uruguay, favor secured creditors over employees' and tax claims (table 6.3). The majority of them give priority to labor over tax claims. Canada, France, Hungary, Portugal, and Spain give priority to employees' claims, at the expense of secured claims and taxes. On average, common-law countries favor secured creditors, whereas countries in the French and socialist legal traditions favor taxes and labor. Lower-income countries are less likely to give priority to secured lenders.

Countries that give secured lenders top priority are also more likely to have efficient insolvency systems that save viable businesses and liquidate bad ones (figure 6.4). Priority creates incentives for all parties to work toward an efficient outcome. Without the assurance that their claims will be paid first, senior

Table 6.3
Priority of Claims—Country Examples

Secured Claims Have Top Priority	Labor Has Top Priority	Taxes Have Top Priority
Bolivia	Brazil	Bosnia
Belgium	France	Chile
Bulgaria	Greece	Egypt
Cambodia	India	Jamaica
Canada	Niger	Lebanon
China	Poland	Taiwan (China)
Finland	Russia	Tanzania
Germany	Thailand	Turkey
Iran, Islamic Rep. of	Vietnam	Uganda
Kenya	Yemen	Uzbekistan

Source: Doing Business database.

creditors are likely to block a company's entry into bankruptcy procedures, even if it would lead to an efficient outcome. Once the company is in insolvency, creditors are encouraged to move the process toward the efficient outcome—if they are confident of the priority of their claim.

Figure 6.4
Priority of Secured Claims Is Associated with Efficient Outcomes

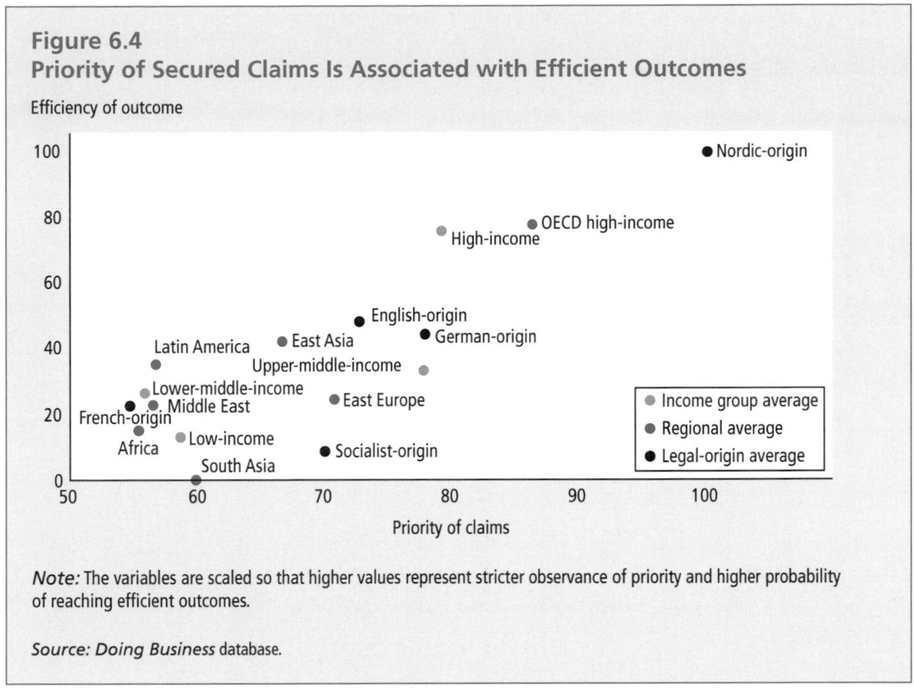

Note: The variables are scaled so that higher values represent stricter observance of priority and higher probability of reaching efficient outcomes.

Source: Doing Business database.

Who Achieves the Goals of Bankruptcy?

Insolvency proceedings thus differ in their length and cost, achievement of an efficient outcome, and preservation of a stable ordering of claims. Which countries manage to achieve all those goals best? Developed countries generally achieve the goals of insolvency, with Canada, Japan, the Netherlands, New Zealand, and Singapore among the top 10 (table 6.4). Ireland and the United States (not shown in the table) are tied for eleventh. The least effective bankruptcy regimes include mainly African countries—Angola, Burundi, Chad, the Democratic Republic of Congo, Ghana, Guinea, Rwanda, and Togo—along with Lao PDR in East Asia, and Honduras in Latin America.

Latvia is perhaps the biggest surprise. A transition economy that only recently revised its bankruptcy laws, Latvia is now among the top ten countries where bankruptcy is effective in achieving the goals of insolvency. Indeed, Latvia adopted its first postsocialist law only in 1996. Subsequent amendments in 2001 defined the power of the insolvency administrator. Lithuania, Moldova, and the Russian Federation have also improved their bankruptcy laws in the past five years.

Transition economies are not the only reformers. With the exception of Norway, all 10 of the best-practice jurisdictions have revised their bankruptcy law since 1990. On average, their current laws are six years old. In the least effective countries, the average bankruptcy law is more than 40 years old.[12]

Table 6.4
Where Is Bankruptcy Most Effective in Achieving the Goals of Insolvency—And Where Least?

Most	Least
Singapore	Angola
Finland	Burundi
Norway	Congo, Dem. Rep.
Netherlands	Guinea
Japan	Rwanda
Canada	Togo
Belgium	Chad
Latvia	Lao PDR
Korea, Rep. of	Ghana
New Zealand	Honduras

Source: Doing Business database.

Reform to improve bankruptcy is not about being friendly to creditors or to debtors. Singapore is extremely creditor-friendly, closely followed by Ireland. Germany, Japan, the Netherlands, and Norway are thought to balance the inte-rests of debtors and creditors. Belgium is very debtor-friendly.[13] Yet all those jurisdictions have quick and cheap bank-ruptcy procedures, reach the efficient outcome, and maintain the absolute priority of claims.

Some countries—especially French-legal-origin jurisdictions in Africa—have attempted to reach the goals of insolvency by giving broader powers to the court. Three powers are especially important. First is the involvement of stakeholders in the appointment and replacement of the insolvency administrator. Second is access to information throughout the insolvency process—specifically, whether the laws require that the bankruptcy administrator submit reports only to the court or also to other stakeholders. Third, in some countries the court adopts a plan for rehabilitating a bankrupt debtor—in others, creditors and other stakeholders are required to accept the plan before it can be implemented.

Expanding court powers in bankruptcy proceedings on those dimensions has not had the desired effects (figure 6.5). Countries with more court power are less likely to achieve the goals of insolvency, even controlling for income. Moreover, higher levels of court power are associated with more corruption—again, even controlling for income. In such jurisdictions, less court involvement is needed, not more. Involving creditors and other stakeholders in the bankruptcy process is important.

Figure 6.5
More Court Power—Less Likely to Achieve the Goals of Insolvency

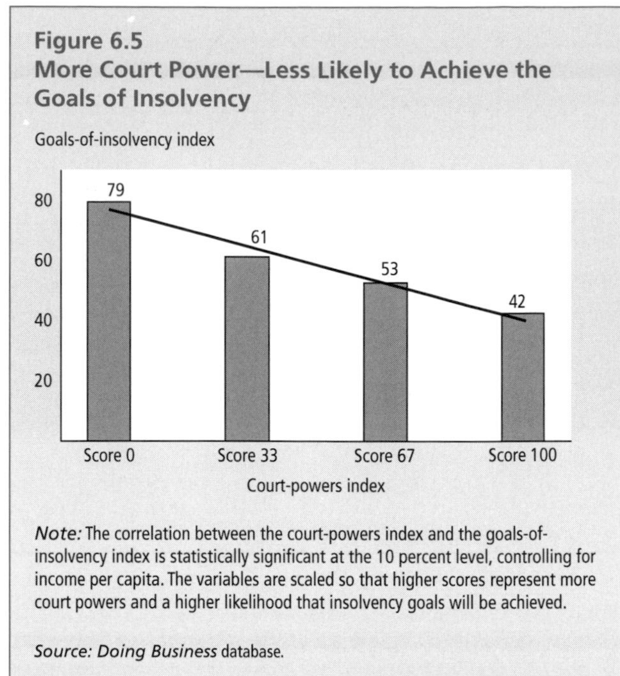

Note: The correlation between the court-powers index and the goals-of-insolvency index is statistically significant at the 10 percent level, controlling for income per capita. The variables are scaled so that higher scores represent more court powers and a higher likelihood that insolvency goals will be achieved.

Source: Doing Business database.

Effects of Good Bankruptcy Laws

The main test of whether bankruptcy laws and judicial procedures are good is whether financially distressed companies and their stakeholders use them. If companies do not see incentives to enter bankruptcy, for example when no rehabilitation procedure exists or when management gets fired automatically, few bankruptcies will take place. Similarly, if creditors find bankruptcy unattractive—for example, if they are left out of the formulation or adoption of a rehabilitation plan, or if there is a prolonged stay on assets—they will find other means of resolving their claims. The result: less chance of maximizing the value of the estate, of achieving the efficient outcome and stability of claims.

That is precisely what happens in many countries with an obsolete bankruptcy regime or with inefficient judicial processes. Interviews with the five largest banks in Mozambique, which account for about 90 percent of bank loans to enterprises, reveal that they never use formal bankruptcy. Instead, each bank has a large debt recovery department that negotiates defaulted loans directly with the customers. Similarly, private banks have very rarely used bankruptcy in Bangladesh, Benin, Burkina Faso, Cameroon, Nepal, Niger, Mali, and Mongolia. When the bankruptcy law is used in those countries, it is generally only to clear a state-owned company from debt—or to liquidate a subsidiary of a foreign company. Though existing on paper, the law is not used in the course of doing business. By contrast, in countries where bankruptcy procedures are efficient, many cases are filed. Belgium, Denmark, Finland, Norway, Sweden, and Switzerland have the highest incidence of filings—on average, 50 companies for every 100,000 citizens each year (figure 6.6).

In countries without efficient bankruptcy procedures, out-of-court negotiations (workouts) are the main mechanism for reorganizing debt (table 6.5).[14] Workouts are usually faster, cheaper, and more predictable than formal bankruptcy. Contracts can be written so as to avoid reference to the bankruptcy law—by using blank promissory notes, writing leasing contracts, or giving power of attorney to creditors.[15] But the contracts may be difficult to enforce in countries where either party has strong rights in

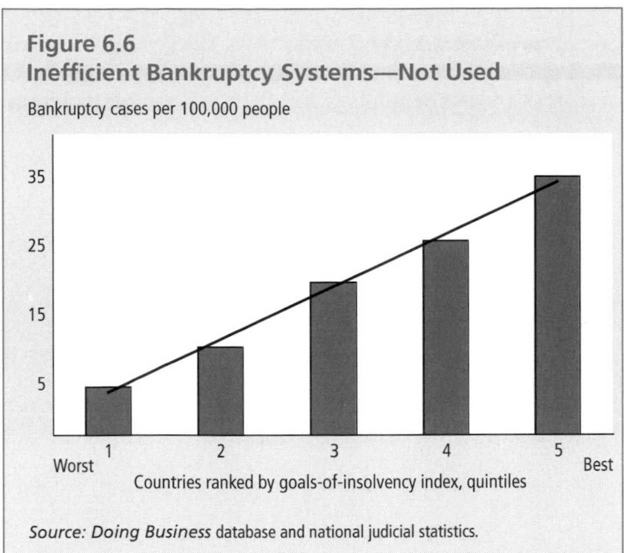

Figure 6.6
Inefficient Bankruptcy Systems—Not Used

Source: Doing Business database and national judicial statistics.

Table 6.5
Private Workouts—the Norm in Countries with Inefficient Bankruptcy

Bangladesh	Ghana	Mauritania
Belarus	Guatemala	Mozambique
Bolivia	Indonesia	Pakistan
Brazil	Iran, Islamic Rep. of	Panama
Costa Rica	Jamaica	Portugal
Egypt	Malaysia	Turkey
Georgia	Mali	Uruguay

Source: Doing Business database.

formal bankruptcy, because that party will hold out and cause delays.

Where the process of bankruptcy is efficient, access to external credit is both easier and cheaper.[16] That is so because creditors can be reasonably sure of collecting on their loans when a firm fails. The enforcement of such rights in bankruptcy is also shown to be associated with deeper private credit markets and smaller interest-rate spreads in developed countries (figure 6.7). Such is not the case in poor countries, suggesting again that financial distress would be better addressed by private negotiations under contract law, without using the bankruptcy law.

What to Reform?

There are several ingredients in an efficient bankruptcy system.[17] One is the choice of appropriate institutions for dealing with bankruptcy, given a country's income. A second ingredient is the involvement of stakeholders rather than the court in business decisions. A third is the availability of well-trained judges and bankruptcy trustees, supported by well-functioning clerical and administrative staff. Some developed countries, such as Italy, are known to have less-than-efficient legal provisions in bankruptcy but a very efficient judicial process. In contrast, many developing countries—such as Côte d'Ivoire, Georgia, and the Philippines—have good bankruptcy laws on the books but an inefficient judicial process. In either case, creditors perceive the bankruptcy system as inefficient and seldom use it.

Choosing Appropriate Institutions

What constitutes good bankruptcy law? The answer depends on the capacity of the judiciary to deal with sophisticated commercial cases. Where judges are well trained and have the support of clerks to do research and manage the workflow, where accounting practices are reliable and the legal profession is experienced in handling business litigation, the law can provide a menu of options—including liquidation and rehabilitation under bankruptcy provisions, as well as enforcement of collateral agreements outside of insolvency under secured-transactions and contracts law. Only high-income countries and a few upper-middle-income countries (Republic of Korea, Latvia) meet those criteria. In lower-middle-income countries, the best bankruptcy law is the one that allows for simple liquidation procedures. Enforcement outside of bankruptcy, under secured-transactions law or private workouts, is another option.

In poor and lower-middle-income countries, ensuring the efficient enforcement of collateral through private mechanisms or summary judgments takes priority. To the extent that they already exist, more-sophisticated bankruptcy procedures may remain in force. But the emphasis for reform should be on creation of simple debt-enforcement mechanisms through improvements in secured-transactions law, commercial codes, and cost reductions. The reason: the judiciary would not have the capacity to administer

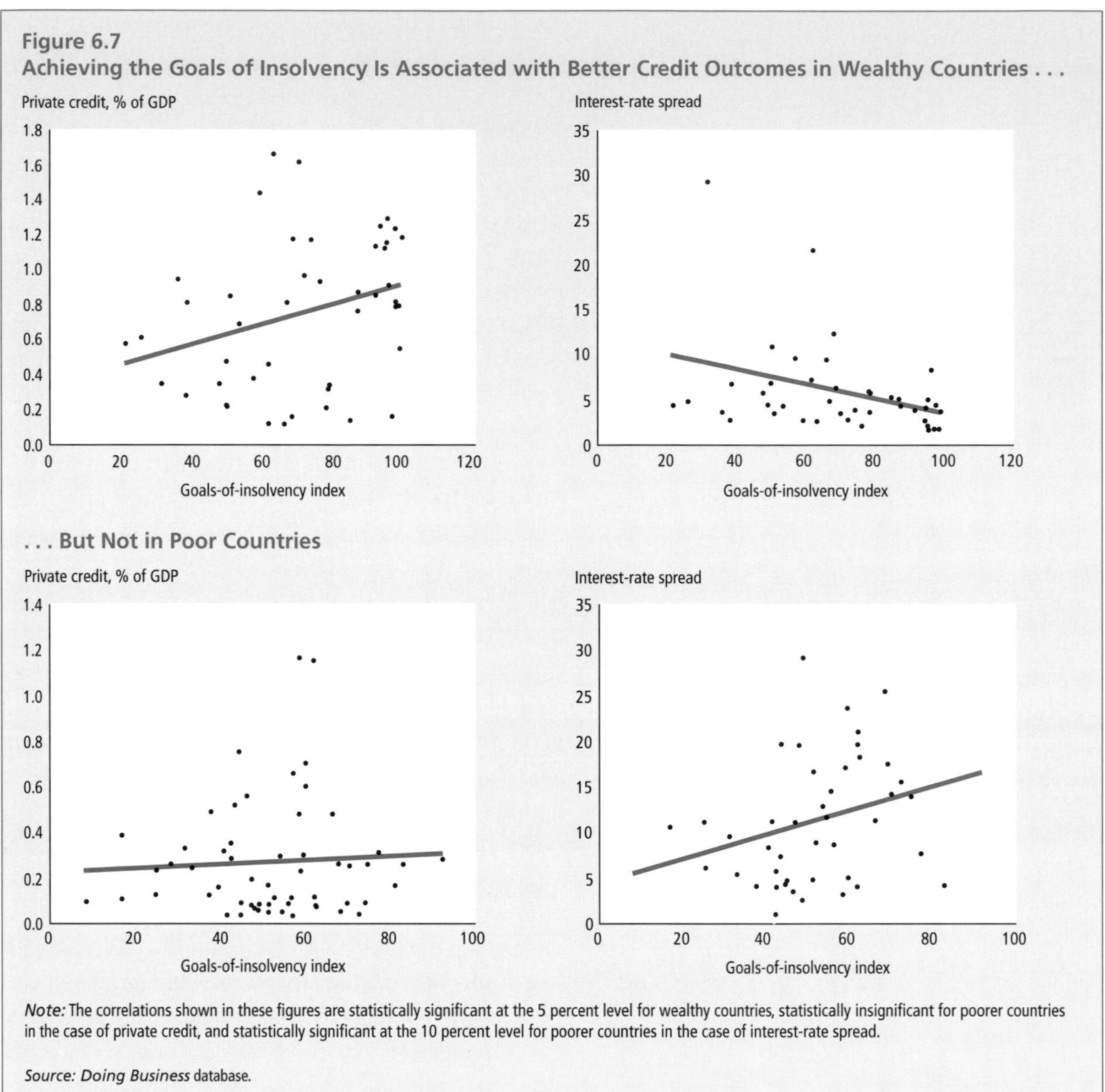

Figure 6.7
Achieving the Goals of Insolvency Is Associated with Better Credit Outcomes in Wealthy Countries . . .

Note: The correlations shown in these figures are statistically significant at the 5 percent level for wealthy countries, statistically insignificant for poorer countries in the case of private credit, and statistically significant at the 10 percent level for poorer countries in the case of interest-rate spread.

Source: Doing Business database.

insolvency provisions efficiently. Furthermore, the credit market is small, and enterprises typically have only one financial institution as the main lender.

The present state of bankruptcy practice around the world is broadly consistent with that pattern. Several poor and lower-middle-income countries—Bosnia and Herzegovina, Egypt, Jordan, Kenya, Nepal, Panama, Syria, Uganda, and Zambia—do not have a rehabilitation procedure. Where one exists, as in Bulgaria or Mozambique, it is rarely used. Of the 1,320 Bulgarian companies that entered bankruptcy in 2000–01, only 37 applied for rehabilitation. In Mozambique, as already mentioned, only state-owned companies have used the rehabilitation procedure. In still other countries, some attractive features of rehabilitation—such as the ability to raise new financing that enjoys priority over existing debt—do not exist. Jordan, Lebanon, Moldova, Ukraine, and the Republic of Yemen do not give priority to new debt (table 6.6). In contrast, Germany and the United Kingdom recently enhanced their rehabilitation procedures.[18]

Table 6.6
Countries Where New Debt Does Not Receive High Priority

Albania	Egypt	Malaysia	Sweden
Austria	Guatemala	Moldova	Syria
Azerbaijan	India	Nicaragua	Turkey
Bolivia	Iran, Islamic Rep.	Nigeria	Ukraine
Bosnia	Jamaica	Pakistan	Uzbekistan
China	Jordan	Singapore	Venezuela
Czech Rep.	Lebanon	Slovak Rep.	Yemen, Rep.

Source: Doing Business database.

In rich countries, a specialized court can improve insolvency procedures, because specialized judges have better training and more expertise and because issues not covered sufficiently in the law are decided swiftly in the profession. The existence of specialized courts is significantly related to insolvency proceedings that are shorter (by almost a year) and cheaper (by a third). But they may be little used in countries with few bankruptcy cases, and they come at the cost of spreading scarce resources more thinly. In such countries, a specialized section or specialized judges within the general court should be the preferred venue for resolving financial distress.

Involving Stakeholders in the Insolvency Process

Reforms of bankruptcy procedures have emphasized more powers for stakeholders, with the judge supervising and facilitating the process—not controlling it. The appointment and replacement of the insolvency administrator is one such area, as with Latvia in its 2001 reforms. Best practice suggests that the court choose at random from a list of licensed administrators. Creditors may request replacement of the administrator in the event of biased or fraudulent behavior.[19] But in many countries, the creditors are not consulted during the administrator's appointment and have no possibility of replacing one, as in Cameroon, Ecuador, France, Lithuania, Paraguay, Poland, and Taiwan (China). Countries where creditors have a say in appointment and replacement have bankruptcy procedures that are significantly cheaper (11 percent of estate value versus 15 percent) and more efficient in achieving the right outcome (65 percent of countries do versus 36 percent that do not).

Creditors also need to be informed about the work of the bankruptcy administrator. The main mechanism of bankruptcy law is to require the administrator to file reports with creditors, during and at the end of the case, on transactions involving the debtor and other decisions made in the course of bankruptcy. In several countries, the administrator is not required to file a report, as in Bolivia, Colombia, India, Korea, Moldova, Spain, Taiwan (China), Thailand, Uruguay, and Vietnam. In yet other countries, a report is filed only with the court and is not accessible to creditors. Such a report would inform the creditors and provide a higher chance of maintaining absolute priority.

Another set of judicial procedures defines the powers of various stakeholders in formulating and adopting a rehabilitation plan. It is hard to justify laws that mandate the formulation of a plan by the court, without effective participation of creditors or management. But such is the case in many countries, including Benin, Mali, Morocco, and Tunisia. During the adoption of the plan, creditors vote individually or by class in most countries, with an acceptance threshold of the majority of claims by value. That method ensures that the will of major creditors is taken into account. But in Azerbaijan, Burkina Faso, Cameroon, Costa Rica, Mali, Moldova, and Niger, the court adopts the plan without considering the views of creditors. Ignoring them is counterintuitive—because one of the goals of bankruptcy is to preserve the value of creditors' claims.

Training Bankruptcy Judges and Administrators

Judicial procedure will improve with qualified judges, and training judges in commercial litigation has become widespread. From Thailand to Ecuador to Nepal to the Dominican Republic, a judicial career depends on going through specialized training, including accounting and business courses. Several countries have recently established institutes to train judges in handling commercial cases (chapter 4).

In most middle-income countries where bankruptcy administrators are responsible for managing a company in insolvency, the profession is still developing.

Progress is needed in two areas: ensuring the proper qualifications for administrators (including the necessary education and business experience prior to receiving a license), and periodically renewing administrators' licenses on the basis of continued training and practice with insolvency cases. Lithuania and the Czech and Slovak Republics do not have education requirements for administrators.[20] Thus, the lists of licensed administrators there are long—more than 800 in Lithuania and more than 1,600 in the Czech Republic. Furthermore, many administrators do not have legal, accounting, or economics education. Nor do the majority of administrators have business experience, which is crucial for managing a company in distress.

In some countries, professional associations have provided training. But continuing education needs to be mandatory, as it is in the accounting and legal professions. In Argentina, insolvency administrators are required to receive a certain number of training credits within a four-year period. If they do not, their licenses are revoked.[21] In lower-middle-income countries, where the trustee profession is still nascent, regulators can consider licensing individual experts, along with consulting, accounting, and law firms, which have an easier time pulling together the required capacity. But stricter licensing should not benefit just one profession, such as lawyers and law firms. Such reform, now being considered in Croatia, alleviates some problems but creates many others—among them, reducing competition in the trustee market and leaving the pool of management skills and accounting competencies deficient.

Notes

1. Bruno 1561.
2. Levinthal 1919.
3. Berglof, Rosenthal, and von Thadden 2002.
4. Rajak 1997.
5. Posner 1992.
6. The survey was conducted in cooperation with Committee J (Insolvency and Creditor Rights). *Doing Business* gratefully acknowledges the leadership of Selinda Melnik, Esq., Chairwoman of Committee J at the time of the survey.
7. For a further description, see the data notes section in the *Doing Business* Indicators tables.
8. Claessens and Klapper (2001) collect data on 35 jurisdictions, primarily in the OECD.
9. The hypothetical case does not address whether a nonviable company is rescued—because the continuing existence of insolvent firms is typically a privilege of large enterprises that have national importance or a large number of employees—firms "too big to fail."
10. Stromberg 2000, Thorburn 2000.
11. Court costs almost always have the top priority. In some countries, postpetition claims take priority over secured claims. A further indicator of priority is the incidence of shareholders' getting paid before secured creditors. Studies of the United States (such as Betker 1995) show that shareholders often get paid in reorganization when creditors have not been fully paid. Such reversals of the order of claims may occur in Argentina, Belarus, the Dominican Republic, Ecuador, Guatemala, Indonesia, Nigeria, Taiwan (China), and Ukraine.
12. Pistor and others (forthcoming) also find that reform of commercial codes is faster in developed countries.
13. Wood 1995.
14. Modigliani and Perotti 2000.
15. Another scenario for the use of private workouts is when the rights of debtors and creditors are balanced and the judicial process is efficient and predictable. For example, private workouts are often used in New Zealand and Switzerland. In contrast, they are seldom used in countries with strong creditors' rights, such as Hong Kong (China), Singapore, and the United Kingdom, or in countries with strong debtors' rights, such as Ireland, Finland, and Spain. This scenario does not arise in developing countries, where the outcome of the bankruptcy process is typically far from predictable.
16. La Porta and Lopez-de-Silanes 2001.
17. See La Porta and others (1997, 1998) for a view from the creditor's perspective, and Hart (2000) and Stiglitz (2001) for an overall review of bankruptcy provisions.
18. Couwenberg 2001.
19. The court approves any request for replacement of the administrator by creditors in all countries except Guatemala, Jamaica, and the United Kingdom.
20. World Bank 2001a, World Bank 2001b, World Bank 2002a, World Bank 2002b.
21. World Bank 2002c.

7 The Practice of Regulation

The pervasiveness of government regulation in business activity raises questions. Which countries regulate the most? Do the activities being regulated or the characteristics of the country influence the choice of regulation? Is the level of regulation an outcome of efficient social choice, or has it persisted because of inertia and a lack of capacity for reform? Is regulation generally good, as the positive correlation between its growth and the growth of income over the last century seems to indicate? Or has business regulation been an obstacle to economic and social progress? What are the main obstacles to regulatory reform? The answers to those questions, presented in this report, have implications for economic theory and public policy.

The analysis reveals three findings concerning the practice of regulation:

- Regulation varies widely around the world.
- Heavier regulation of business activity generally brings bad outcomes, while clearly defined and well-protected property rights enhance prosperity.
- Rich countries regulate business in a consistent manner. Poor countries do not.

Regulation Varies Widely around the World

Belarus, Chad, and Colombia have the most procedures to start a business: 19. Algeria, Bolivia, Paraguay, and Uganda come next, each with more than 15 procedures. Burundi has the most procedures to enforce contracts through the courts: 62. Angola, Benin, Bolivia, Cameroon, El Salvador, Kazakhstan, the Kyrgyz Republic, Mexico, Panama, Paraguay, Sierra Leone, and Venezuela come next, with more than 40 each. Costa Rica and Guatemala have the most complex contract enforcement processes. In employment regulation, Ethiopia has the most generous paid-vacation allowance of any country, at 39 working days a year. Panama has the most restrictive regulations on part-time and fixed-term employment contracts. Bolivia and Nicaragua have the longest minimum daily rest for workers. Angola, Belarus, and Paraguay place the most restrictions on firing. The powers of the judge in deciding the course of insolvency proceedings are greatest in Benin, Bolivia, Burkina Faso, Cameroon, Côte d'Ivoire, and the Philippines.

In contrast, Australia has the fewest entry procedures: 2. Canada, Ireland, New Zealand, and Sweden come next. Australia has the fewest procedures to enforce a contract through the courts, with 11. Norway and the United Kingdom come next, with 12. With respect to labor regulations, Singapore makes the dismisal of workers the easiest. Denmark, Hong Kong (China), New Zealand, Sweden, and the United States are among the countries with the most flexible labor regulations overall. The powers of the judge in deciding the course of bankruptcy proceedings are the weakest in Australia, Finland, New Zealand, the United Kingdom, and the United States.

Rich countries regulate less on all aspects of business activity covered in this report (figure 7.1). The average number of procedures to start a new business is 7 in high-income countries, 10 in upper-middle-income countries, 12 in lower-middle-income countries, and 11 in low-income countries. The employment regulation index has an average value of 43 in high-income

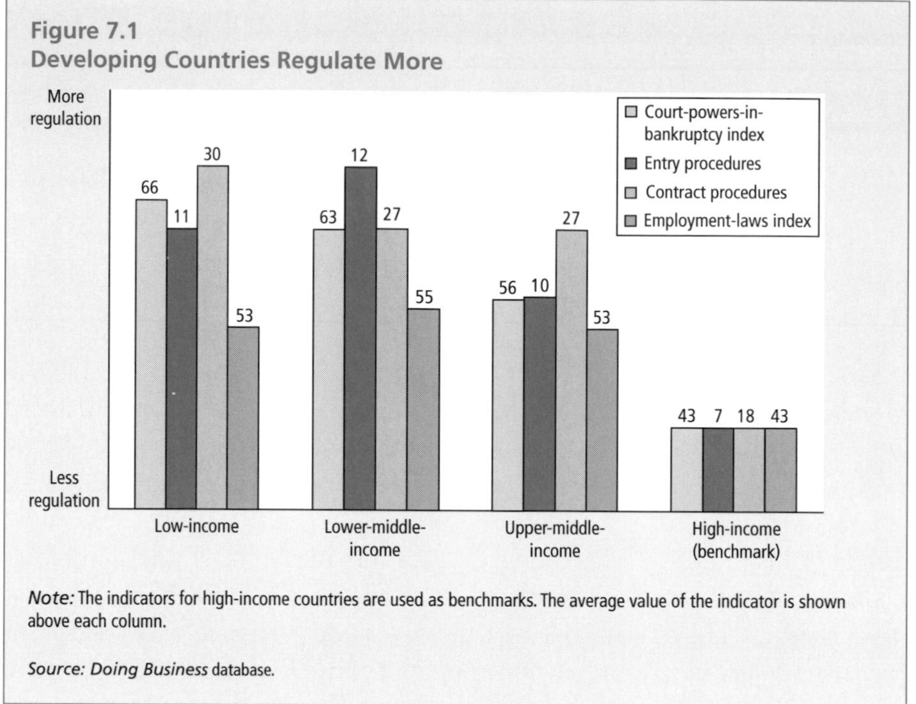

Figure 7.1
Developing Countries Regulate More

Note: The indicators for high-income countries are used as benchmarks. The average value of the indicator is shown above each column.

Source: Doing Business database.

countries, 53 in upper-middle-income countries, 55 in lower-middle-income countries, and 53 in low-income countries. The average number of procedures to enforce a contract is 18 in high-income countries, 27 in middle-income countries, and 30 in low-income countries. The index of court powers in bankruptcy has an average value of 43 in high-income countries, 56 in upper-middle-income countries, 63 in lower-middle-income countries, and 66 in low-income countries, where higher index scores reflect more regulation.

Income is not the only important factor determining differences in regulation. The regulatory regimes of most developing countries are not indigenous—they are shaped by their colonial heritage. When the Dutch, English, French, Germans, Spaniards, and Portuguese colonized much of the world, they brought with them their laws and institutions. After independence, many countries revised their legislation, but in only a few cases have they strayed far from the original.

Regulation in developed countries varies systematically, shaped by their history over the last millennium.[1] England developed a common-law tradition, characterized by independent judges and juries, the low importance of regulation, and a preference for private litigation as a means of addressing social problems. France, following the Romans, developed a civil-law tradition, characterized by state-employed judges, emphasis on legal and procedural codes, and a preference for state regulation over private litigation. Germany and the Nordic countries developed their own civil-law traditions, also based on Roman law.

Napoleon exported the French legal system, after his conquests, to Spain, Portugal, and Holland. Through his and subsequent colonial conquests, the French legal system was transplanted to all of Latin America, Quebec, large parts of Europe, North and West Africa, parts of the Caribbean, and parts of Asia.[2] The common-law tradition was transplanted by England to the United States, Canada (except for Quebec), Australia, New Zealand, East Africa, large parts of Asia (including India), and most of the Caribbean. The German legal system was adopted voluntarily in Japan, and through Japan it influenced the legal systems of the Republic of Korea, Taiwan (China), and China. Austria and Switzerland were also influenced by German legal scholarship. Through the Austro-Hungarian Empire, much of today's central and eastern Europe inherited German commercial laws. Finally, the Soviet Union instituted its socialist legal system in the 15 republics, and influenced commercial law in Mongolia (figure 7.2).

Those channels of transplantation suggest the existence of systematic variations in regulation that are not a consequence of either domestic political choice or pressures toward regulatory efficiency. The data agree. Nordic and common-law countries regulate the least (figure 7.3). This finding is especially striking for the common-law group, which includes poor countries like Ethiopia, Ghana, Nigeria, Sierra Leone,

Figure 7.2
World Map of Legal Origin

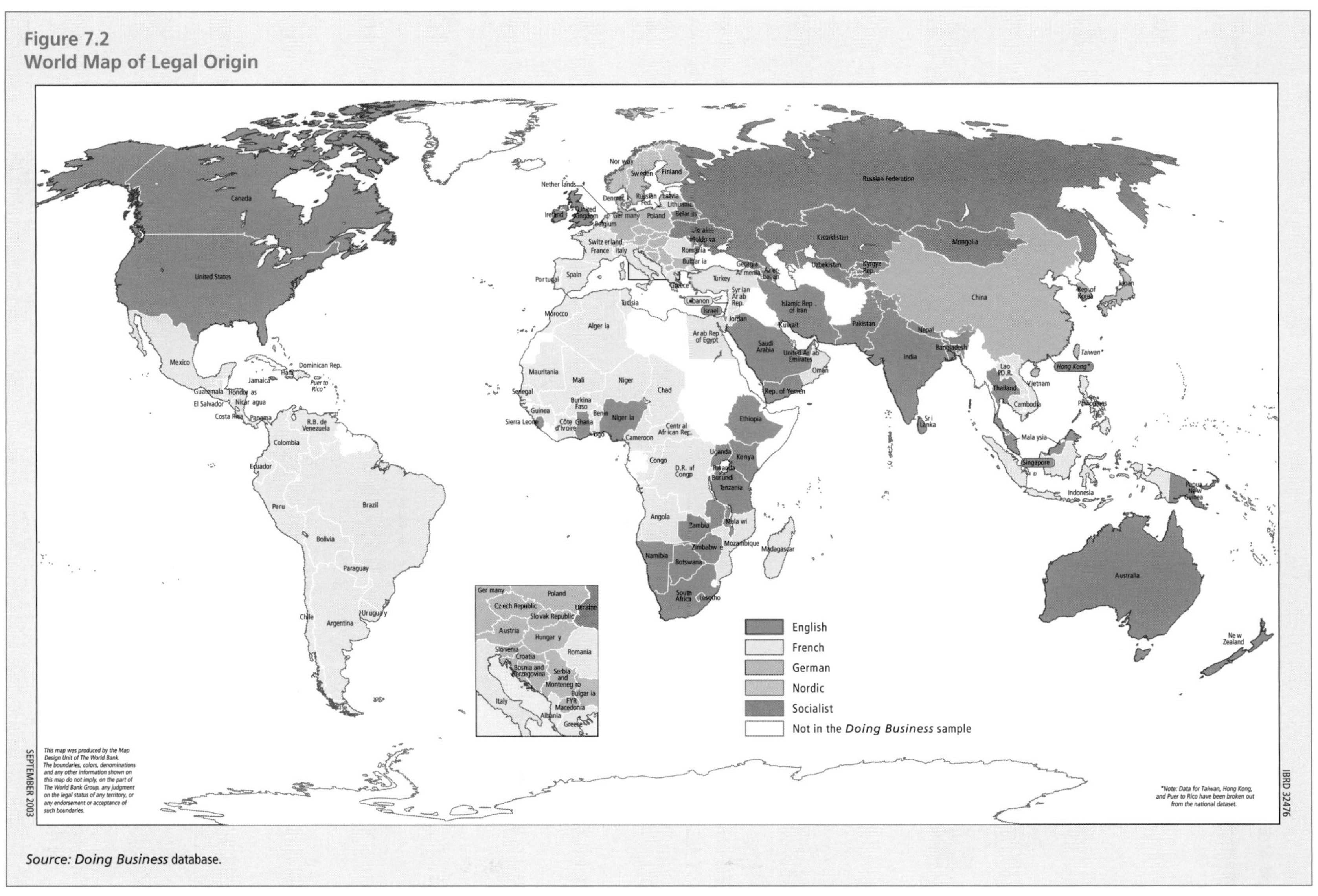

Source: Doing Business database.

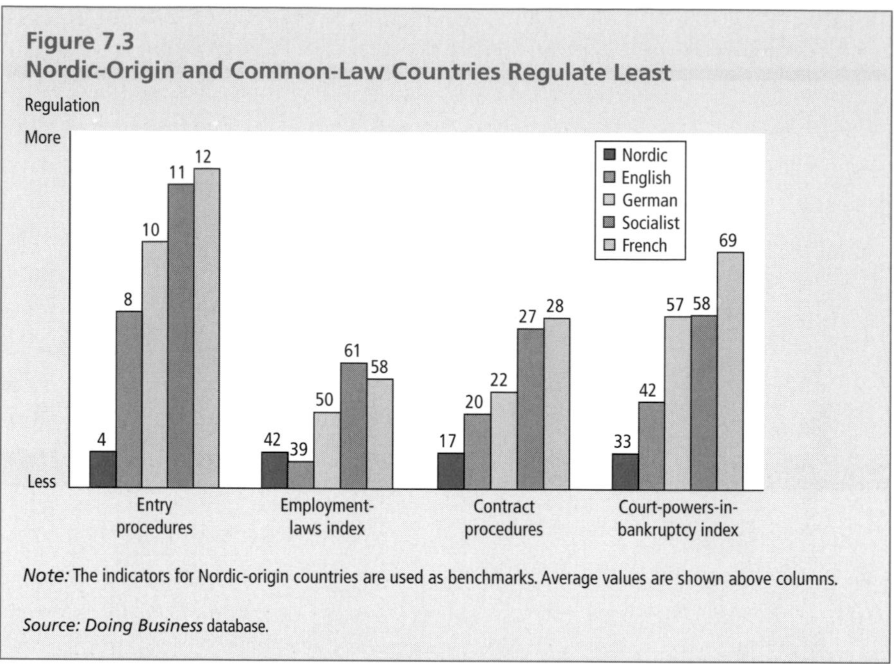

Figure 7.3
Nordic-Origin and Common-Law Countries Regulate Least

Note: The indicators for Nordic-origin countries are used as benchmarks. Average values are shown above columns.

Source: Doing Business database.

more likely to be captured by incumbent businesses and to have regulation aimed at maximizing benefits of an elite group.[3]

Regulation is lighter in countries with more-representative governments, more openness to competition, and greater political rights and media freedoms, even controlling for income per capita and legal origin.[4] Regulation is heavy in Belarus and Syria, light in Canada, Latvia, and Norway. The countries with the heaviest employment regulations in Europe—Portugal and Spain—inherited them from the dictatorships of António Salazar and Generalissimo Franco.

But might not other institutional determinants be at work?

- Democracy is more difficult to maintain in countries with ethnolinguistic differences, religious divisions, and low levels of human capital. And the association between stricter regulation and democracy could be driven by Latin America or Africa, the two continents with the most-checkered history of repressive governments.
- Geography might influence institutional development in other ways—for example, it has been argued that the environment in Latin America was relatively suitable to large-scale production technologies, which in turn led to significant inequalities and poor-quality institutions.[5]
- Inhospitable environments for European settlers, as measured by mortality rates, may have shaped institutional development.[6] The lack of investment in public administration capacity in the Congo under King Leopold of Belgium is one example.[7]
- Finally, the openness of countries to trade could encourage institutional development.

and Zimbabwe. Because Nordic laws and regulation have not been transplanted to other parts of the world, it is not known whether they would be as effective in poor countries, where the inclination to regulate is greater, as in, say, Finland or Norway.

Across all sets of indicators, income and legal origin are the most important variables for explaining different levels of regulatory intervention, together accounting for more than 60 percent of the variation in regulation among the 133 *Doing Business* countries. However, heritage is not destiny. Tunisia, a lower-middle-income country in the French legal tradition, is among the world's best at contract enforcement. Uruguay has one of the world's most flexible regulations on firing, standing alone among Latin American countries. And France is among the richest countries despite heavier regulatory intervention in relation to its peers.

The effect of other factors is weaker and less systematic. Of particular importance: the political system. If regulations were put in place to remedy market failures, the level of regulation should be higher in countries with political systems characterized by the convergence of policy choices and social preferences—countries with more-representative governments. In contrast, less-democratic regimes are

The Netherlands, an early free-trader, developed credit registries to help its merchant class extend business to new places.

Analysis controlling for all the above factors shows that heavy regulation on the dimensions measured here is strongly and consistently associated with lower incomes and with French and socialist legal origins. It is sometimes associated with less democracy and with tropical climates. Other factors are not significantly associated with the extent of regulation.

Heavier Regulation Brings Bad Outcomes

Heavier regulation is generally associated with greater inefficiency of public institutions (see, for example, figures 4.7 and 4.8) and more corruption (see, for example, figures 2.6 and 4.4)—but not with better quality of private or public goods. The countries that regulate the most—the poor countries—have the least enforcement capacity and the fewest checks and balances to ensure that regulatory discretion is not used to abuse businesses and extract bribes.

Regulation has a perverse effect on the people it is meant to protect. Faced with a large regulatory burden and few incentives to become formal, entrepreneurs in many developing countries choose to operate in the unofficial economy (see figure 2.5). Bad institutions—cumbersome entry procedures, rigid employment laws, weak creditor rights, inefficient courts, and overly complex bankruptcy laws—simply do not get used. Instead, businesses use informal institutions—an improvement but a poor substitute for good-practice regulation.

In Bolivia, one of the most heavily regulated economies, an estimated 82 percent of the business activity takes place in the informal sector. There, workers enjoy no paid vacations or maternity leave. It is hard for businesses to get credit or resolve disputes through formal institutions, such as courts. Growth is inhibited because transactions take place only within a narrow group of established business relationships. The resources for delivering basic infrastructure are reduced because businesses do not pay taxes. There is no quality control of products. And entrepreneurs keep their operations small, below an efficient production size, for fear of inspectors and the police.

The results: poor economic outcomes, a reduced tax base, a large group of entrepreneurs and businesses never entering the formal sector, and a general failure of the state to provide for its citizens. It is in the most heavily regulated countries that investment and productivity are low, and unemployment is high (figure 7.4).[8]

It might be argued that having less regulation would result in lower quality products, an inability to resolve disputes, poor protection of worker rights, and, ultimately, social unrest. That democratic countries regulate less and that regulatory countries do not differ from nonregulatory ones in social capital (religion, ethnolinguistic divisions) suggest otherwise. In the absence of many burdensome regulations, businesses in poor countries would rely on private reputation mechanisms—as they have for centuries.[9]

Instead of imposing burdensome regulations on business, a government may focus on better defining the property rights of its citizens and protecting

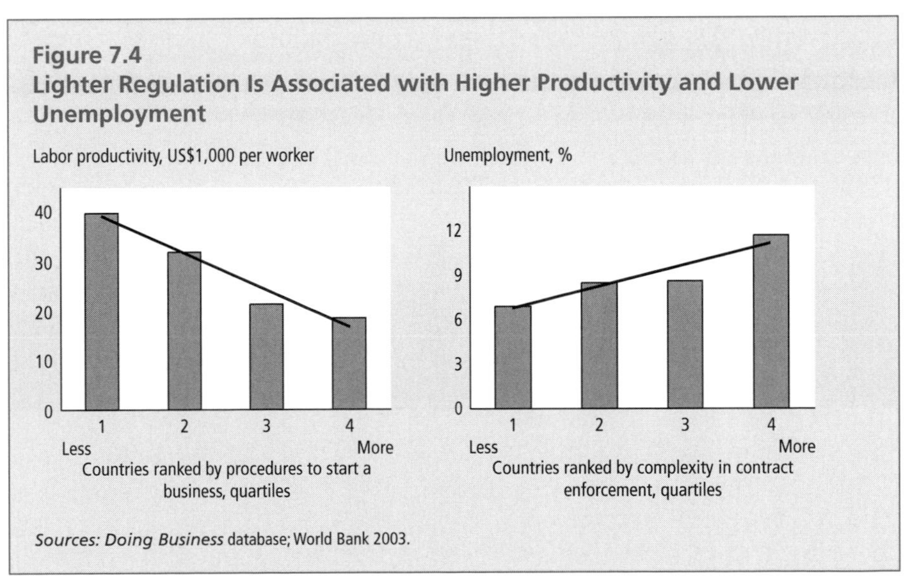

Figure 7.4
Lighter Regulation Is Associated with Higher Productivity and Lower Unemployment

Sources: Doing Business database; World Bank 2003.

Figure 7.5
Efficient Courts and Creditor Rights Are Associated with Deeper Credit Markets

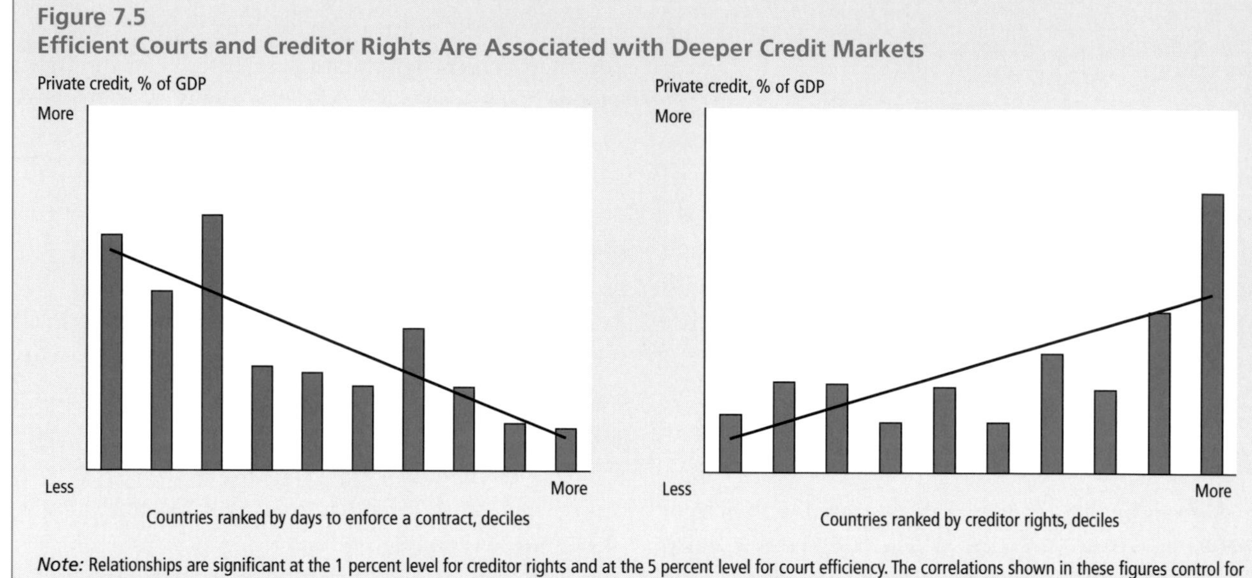

Note: Relationships are significant at the 1 percent level for creditor rights and at the 5 percent level for court efficiency. The correlations shown in these figures control for income, GDP growth, inflation, credit information, creditor rights, and number of days to enforce a contract.

Source: Doing Business database.

them against injury from other citizens and the state. Two examples are creditor rights—the legal rights of lenders to recover their investment if a borrower defaults—and the efficient enforcement of property rights in court. Countries that protect such rights achieve better economic and social outcomes. Assuring lenders of a fair return on their investments is associated with depth of credit markets, even controlling for income, growth, inflation, and credit information (figure 7.5). It also democratizes access to markets, because lenders will be willing to extend credit beyond large, well-connected firms if they know that their rights to recover loans are secure.

The fact that the governments best at defining and protect property rights do so by using little regulation suggests a trade-off between regulatory intervention and a narrow focus on achieving the main purpose of government. For example, countries with stronger creditor rights—a subset of property rights—regulate employment relations lightly (figure 7.6). Rather than spend resources on costly (and often ineffective) regulation, good governments channel their energies into enhancing prosperity.

Figure 7.6
Countries with Stronger Property Rights Regulate Employment Lightly

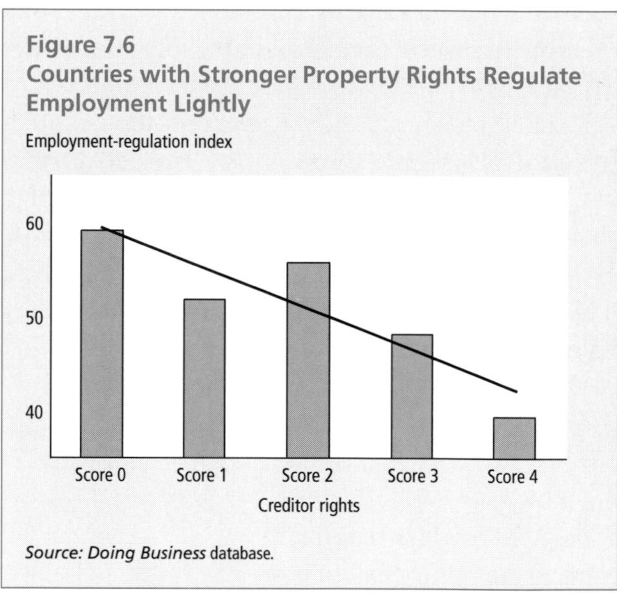

Source: Doing Business database.

Rich Countries Regulate Business in a Consistent Manner

In the well-known opening to *Anna Karenina*, Tolstoy pronounced: “All happy families are alike; each unhappy family is unhappy in its own way.” The way governments regulate business is similar. Rich countries tend to regulate consistently on all

Table 7.1
Correlations of Regulation Indicators for Rich Countries

	Entry Procedures	Employment-Regulation Index	Contract Procedures	Procedural-Complexity Index	Creditor-Rights Index
Employment-laws index	0.50***				
Contract procedures	0.23*	0.47***			
Procedural-complexity index	0.59***	0.56***	0.36***		
Creditor-rights index	–0.23*	–0.29**	–0.15	–0.08	
Court-powers-in-bankruptcy index	0.30**	0.36***	0.23	0.41***	–0.27*

Note: *** The correlation is significant at the 1 percent level. ** The correlation is significant at the 5 percent level. * The correlation is significant at the 10 percent level. The sample includes 49 countries classified as high-income and upper-middle-income by the World Bank.

Source: Doing Business database.

dimensions of business regulation and property-rights protection. Those that encourage business entry by means of fewer and simpler regulations also permit more-flexible hiring and firing, protect creditors, and have less regulation in their courts and insolvency systems (table 7.1).

Common-law countries (Australia, Canada, Hong Kong [China], New Zealand, the United Kingdom, and the United States) and Nordic countries (Denmark, Finland, Norway, and Sweden) offer the best practices in business regulation. Japan, the Republic of Korea, the Netherlands, and Singapore also figure among the best-practice regulators (table 7.2). Regardless of how the indices are constructed, those countries regulate the least and protect property rights the most. By combining modest levels of regulation with property rights that are clearly defined and well protected, the countries achieve what many others strive to do: have regulators act as public servants and not public masters.

The significant correlations across the indicators suggest that all governments have a general regulatory stance toward more or less intervention. Does that fact mean that the indicators capture one underlying variable? Apparently not, at least for the topics covered in this report. The indicators have distinctly different explanatory power over specific economic outcomes, as theory predicts. The creditor-rights index helps explain the depth of credit markets, but the employment-laws index and entry-regulations measures do not. And though the employment-laws index helps explain unemployment levels, the creditor-rights and court-powers-in-insolvency indices bear no relation to unemployment.

Although rich countries converge, there is much more variation among poor countries. Correlations across the indicators are much less significant (table 7.3). Some countries have reformed one or two areas of business regulation while maintaining heavy control in others. Over the last decade, Latvia, Serbia

Table 7.2
The Ten Least-Regulated Countries across *Doing Business* Indicators

Regulation[1]	Regulatory Outcomes[2]	Regulation and Its Outcomes[3]
Australia	Canada	Australia
Canada	Ireland	Canada
Denmark	Japan	Denmark
Hong Kong (China)	Netherlands	Netherlands
Jamaica	New Zealand	New Zealand
Netherlands	Norway	Norway
New Zealand	Republic of Korea	Singapore
Singapore	Singapore	Sweden
Sweden	Sweden	United Kingdom
United Kingdom	United Kingdom	United States

Notes: [1]Entry procedures, contract-enforcement procedures, procedural-complexity index, employment-regulation index, court-powers-in-bankruptcy index. [2]Business-entry days and cost, contract-enforcement days and cost, bankruptcy days and cost. [3]Entry procedures, contract-enforcement procedures, procedural-complexity index, employment-regulation index, court-powers-in-bankruptcy index, business-entry days and cost, contract-enforcement days and cost, bankruptcy days and cost. Combining the indicators in different ways can change country rankings. Aggregated *Doing Business* indicators will be further analyzed in future reports as the coverage of business environment topics expands.

Source: Doing Business database.

Table 7.3
Correlations of Regulation Indicators for Poor Countries

	Entry Procedures	Employment-Regulation Index	Contract Procedures	Procedural-Complexity Index	Creditor-Rights Index
Employment-laws index	0.48***				
Contract procedures	0.14	0.32***			
Procedural-complexity index	0.28*	0.27**	0.35***		
Creditor-rights index	–0.09	–0.21*	–0.03	0.03	
Court-powers-in-bankruptcy index	0.13	0.04	0.13	0.37***	–0.14

Note: *** The correlation is significant at the 1 percent level. ** The correlation is significant at the 5 percent level. * The correlation is significant at the 10 percent level. The sample includes 70 countries classified as lower-middle-income and low-income by the World Bank.

Source: Doing Business database.

and Montenegro, South Africa, Thailand, Tunisia, and Vietnam have achieved great improvements in some areas of regulatory efficiency. In 1996, Tunisia reformed its judicial procedures to allow summary execution of judgment in commercial cases, and today it is one of the most efficient countries at resolving commercial disputes. South Africa implemented a similar reform in 1999. Latvia, Serbia and Montenegro, and Vietnam have all reformed business entry regulations, making them among the most efficient in that area. In Thailand, the 1999 bankruptcy reforms have achieved great success—but whether they will extend to other areas of business regulation remains to be seen.

Those developments are grounds for optimism, because they suggest that partial reforms have already been undertaken in many developing countries. Further reforms, in other areas of business regulation, are now necessary. With more-limited capacity in their public administration, developing countries are less equipped to do comprehensive reforms. However, the practice shows that small steps in some reforms have made larger reforms possible elsewhere. Only a handful of countries—for example, Angola, Bolivia, Guatemala, Mozambique, and Paraguay—have heavy regulation in all aspects of business activity. For those countries, comprehensive reforms may be necessary.

What Do These Findings Mean for Economic Theory?

During the 20th century, economists have come up with several ways of thinking about government regulation.[10] The three main theories are the public-interest theory of regulation associated with Arthur Pigou,[11] the contracting theory associated with Ronald Coase,[12] and the capture theory of George Stigler.[13] The data and analysis in this report have implications for all three—and provide the empirical foundations for new theoretical work.

The public-interest theory of regulation holds that unregulated markets exhibit frequent failures. A government that pursues social efficiency protects the public by means of regulation. As applied to business entry, this theory says that governments should screen new entrants to make sure that consumers buy high-quality products from "desirable" sellers. In addition, governments control prices to prevent natural monopolies from overcharging, impose safety standards to prevent accidents such as fires or food poisonings, regulate labor markets to counter employers' power over employees, regulate bankruptcy procedures to ensure that stakeholders are not cheated, and so on.[14] Joseph Stiglitz takes the theory further by arguing that developing countries have more market failures—and thus a greater need for government regulation.[15]

The theory has been subject to three criticisms. The first critique blames public-interest theory for exaggerating the extent of market failure and for not recognizing the ability of competition to address many of the alleged problems. For example, competition for labor ensures that employers provide good working conditions for employees. If an employer failed to do so, competitors would offer better packages and attract workers. Similarly, private markets ensure efficient safety levels in a variety of products and services, such as food, houses, and cars. Sellers who fail to deliver those levels lose market share to competitors who sell unspoiled food, build safer houses, or produce safer cars. As discussed in chapter 2, the data show that stricter entry regulation is not associated with better consumer protection.

Even when competitive forces are not strong enough, private orderings work to address potential market failures. Neighbors resolve disputes among themselves, without government intervention, because they need to get along with each other over long stretches of time.[16] Private credit information bureaus are established to protect lenders from extending credit to bad borrowers. Professional associations of accountants, exporters, and teachers impose standards on their members to guarantee quality and penalize cheaters so that, in the long run, customers continue their patronage.[17]

The second critique, originating in the work of Coase, maintains that where competition and private orderings do not address market failures, impartial courts can do so. Employers can offer workers employment contracts that specify what happens in the event of an accident. Security issuers can voluntarily disclose information to potential investors and guarantee its accuracy. And so on. With well-functioning courts enforcing property rights and contracts, the scope for desirable regulation is reduced. The data agree. As shown in chapter 4, countries with more efficient courts tend to regulate lightly.

Private orderings do work well in some situations, but they also degenerate into anarchy, wherein the strong—not the just—win the day. Moreover, the empirical evidence suggests that courts around the world are often inefficient. Courts in Guatemala take more than 4 years to resolve a simple dispute, those in Brazil take more than 5 to collect collateral, and those in India take more than 10 to close down an unviable business.

The third critique of regulation questions the assumption that a government is benevolent and competent, the essence of Stigler's theory.[18] First, incumbent business interests typically capture the process of regulation. Regulation not only fails to counter monopoly pricing—it sustains it. Second, even where regulators try to promote social welfare, they lack the capacity to do so, and regulation makes things even worse. Empirical evidence provides support for this conclusion. Bureaucratic entry is associated with more corruption, and heavy regulation of court procedures leads to less impartiality and longer delays. However, today we also live in a much richer but also more regulated society, and we are generally happy as consumers with many of the regulations that protect us. A more nuanced theory—which recognizes the benefits of public involvement in at least some activities—is clearly needed to keep theory and facts together.

To be effective, regulation needs to be enforced. The nature of the regulation and the activity regulated determines the success in enforcement. In the background research for this report, conducted with Professor Andrei Shleifer and his colleagues, that view is called "the enforcement theory."[19] Its premise is a basic tradeoff between two social costs: the cost of private injury and the cost of state intervention. Private injury refers to the ability of private agents to harm others—to steal, cheat, overcharge, or impose external costs. State intervention refers to the ability of government officials to expropriate private agents through bureaucratic hassle or the confiscation of property. As one moves from private orderings to private litigation to regulation to public ownership, the powers of the government rise and those of private agents fall. The social losses from private injury decline as those from state intervention increase.

The theory implies that the appropriate choice of government intervention—from market discipline, to reliance on courts and litigation, to regulation, to state ownership—depends on the type of activity and on country circumstances, such as administrative capacity.

The main strength of market discipline as a method of enforcement is that it is free of public enforcers. When market discipline can control private injury, it is the best approach, because it has the lowest social costs of state intervention—particularly in developing countries, where government capacity in the courts and public administration is low.

But market discipline may not be enough. Employers may underinvest in safety and blame accidents on an injured worker's own carelessness. In such instances, societies efficiently accept a higher level of government intervention by relying on enforcement through private litigation. Injured employees can sue their employers for damages. A judge would recognize whether employees had not been adequately protected and award damages to compensate them for their losses. But in many countries, even simple litigation can take years to resolve and may incur substantial costs.

Compared with court enforcement, regulation has advantages. Unlike judges, public regulators can be expert and motivated to pursue social objectives in specific areas. Indeed, this has been the main argument for public regulation of securities markets.[20] This combination of expertise and incentives makes public enforcement more efficient, in some circumstances, than private or court enforcement.[21]

Alas, public regulation has problems. The key problem is the risk of an official's abuse of market participants. Overzealous enforcement is a particular problem in developing countries, where officials sometimes hassle businesses for bribes,[22] thereby pushing them into the informal economy. This situation suggests that regulation is a more attractive option in richer countries, where the checks on government regulators are stronger. Heavy regulatory intervention is generally a bad idea in developing countries and in countries with undemocratic governments, where the risks of abuse are the greatest.

In some situations, nothing short of government ownership can foster a good business environment. If monopolies cannot be restrained through regulation, if quality cannot be assured except with full state control, if public safety is jeopardized, there is a case for state ownership. For example, the police function needs to be state-controlled if it is to protect businesses from injury by others. Otherwise businesses would have an incentive to support private police. The more powerful the business, the more likely it is that its police will dominate the others. Commercial disputes would then be resolved in favor of the powerful.[23]

More mundanely, in countries with underdeveloped and concentrated financial markets, public ownership of credit registries may be the only way to increase the sharing of credit information in the short term. Why? Because there would not be enough profit opportunities for a private business to enter before the credit market was sufficiently developed. And because banks that already control a large customer base would not voluntarily divulge information. Public registries perform an admirable job in countries as diverse as Mozambique and Nepal.

Enforcement theory predicts that regulation may best be limited in countries with insufficient enforcement capacity and in undemocratic countries, because heavy regulation would result in inferior social outcomes.

Principles of Good Regulation

In the regulation of business activity, two principles apply. First, regulate only when private ordering or litigation are not sufficient to induce good conduct. Second, regulate only if there is capacity to enforce. Countries that perform well have common elements in their approach to regulation:

- Simplify and deregulate in competitive markets.
- Focus on enhancing property rights.
- Expand the use of technology.
- Reduce court involvement in business matters.
- Make reform a continuous process.

Simplify and Deregulate in Competitive Markets

There is too much regulation in countries, particularly in developing countries, where other means would suffice and where its complexity and volume cannot be enforced. Rather than inducing good conduct, such regulation puts businesses at the discretion of government inspectors and officials, who sometimes abuse their powers to extract bribes. So there is less business activity, and much of it remains informal.

Several areas for deregulation stand out. Where there is enough competition in business and in labor markets, markets would be enhanced if fewer regulations were imposed on the participants. If newly established firms produce inferior products, they will soon be driven out of business. And if a business does not provide its workers with adequate conditions of employment, other companies will attract the workers.

In most developing countries, government lacks the capacity to enforce complex regulation, as in bankruptcy. But out-of-court resolution of insolvency can be sought—say, through private contracts and efficient collateral enforcement. If reform is not pursued, regulation imposes high costs and breeds corruption, thereby encouraging businesses to operate in the informal economy. There, workers have no protections, and entrepreneurs live in constant fear of the tax administrator and the police. Firms do not grow to their efficient size, thus reducing the number of productive jobs and severely diminishing the opportunities for growing out of poverty.

Some regulations—such as those for commercial dispute resolution—are necessary, but need simplification and fewer formalities to be effective. There is no reason to believe that Benin needs or can enforce 44 procedures to resolve disputes in the courts if France, whose laws Benin adapted, has only 21. And there is no reason for Angola to have one of the most rigid employment laws if Portugal, whose laws Angola adapted, has already revised them twice to make the labor market more flexible. In both Angola and Benin, greatly simplifying the regulatory process is advisable.

Focus on Enhancing Property Rights

Much of the evidence in this report shows that in most countries government intervention is excessive and that it hurts business. There is also evidence that governments do too little to protect property rights. The best-practice countries build efficient courts and support laws and institutions that define the rights of citizens and businesses to their property. This year *Doing Business* has constructed indicators on two aspects of property rights: court efficiency and the legal rights of creditors. On these dimensions, high regulatory intervention is associated with less protection of property rights, not more. Ironically, the institutions that define and enforce property rights in many developing countries—the court system, property registries, and law enforcement agencies—are often the least modern and least funded of all public institutions.

Better protection of property rights benefits everyone, especially the poor. One example comes from Peru, where in the last decade the government has issued property titles to 1.2 million urban squatter households. As a result, there has been a substantial increase, of almost 20 percent, in the number of work hours away from home, and a nearly 30 percent reduction in the incidence of child labor. Secure property rights have enabled parents to leave their homes and find jobs instead of staying in to protect the property. The main beneficiaries are their children, who can now go to school.[24]

Expand the Use of Technology

For all areas of regulation covered in this report, the use of technology is improving efficiency, increasing information, and reducing opportunities for bureaucratic discretion. In the best-practice countries, modern technology minimizes the regulatory burden on business. With Internet-based business registration systems in Canada and Australia, application-processing time is the fastest in the world. And because entrepreneurs never have to face a bureaucrat, there are no opportunities to extract bribes. Electronic information systems in the Slovak Republic have dramatically improved court efficiency.

In credit markets, technology enables developing countries to leapfrog levels of institutional development, the spread of credit registries to poorer countries being spurred by falling costs and easier access to new technologies. For a few hundred thousand dollars, Albania is establishing a comprehensive electronic registry with access in real time, similar to the systems of many OECD countries. Technology also helps create regional markets, enabling small countries to realize faster and cheaper access to best-practice institutions. In southern Africa, the private credit bureau in South Africa has recently expanded its services to cover businesses in Botswana, Namibia, and Swaziland. Similarly, the private credit registry in Fiji operates from a server based in New Zealand.

Such technology has other positive effects. In Malaysia, one benefit of the credit bureau is the capability to validate records and detect fraud. Lenders are able to identify multiple charges of the same collateral for different loans.

Reduce Court Involvement in Business Matters

One of the major simplifications in many areas of regulation is to reduce the involvement of courts. For business entry, taking registration out of the courts and making it an administrative process radically reduces registration time and eases the backlog of commercial cases in the judiciary. Permitting private enforcement of collateral, with recourse to the courts only for disputes, substantially reduces enforcement time and encourages lending. For contract enforcement and bankruptcy proceedings, reducing the involvement of courts can open the way for specialized sections in the general jurisdiction courts or for specialized commercial courts, which can use streamlined processes in notification, evidence collection, and judgments, especially in countries with greater administrative capacity. Alternatively, the role of courts can be reduced by introducing summary procedures for commercial disputes, thereby limiting the time for judgment. One example comes from Nicaragua, where the summary procedure to collect debt takes about four months on average. In contrast, the normal civil procedure for resolving commercial disputes lasts more than five years.

Make Reform a Continuous Process

Countries that consistently perform well across the *Doing Business* indicators do so because of continuous reform. Denmark revised its business entry regulations in 1996 by removing several procedures, making the process electronic, and eliminating all fees. Australia has built in regulatory reform by including "sunset provisions" in new regulations, which automatically expire after a certain period if not renewed by parliament. And Sweden introduced a "guillotine" approach for regulatory reform, with hundreds of obsolete regulations being canceled after the government required regulatory agencies to register all essential regulations.[25] On average, laws in wealthy countries have been enacted or amended much more recently than those in developing countries, which often date to colonial times. This situation makes the often-heard complaint of "reform fatigue" in developing countries difficult to fathom.

Over the last decade, several countries have introduced regulatory impact assessments, which are carried out when new regulation is proposed. Requiring government agencies and ministries to engage in cost-benefit analyses has proven to be an effective tool in winnowing out burdensome, poorly designed, and socially costly regulations and in improving those that are necessary. Regulatory impact assessments are a standard feature of new business regulation in the European Union and have been adopted in many accession candidates, such as Hungary and Poland.

Continuous reforms require political will but not necessarily large resources. Indeed, if properly implemented, deregulation can save the government money and permit allocations to meet the needs of poor people. Other reforms, such as introducing administrative registration of businesses and creating credit registries, can pay for themselves in two to three years, as Serbia and Montenegro and Malaysia show. In the Netherlands, for example, administrative costs are reduced by an independent agency, ACTAL. ACTAL, which has only nine staff members, is empowered to advise on all proposed laws and regulations. To date, simplification of administrative procedures has been achieved in the areas of corporate taxation, social security, environmental regulations, and statistical requirements. The estimated savings are $600 million in streamlining of the tax requirements alone.

Other reforms, such as revising commercial codes and company laws, require big investments and take several years. In the interim, the public needs to be consulted, and the costs and benefits of the new legislation must be evaluated. Once in place, however, such reforms have enormous impact on private business. Vietnam's new enterprise law is but one example.

Notes

1. Glaeser and Shleifer 2002.
2. Thailand is the only East Asian country never to have be colonized. In the Middle East, several present-day countries had British rule after World War I.
3. Olson 1991; De Long and Shleifer 1993.
4. Also see Djankov and others 2002, 2003.
5. Engerman and Sokoloff 2002.
6. Acemoglu, Johnson, and Robinson 2001.
7. Hochschild 1998.
8. Alesina and others 2003; Dollar and others 2003.
9. Greif 1989; MacMillan and Woodruff 1999a, b.
10. This section is based on Djankov and others (forthcoming).
11. Pigou 1938.
12. Coase 1960.
13. Stigler 1971.
14. Allais 1947; Meade 1948; Lewis 1949.
15. Stiglitz 1989.
16. Ellickson 1991.
17. Greif 1989; Bernstein 1992.
18. See also Posner 1974.
19. See Djankov and others (forthcoming).
20. Landis 1938; Acemoglu and others 2001; Pistor and Xu 2002.
21. Along these lines, Glaeser and Shleifer (2003) argue that "The Rise of the Regulatory State" in the United States during the Progressive Era at the beginning of the twentieth century was a response to the growing problems of subversion of courts by robber barons.
22. Hellman, Jones, and Kaufmann (forthcoming).
23. Hart, Shleifer, and Vishny (1997) make the case for state ownership of prisons.
24. Field 2002.
25. Jacobs 2002.

References

Chapter 1

Batra, Geeta. 2003. "Investment Climate Measurement: Pitfalls and Possibilities." Discussion Paper. Investment Climate Unit, Private Sector Vice-Presidency, World Bank, Washington, D.C.

Batra, Geeta, Daniel Kaufmann, and Andrew Stone. 2003. *Investment Climate Around the World: Voices of the Firms from the World Business Environment Survey.* Washington, D.C.: World Bank.

BERI (Business Environment Risk Intelligence). 2002. *User Guide.* Friday Harbor, Wash.

Bertrand, Marianne, and Sendhil Mullainathan. 2002. "Do People Mean What They Say? Implications for Subjective Survey Data." *American Economic Review* 91(2): 67–72.

Bishop, George, Robert Oldendick, and Alfred Tuchfarber. 1986. "Opinions on Fictitious Issues: The Pressure to Answer Survey Questions." *Public Opinion Quarterly* 50(3): 240–50.

Botero, Juan, Simeon Djankov, Rafael La Porta, Florencio Lopez-de-Silanes, and Andrei Shleifer. 2003. "The Regulation of Labor." Working Paper 9756. National Bureau of Economic Research, Cambridge, Mass.

Cornelius, Peter, Michael Porter, and Klaus Schwab. 2003. *The Global Competitiveness Report 2002–2003.* London: Oxford University Press.

Djankov, Simeon, Oliver Hart, Tatiana Nenova, and Andrei Shleifer. 2003. "The Efficiency of Bankruptcy." Working Paper. Department of Economics, Harvard University, Cambridge, Mass.

Djankov, Simeon, Rafael La Porta, Florencio Lopez-de-Silanes, and Andrei Shleifer. 2002. "The Regulation of Entry." *Quarterly Journal of Economics* 117(1): 1–37.

———. 2003. "Courts." *Quarterly Journal of Economics* 118(2): 342–87.

Djankov, Simeon, Caralee McLiesh, and Andrei Shleifer. 2003. "Remedies in Credit Markets." Working Paper. Department of Economics, Harvard University, Cambridge, Mass.

Ernst&Young. 2003. *Corporate Tax Guide.* Washington, D.C.

Friedman, Eric, Simon Johnson, Daniel Kaufman, and Pablo Zoido-Lobaton. 2000. "Dodging the Grabbing Hand: Determinants of Unofficial Activity in 69 Countries." *Journal of Public Economics* 76(3): 459–93.

Hammergren, Linn. 2003. "Uses of Empirical Research in Refocusing Judicial Reforms: Lessons from Five Countries." Latin America and Caribbean Region, World Bank, Washington, D.C.

Jentzsch, Nicola. 2003. "The Regulatory Environment for Business Information Sharing." Working Paper. Monitoring, Analysis and Policy Unit, Private Sector Development Vice Presidency, World Bank, Washington, D.C.

Kaufmann, Daniel, Aart Kraay, and Massimo Mastruzzi. 2003. "Governance Matters III: Governance Indicators for 1996–2002." Policy Research Working Paper 3106. World Bank, Washington, D.C.

La Porta, Rafael, Florencio Lopez-de-Silanes, Andrei Shleifer, and Robert Vishny. 1998. "Law and Finance." *Journal of Political Economy* 106(6): 1113–55.

Meade, James, and Richard Stone. 1941. *An Analysis of the Sources of War Finance and an Estimate of the National Income and Expenditure in 1938 and 1940.* United Kingdom, Treasury, H.M.S.O., London.

Petty, Sir William. 1691. *Verbum Sapienti.* Published with *The Political Anatomy of Ireland.* London: Brown and Rogers. Reprinted in *The Economic Writings of Sir William Petty.* C.H. Hull, ed., 2 vols. Cambridge University Press, 1899.

Schneider, Friedrich. 2002. "Size and Measurement of the Informal Economy in 110 Economies Around the World." Discussion Paper. Monitoring, Analysis and

Policy Unit, Private Sector Vice-Presidency, World Bank, Washington, D.C.

Schwarz, Norbert, Fritz Strack, and Hans-Peter Mai. 1991. "Assimilation and Contrast Effects in Part-Whole Question Sequence: A Conversational Logic Analysis." *Public Opinion Quarterly* 55(1): 3–23.

Schwarz, Norbert, Hans J. Hippler, Brigitte Deutsch, and Fritz Strack. 1985. "Response Categories: Effects on Behavioral Reports and Comparative Judgments." *Public Opinion Quarterly* 49(2): 388–95.

The Heritage Foundation. 2002. *2003 Index of Economic Freedom*. Washington, D.C.

World Bank. 2002a. *Investment Climate Assessment: India.* Private Sector Advisory Services. Washington, D.C.

———. 2002b. *The Juicio Executivo Mercantil in the Federal District Courts of Mexico*. Latin America and the Caribbean Region, Poverty Reduction and Economic Management Unit. Washington, D.C.

———. 2002c. *The Investment Climate for Foreign Companies in Turkey*. Monitoring, Analysis and Policy Unit, Private Sector Vice-Presidency. Washington, D.C.

———. 2003. *Informe Final de la Investigacion Sobre el Uso de la Justicia Civil en el Ecuador*. Latin America and the Caribbean Region, Poverty Reduction and Economic Management Unit, World Bank, Washington, D.C.

Chapter 2

Alesina, Alberto, Silvia Ardagna, Guiseppe Nicoletti, and Fabio Schiantarelli. 2003. "Regulation and Investment." Working Paper 9560. National Bureau of Economic Research, Cambridge, Mass.

Batra, Geeta, Daniel Kaufmann, and Andrew Stone. 2003. *Investment Climate Around the World: Voices of the Firms from the World Business Environment Survey.* Washington, D.C.: World Bank.

Berle, Adolf, and Gardiner Means. 1932. *The Modern Corporation and Private Property.* New York: Macmillan.

Bertrand, Marianne, and Francis Kramarz. 2002. "Does Entry Regulation Hinder Job Creation? Evidence from the French Retail Industry." *Quarterly Journal of Economics* 117(4): 1369–1413.

Blumberg, David. 1986. "Limited Liability and Corporate Groups." *Journal of Corporation Law* 11(4): 573–631.

DeLong, J. Bradford, and Andrei Shleifer. 1993. "Princes and Merchants: European City Growth Before the Industrial Revolution." *Journal of Law and Economics* 36(2): 671–702.

De Soto, Hernando. 1989. *The Other Path.* New York: Harper and Row.

Diamond, Aubrey. 1982. "Corporate Personality and Limited Liability." In *Limited Liability and the Corporation*, ed. Tony Orhnial. London and Canberra: Croom Helm.

Djankov, Simeon, Rafael La Porta, Florencio Lopez-de-Silanes, and Andrei Shleifer. 2002. "The Regulation of Entry." *Quarterly Journal of Economics* 117(1): 1–37.

European Commission. 2002. *Benchmarking the Administration of Business Start-Ups*. Center for Strategy and Evaluation Services, Enterprise Directorate General, Brussels.

FIAS (Foreign Investment Advisory Service). 2003. *Administrative Barriers in Latvia*. World Bank, Washington, D.C.

Friedman, Eric, Simon Johnson, Daniel Kaufman, and Pablo Zoido-Lobaton. 2000. "Dodging the Grabbing Hand: Determinants of Unofficial Activity in 69 Countries." *Journal of Public Economics* 76(3): 459–93.

Hoekman, Bernard, Hiao Looi Kee, and Marcelo Olarreaga. 2001. "Mark-ups, Entry Regulation and Trade: Does Country Size Matter?" Policy Research Working Paper 2662. World Bank, Washington, D.C.

Horvath, Michael, and Michael Woywode. 2003. "The Entrepreneurial Choice of Limited Liability." Discussion Paper, University of Mannheim, Germany.

Jacobs, Scott. 2002. "Reforming Business Registration in Serbia." Jacobs and Associates, Washington, D.C.

Japanese Association of Small Business. 1999. *Constraints to Business Entry*. Tokyo.

Jordan, Cally. 1996. "A Comparative Survey of Companies' Laws in Selected Jurisdictions." McGill University, Law Faculty, Montreal.

Kaufmann, Daniel, Aart Kraay, and Massimo Mastruzzi. 2003. "Governance Matters III: Governance Indicators for 1996–2002." Policy Research Working Paper 3106. World Bank, Washington, D.C.

La Porta, Rafael, Florencio Lopez-de-Silanes, Andrei Shleifer, and Robert W. Vishny. 1999. "The Quality of Government." *Journal of Law, Economics, and Organization* 15(1): 222–79.

Morisset, Jacques, and Olivier Lumenga Neso. 2002. "Administrative Barriers to Foreign Investment in Developing Countries." Policy Research Working Paper 2848. World Bank, Washington, D.C.

Olson, Mancur. 1991. "Autocracy, Democracy, and Prosperity." In *Strategy of Choice*, ed. Richard Zeckhauser. Cambridge, Mass: MIT Press.

Pistor, Katharina, and Daniel Berkowitz. 2003. "Of Legal Transplants, Legal Irritants, and Economic Development." In *Corporate Governance and Capital Flows in a Global Economy*, eds. Peter Cornelius and Bruce Kogut. London: Oxford University Press.

Sader, Frank. 2002. *A View on One-Stop Shops*. Working Paper. Foreign Investment Advisory Service, Investment Climate Department, Private Sector Vice-Presidency, World Bank, Washington, D.C.

Shannon, Herbert. 1931. "The Coming of General Limited Liability." *Economic History* 2(6): 267–91.

Smith, Adam. 1776. [1937]. *The Wealth of Nations*. New York: Random House.

Trang, Nguyen Phuong Quynh, Bui Tuong Anh, Han Manh Tien, Hoang Xuan Thanh, and Nguyen Thi Hoai Thu. 2001. "Doing Business Under the New Enterprise Law: A Survey of Newly Registered Companies." Private Sector Discussion Paper 12. Mekong Project Development Facility, Mekong, Vietnam.

Zhuravskaya, Ekaterina. 2003. "Assessing Russia's Business Climate." Working Paper. Department of Economics, Princeton University, Princeton, NJ.

Chapter 3

Abowd, John, Francis Kramarz, and David Margolis. 1999. "Minimum Wages and Employment in France and the United States." Working Paper 6996. National Bureau of Economic Research, Cambridge, Mass.

Aidt, Toke, and Zafiris Tzannatos. 2002. *Unions and Collective Bargaining: Economic Effects in a Global Environment*. Washington, D.C.: World Bank.

Batra, Geeta, Daniel Kaufmann, and Andrew Stone. 2003. *Voices of the Firms: Investment Climate and Governance Findings of the World Business Environment Survey*. World Bank, Washington, D.C.

Becker, Gary. 1971. *The Economics of Discrimination*. Chicago: University of Chicago Press.

Besley, Timothy, and Robin Burgess. 2003. "Can Labor Regulation Hinder Economic Performance? Evidence from India." Discussion Paper. Department of Economics, London School of Economics, London.

Betcherman, Gordon. 2002. "An Overview of Labor Markets World-Wide: Key Trends and Major Policy Issues." Discussion Paper 205. Social Protection Discussion Series. World Bank, Washington, D.C.

Betcherman, Gordon, Amy Luinstra, and Makoto Ogawa. 2001. "Labor Market Regulation: International Experience in Promoting Employment and Social Protection." Discussion Paper 128. Social Protection Discussion Series, World Bank, Washington, D.C.

Blanchard, Olivier. 2002. "Designing Labor Market Institutions." MIT, Department of Economics, Cambridge, Mass.

Blanchard, Olivier, and Augustin Landier. 2000. "The Perverse Effects of Partial Labor Market Reform: Fixed Duration Contracts in France." MIT, Department of Economics, Cambridge, Mass.

Blanchard, Olivier, and Pedro Portugal. 2001. "What Hides Behind an Unemployment Rate: Comparing Portuguese and U.S. Labor Markets." *American Economic Review* 91 (1): 187–207.

Booth, Alison. 1995. *The Economics of the Trade Union*. Cambridge: Cambridge University Press.

Botero, Juan, Simeon Djankov, Rafael La Porta, Florencio Lopez-de-Silanes, and Andrei Shleifer. 2003. "The Regulation of Labor." Working Paper 9756. National Bureau of Economic Research, Cambridge, Mass.

Dolado, Juan, Florentino Felgueroso, and Juan Jimeno. 1997. "The Effects of Minimum Bargained Wages on Earnings: Evidence from Spain." *European Economic Review* 41 (3–5): 713–21.

Fallon, Peter, and Robert Lucas. 1991. "The Impact of Changes in Job Security Regulations in India and Zimbabwe." *World Bank Economic Review* 5 (3): 395–413.

Fields, Gary, and Henry Wan. 1989. "Wage-Setting Institutions and Economic Growth." *World Development* 17 (4): 1471–83.

Fishback, Price. 1998. "Operations of 'Unfettered' Labor Markets: Exit and Voice in American Labor Markets at the Turn of the Century." *Journal of Economic Literature* 36 (2): 722–65.

Gill, Indermit, Claudio Montenegro, and Dorte Domeland, editors. 2002. *Crafting Labor Policy: Techniques and Lessons from Latin America*. Washington, D.C.: World Bank.

Heckman, James, and Carmen Pages. 2000. "The Cost of Job Security Regulation: Evidence from Latin American Labor Markets." *Economia* 6 (2): 109–54.

Holmes, Thomas. 1998. "The Effect of State Policies on the Location of Manufacturing: Evidence from State Borders." *Journal of Political Economy* 106 (4): 667–705.

Hopenhayn, Hugo. 2001. "Labor Market Policies and Employment Duration: The Effects of Labor Market Reform in Argentina." Working Paper 407. Inter-American Development Bank, Research Department, Washington, D.C.

International Labor Organization. 1998. *ILO Declaration on Fundamental Principles and Rights at Work.* Geneva, Switzerland.

———. 2000. *Your Voice at Work.* Geneva, Switzerland.

———. 2001. *Stopping Forced Labor.* Geneva, Switzerland.

———. 2002. *A Future Without Child Labor.* Geneva, Switzerland.

———. 2003. *Time for Equality at Work.* Geneva, Switzerland.

Kugler, Adriana. 2000. "The Incidence of Job Security Regulations on Labor Market Flexibility and Compliance in Colombia: Evidence from the 1990 Reform." Working Paper 393. Inter-American Development Bank, Research Department, Washington, D.C.

Lazear, Edward. 1990. "Job Security Provisions and Employment." *Quarterly Journal of Economics* 105 (3): 699–727.

Maloney, William, and Juan Nunez. 2000. "Minimum Wages in Latin America." World Bank, Washington, D.C.

Montenegro, Claudio, and Carmen Pages. 2003. "Who Benefits from Labor Market Regulations: Chile 1960–1988." Working Paper 9850. National Bureau of Economic Research, Cambridge, Mass.

Mulligan, Casey, and Xavier Sala-i-Martin. 2000. "Measuring Aggregate Human Capital." *Journal of Economic Growth* 5 (2): 215–52.

Nicoletti, Giuseppe, and Stefano Scarpetta. 2003. "Regulation, Productivity and Growth: The OECD Evidence." Policy Research Working Paper 2944. World Bank, Washington, D.C.

Nicoletti, Giuseppe, Andrea Bassanini, Ekkehard Ernst, Sebastien Jean, Paulo Santiago, and Paul Swain. 2001. "Product and Labor Market Interactions in OECD Countries." Economics Department Working Paper 312. Organisation of Economic Co-operation and Development, Paris.

OECD (Organisation of Economic Co-operation and Development). 1998. *The OECD Employment Outlook.* Paris.

Saavedra, Jaime, and Maximo Torrero. 2000. "Labor Market Reforms and Their Impact on Formal Labor Demand and Job Market Turnover: The Case of Peru." Research Network Working Paper 394. Inter-American Development Bank, Research Department, Washington, D.C.

Scarpetta, Stefano. 1996. "Assessing the Role of Labor Market Policies and Institutional Settings on Unemployment: A Cross-Country Study." *OECD Economic Studies* 26 (1): 43–98.

Scarpetta, Stefano, and Thierry Verdier. 2002. "Productivity and Convergence in a Panel of OECD Industries: Do Regulations and Institutions Matter?" Economics Department Working Paper 28/2002. Organisation of Economic Co-operation and Development, Paris.

Schneider, Friedrich. 2002. "Size and Measurement of the Informal Economy in 110 Economies around the World." Discussion Paper. Monitoring, Analysis and Policy Unit, Private Sector Vice-Presidency, World Bank, Washington, D.C.

Turin, Stepan. 1934. *From Peter the Great to Lenin.* London: King and Sons.

Valenze, David. 1985. *The First Industrial Woman.* London: Oxford University Press.

Van Audenrode, Marc. 1994. "Short-Time Compensation, Job Security, and Employment Contracts: Evidence from Selected OECD Countries." *Journal of Political Economy* 102 (1): 76–102.

World Bank. 2002a. *The Russian Labor Market: Moving from Crisis to Recovery.* Europe and Central Asia Region, Human Development Sector Unit, Washington, D.C.

———. 2002b. *Brazil Jobs Report.* Brazil Country Management Unit, Latin America and the Caribbean Region, Washington, D.C.

———. 2003a. *Republic of Tunisia: Employment Strategy.* Washington, D.C.

———. 2003b. *World Development Indicators.* Washington D.C.

Chapter 4

Bannerjee, Abhijit, and Esther Duflo. 2000. "Reputation Effects and the Limits of Contracting: A Study of the Indian Software Industry." *Quarterly Journal of Economics* 115 (3): 989–1017.

Batra, Geeta, Daniel Kaufmann, and Andrew Stone. 2003. *Voices of the Firms: Investment Climate and*

Governance Findings of the World Business Environment Survey. World Bank, Washington, D.C.

Bigsten, Arne, Paul Collier, Stefan Dercon, Marcel Fafchamps, Bernard Gauthier, Jan Willem Gunning, Abena Oduro, Remco Oostendorp, Cathy Patillo, Mans Soderbom, Francis Teal, and Albert Zeufack. 2000. "Contract Flexibility and Dispute Resolution in African Manufacturing." *Journal of Development Studies* 36 (4): 1–17.

Botero, Juan, Rafael La Porta, Florencio Lopez-de-Silanes, Andrei Shleifer, and Alexander Volokh. 2003. "Judicial Reform." *World Bank Research Observer* 18 (1): 61–88.

Clay, Karen. 1997. "Trade without Law: Private Order Institutions in Mexican California." *Journal of Law, Economics, and Organization* 13 (1): 202–31.

Dakolias, Maria. 1999. "Court Performance around the World: A Comparative Perspective." Technical Paper 430. World Bank, Washington, D.C.

Dawson, John. 1960. *A History of Lay Judges*. Cambridge, Mass: Harvard University Press.

Djankov, Simeon, Rafael La Porta, Florencio Lopez-de-Silanes, and Andrei Shleifer. 2003. "Courts." *Quarterly Journal of Economics* 18 (2): 453–517.

Dollinger, Philippe. 1970. *The German Hansa*. Stanford: Stanford University Press.

Finnegan, David. 2001. "Observations on Tanzania's Commercial Court: A Case Study." Paper presented at the World Bank Conference on Empowerment, Security, and Opportunity through Law and Justice, July 8–12, St. Petersburg, Russia.

Goethe, Johann Wolfgang von. 1969. *The Autobiography of Johann Wolfgang von Goethe*. Trn. John Oxenford. New York: Horizon Press.

Greif, Avner. 1993. "Contract Enforceability and Economic Institutions in Early Trade: The Maghribi Traders' Coalition." *American Economic Review* 83 (3): 525–548.

Hammergren, Linn. 2000. "Diagnosing Judicial Performance: Toward a Tool to Help Guide Judicial Reform Programs." Latin America and the Caribbean Region, World Bank, Washington, D.C.

———. 2003. "Uses of Empirical Research in Refocusing Judicial Reforms: Lessons from Five Countries." Latin America and the Caribbean Region, World Bank, Washington, D.C.

McMillan, John, and Christopher Woodruff. 1999. "Dispute Prevention without Courts in Vietnam." *Journal of Law, Economics, and Organization* 15 (3): 637–58.

———. 2000. "Private Order under Dysfunctional Public Order." *Michigan Law Review* 98(8): 101–38.

Milgrom, Paul, Douglass North, and Barry Weingast. 1990. "The Role of Institutions in the Revival of Trade: The Law Merchant, Private Judges, and Champagne Fairs." *Economics and Politics* 2(1): 1–23.

Muldrew, Craig. 1993. "Credit and the Courts: Debt Litigation in a Seventeenth-Century Urban Community." *Economic History Review* XLVI (1): 23–38.

Murrell, Peter. 2003. "Firms Facing New Institutions: Transactional Governance in Romania." *Journal of Comparative Economics* 31(4): 768–791.

PriceWaterhouseCoopers. 2001. *Report on Debt Recovery Tribunals in India*. Washington, D.C.

Sinare, Hanwa. 2000. "An Advocate's Overview of the Commercial Court Performance." The High Court, Dar es Saalam, Tanzania.

Woodruff, Christopher. 1998. "Contract Enforcement and Trade Liberalization in Mexico's Footwear Industry." *World Development* 26 (5): 979–91.

World Bank. 2001a. "Administration of Justice and the Legal Profession in Slovakia." Europe and Central Asia Region, Poverty Reduction and Economic Management Unit, Washington, D.C.

———. 2001b. *Argentina: Legal and Judicial Sector Assessment*. Legal Vice Presidency, Washington, D.C.

———. 2002a. *The World Bank Legal Review: Law and Justice for Development*. Legal Vice Presidency, Washington, D.C.

———. 2002b. *Initiatives in Legal and Judicial Reform*. Legal Vice Presidency, Washington, D.C.

Zweigert, Konrad. 1983. *The International Encyclopedia of Comparative Law: Specific Contracts*. J.C.B. Mohr. Tubingen, Germany.

Chapter 5

Barron, John, and Michael Staten. 2003. "The Value of Comprehensive Credit Reports: Lessons from the U.S. Experience." In *Credit Reporting Systems and the International Economy*, ed. Margaret Miller. Boston: MIT Press.

Besley, Timothy. 1995. "Nonmarket Institutions for Credit and Risk Sharing in Low-Income Countries." *Journal of Economic Perspectives* 9 (3): 115–27.

Bolton, Patrick, and David Scharfstein. 1996. "Optimal Debt Structure and the Number of Creditors." *Journal of Political Economy* 101 (1): 1–25.

Campbell, Tim, and William Kracaw. 1980. "Information Production, Market Signalling and the Theory of Financial Intermediation." *Journal of Finance* 25 (4): 853–82.

Demirguc-Kunt, Asli, and Vojislav Maksimovic. 1998. "Law, Finance, and Firm Growth." *Journal of Finance* 53(6): 2107–37.

Diamond, Douglas. 1984. "Financial Intermediation and Delegated Monitoring." *Review of Economic Studies* 51(3): 393–414.

Djankov, Simeon, Caralee McLiesh, and Andrei Shleifer. 2003. "Remedies in Credit Markets." Working Paper, Department of Economics, Harvard University, Cambridge, Mass.

Fuentes, Rodrigo, and Carlos Maquieira. 2001. "Why Borrowers Repay: Understanding High Performance in Chile's Financial Market." In *Defusing Default: Incentives and Institutions*, ed. Marco Pagano. Washington, D.C.: Inter-American Development Bank.

Galindo, Arturo, and Margaret Miller. 2001. "Can Credit Registries Reduce Credit Constraints? Empirical Evidence on the Role of Credit Registries in Firm Investment Decisions." Paper prepared for the annual meetings of the Inter-American Development Bank, Santiago, Chile, March.

Garro, Alejandro. 1998. "Difficulties in Obtaining Secured Lending in Latin America: Why Law Reform Really Matters." In Norton, Joseph and Mads Andenas, eds., *Emerging Financial Markets and Secured Transactions.* London: Klewer Law International.

Greif, Avner, Paul Milgrom, and Barry Weingast. 1994. "Coordination, Commitment and Enforcement: The Merchant Guild as a Nexus of Contracts." *Journal of Political Economy* 102 (5): 745–76.

Hadlow, John. 2003. *How a Credit Bureau Enhances the Credit Approval and Risk Management Process.* Paper presented at The Central Asian Credit Bureau Conference, held January 29-31 in Almaty, Kazakhstan.

Hill, Claire. 2002. "Is Secured Debt Efficient?" *Texas Law Review* 80 (3): 1117–1173.

Hoffman, Philip T., Gilles Poste-Vinay, and Jean-Laurent Rosenthal. 1998. "What Do Notaries Do? Overcoming Asymetric Information in Financial Markets: The Case of Paris, 1751." *Journal of Institutional and Theoretical Economics* 154 (3): 499–530.

Jappelli, Tullio, and Marco Pagano. 1993. "Information Sharing in Credit Markets." *The Journal of Finance* 43 (5): 1693–718.

———. 2000. "Information Sharing in Credit Markets: The European Experience." Center for Studies in Economics and Finance Working Paper No. 35, The University of Salerno, Italy.

———. 2002. "Information Sharing, Lending and Defaults: Cross-Country Evidence." *Journal of Banking and Finance* 26 (10): 2017–45.

Jentzsch, Nicola. 2003a. "The Regulatory Environment for Business Information Sharing." Working Paper. Monitoring, Analysis and Policy Unit, Private Sector Development Vice Presidency, World Bank, Washington D.C.

Jentzsch, Nicola. 2003b. "The Regulation of Financial Privacy: A Comparative Study of the United States and Europe." Research Report 5, European Credit Research Institute, Brussels.

Keinan, Yoram. 2001. "The Evolution of Secured Transactions." *World Development Report 2002,* Background Paper. World Bank, Washington, D.C.

King, Robert, and Ross Levine. 1993. "Finance and Growth: Schumpeter Might be Right." *Quarterly Journal of Economics* 108(3): 717–37.

La Porta, Rafael, Florencio Lopez-de-Silanes, Andrei Shleifer, and Robert Vishny. 1998. "Law and Finance." *Journal of Political Economy* 106 (6): 1113–55.

Levine, Ross. 1997. "Financial Development and Economic Growth: Views and Agenda." *Journal of Economic Literature* 35 (2): 688–726.

———. 1998. "The Legal Environment, Banks, and Long-Run Economic Growth." *Journal of Money, Credit, and Banking* 30 (3): 596–613.

Love, Inessa, and Nataliya Mylenko. 2003. "Credit Reporting and Financing Constraints." Working Paper. World Bank, Washington, D.C.

Mann, Ronald. 1997. "Explaining the Pattern of Secured Credit." *Harvard Law Review* 110 (2): 625–668.

Miller, Margaret. 2003. "Credit Reporting Systems Around the Globe: The State of the Art in Public and Private Credit Registries." In *Credit Reporting Systems and the International Economy,* ed. Margaret Miller. Boston: MIT Press.

Olegario, Rowena. 2003. "Credit-Reporting Agencies: A Historical Perspective." In *Credit Reporting Systems and the International Economy,* ed. Margaret Miller. Boston: MIT Press.

Rajan, Raghuram, and Luigi Zingales. 1998. "Financial Dependence and Growth." *American Economic Review* 88 (3): 559–86.

Stiglitz, Joseph, and Andrew Weiss. 1988. "Banks as Social Accountants and Screening Devices for the Allocation of Credit." Working Paper 2710. National Bureau of Economic Research, Cambridge, Massachusetts.

Chapter 6

Berglof, Erik, Howard Rosenthal, and Ernst-Ludwig von Thadden. 2002. "The Formation of Legal Institutions for Bankruptcy: A Comparative Study of the Legislative History." Princeton University, Department of Politics, Princeton, N.J.

Betker, Brian. 1995. "Power and Deviations from Absolute Priority in Chapter 11 Bankruptcies." *Journal of Business* 68 (2): 161–183.

Bruno, Matteo. 1561. Tractatus Matthaei Bruni Arimineni de Cessione Bonoru, Apud Haeredes Aloysij Valuassoris & Ioannem Dominicum Michaelem. Vol. 115. Venice.

Claessens, Stijn, and Leora Klapper. 2001. "Bankruptcy Around the World: Explanations for Its Relative Use." University of Amsterdam, Department of Finance, The Netherlands.

Couwenberg, Oscar. 2001. "Survival Rates in Bankruptcy Systems: Overlooking the Evidence." Working Paper. University of Groningen, Department of Economics, The Netherlands.

Hart, Oliver. 2000. "Different Approaches to Bankruptcy." Working Paper 7921. National Bureau of Economic Research, Cambridge, Mass.

La Porta, Rafael, and Florencio Lopez-de-Silanes. 2001. "Creditor Protection and Bankruptcy Law Reform." In *Resolution of Financial Distress,* eds. Stijn Claessens, Simeon Djankov, and Ashoka Mody. Washington, D.C.: World Bank.

La Porta, Rafael, Florencio Lopez-de-Silanes, Andrei Shleifer, and Robert Vishny. 1997. "Legal Determinants of External Finance." *Journal of Finance* 52 (3): 1131–50.

———. 1998. "Law and Finance." *Journal of Political Economy* 106 (6): 1113–55.

Levinthal, Louis. 1919. "The Early History of English Bankruptcy." *University of Pennsylvania Law Review* 1: 17–81.

Modigliani, Franco, and Enrico Perotti. 2000. "Security Markets versus Bank Finance: Legal Enforcement and Investor Protection." *International Review of Finance* 1(2): 81–96.

Pistor, Katharina, Yoram Keinan, Jan Kleinheisterkamp, and Mark D. West. Forthcoming. "Innovation in Corporate Law." *Journal of Comparative Economics.*

Posner, Richard. 1992. *Economic Analysis of Law.* Boston: Little Brown.

Rajak, Harry. 1997. "Rescue versus Liquidation in Central and Eastern Europe." *Texas International Law Journal* 33 (4): 157–72.

Stiglitz, Joseph. 2001. "Bankruptcy Laws: Basic Economic Principles." In *Resolution of Financial Distress,* eds. Stijn Claessens, Simeon Djankov, and Ashoka Mody. Washington, D.C.: World Bank.

Stromberg, Per. 2000. "Conflicts of Interest and Market Illiquidity in Bankruptcy Auctions: Theory and Tests." *Journal of Finance* 55 (4): 2641–92.

Thorburn, Karin. 2000. "Bankruptcy Auctions: Costs of Debt Recovery and Firm Survival." *Journal of Financial Economics* 58 (3): 337–68.

Wood, Philip. 1995. "Principles of International Insolvency." *International Insolvency Review* 4: 94–138.

World Bank. 2001a. "The Insolvency and Creditor Rights System in the Czech Republic." Report on Observance of Standards and Codes. Legal Vice Presidency, Private Sector Development Law, Washington, D.C.

———. 2001b. "Principles and Guidelines for Effective Insolvency and Creditor Rights Systems." Legal Vice Presidency, Private Sector Development Law, Washington, D.C.

———. 2002a. "The Insolvency and Creditor Rights System in the Slovak Republic." Report on Observance of Standards and Codes. Legal Vice Presidency, Private Sector Development Law, Washington. D.C.

———. 2002b. "The Insolvency and Creditor Rights System in Lithuania." Report on Observance of Standards and Codes. Legal Vice Presidency, Private Sector Development Law, Washington, D.C.

———. 2002c. "The Insolvency and Creditor Rights System in Argentina." Report on Observance of Standards and Codes. Legal Vice Presidency, Private Sector Development Law, Washington, D.C.

Chapter 7

Acemoglu, Daron, James Robinson, and Simon Johnson. 2001. "The Colonial Origins of Comparative Development: An Empirical Investigation." *American Economic Review* 91 (5): 1369–1401.

Alesina, Alberto, Silvia Ardagna, Guiseppe Nicoletti, and Fabio Schiantarelli. 2003. "Regulation and Investment." Working Paper 9560. National Bureau of Economic Research, Cambridge, Mass.

Allais, Maurice. 1947. "Le Probleme de la Planification Economique dans une Economie Collectiviste." *Kyklos* II: 48–71.

Bernstein, Lisa. 1992. "Opting Out of the Legal System: Extralegal Contractual Relations in the Diamond Industry." *Journal of Legal Studies* 21(1): 115–57.

Coase, Ronald. 1960. "The Problem of Social Cost." *Journal of Law and Economics* 3(1): 1–44.

DeLong, J. Bradford, and Andrei Shleifer. 1993. "Princes and Merchants: European City Growth Before the Industrial Revolution." *Journal of Law and Economics* 36(2): 671–702.

Djankov, Simeon, Rafael La Porta, Florencio Lopez-de-Silanes, and Andrei Shleifer. 2002. "The Regulation of Entry." *Quarterly Journal of Economics* 117 (1): 1–37.

———. 2003. "Courts." *Quarterly Journal of Economics* 118 (2): 452–517.

———. Forthcoming. "The New Comparative Economics." *Journal of Comparative Economics.*

Dollar, David, Mary Hallward-Driemeier, and Taye Mengistae. 2003. "Investment Climate and Firm Performance in Developing Countries." Working Paper, Development Research Department, World Bank, Washington D.C.

Ellickson, Robert. 1991. *Order Without Law.* Cambridge, Mass: Harvard University Press.

Engerman, Stanley, and Kenneth Sokoloff. 2002. "Factor Endowments, Inequality, and Paths of Development among New World Economies." *Journal of the Latin American and Caribbean Economic Association* 3 (1): 41–88.

Field, Erica. 2003. "Entitled to Work: Urban Property Rights and Labor Supply in Peru." Working paper. Department of Economics, Harvard University, Cambridge, Mass.

Glaeser, Edward, and Andrei Shleifer. 2002. "Legal Origins." *Quarterly Journal of Economics* 117 (4): 1193–229.

———. 2003. "The Rise of the Regulatory State." *Journal of Economic Literature* 41 (2): 401–425.

Greif, Avner. 1989. "Reputation and Coalitions in Medieval Trade: Evidence on the Maghribi Traders." *Journal of Economic History* 49(4): 857–882.

Hart, Oliver, Andrei Shleifer, and Robert Vishny. 1997. "The Proper Scope of Government." *Quarterly Journal of Economics* 112 (4): 1127–1161.

Hellman, Joel, Geraint Jones, and Daniel Kaufmann. Forthcoming. "Seize the State, Seize the Day: State Capture and Influence in Transition Economies." *Journal of Comparative Economics.*

Hochschild, Adam. 1998. *King Leopold's Ghost: A Story of Greed, Terror and Heroism in the Congo.* Boston, Mass: Houghton Mifflin.

Jacobs, Scott. 2002. "Reforming Business Registration in Serbia." Jacobs and Associates, Washington, D.C.

Landis, James. 1938. *The Administrative Process.* New Haven, Conn: Yale University Press.

Lewis, Arthur. 1949. *The Principles of Economic Planning.* London: George Allen and Unwin Ltd.

McMillan, John, and Christopher Woodruff. 1999a. "Dispute Prevention without Courts in Vietnam." *Journal of Law, Economics, and Organization* 15 (3): 637–58.

———. 1999b. "Interfirm Relationships and Informal Credit in Vietnam." *Quarterly Journal of Economics* 114 (4): 1285–1320.

Meade, James. 1948. *Planning and the Price Mechanism: the Liberal Socialist Solution.* London: George Allen and Unwin Ltd.

Olson, Mancur. 1991. "Autocracy, Democracy, and Prosperity." In *Strategy of Choice,* ed. Richard Zeckhauser. Cambridge, Mass: MIT Press.

Pigou, Alfred. 1938. *The Economics of Welfare.* 4th edition. London: Macmillan.

Pistor, Katharina, and Chenggang Xu. 2002. "Law Enforcement under Incomplete Law: Theory and Evidence from Financial Market Regulation." Columbia School of Law, New York.

Posner, Richard. 1972. "A Theory of Negligence." *Journal of Legal Studies* 1 (1): 29–96.

———. 1974. "Theories of Economic Regulation." *Bell Journal of Economics* 5 (2): 335–58.

Stigler, George. 1971. "The Theory of Economic Regulation." *Bell Journal of Economics* 2: 3–21.

Stiglitz, Joseph. 1989. "Markets, Market Failures, and Development." *American Economic Review* 79 (2): 197–203.

World Bank. 2003. *World Development Indicators.* Washington, D.C.

The indicators presented and analyzed in *Doing Business* focus on government regulation and its effect on businesses. Based on assessment of laws and regulations, with verification and input from local government officials, lawyers, business consultants, and other professionals administering or advising on legal and regulatory requirements, this methodology offers several advantages. It is based on factual information and allows multiple interactions with the respondent, ensuring accuracy by clarifying possible misinterpretations of the survey questions. It is inexpensive, so data can be collected in a large sample of countries. And because the same standard assumptions are applied in the data collection, which is transparent and easily replicable, comparisons and benchmarks are valid across countries.

The *Doing Business* methodology has three limitations that should be considered when interpreting the data. First, in many cases the collected data refer to businesses in the country's most populous city and may not be representative of regulatory practices in other parts of the country. Second, the data often focus on a specific business form—limited liability company—and may not be representative of the regulation on other businesses, for example sole proprietorships. Finally, some indicators—e.g. on time—involve an element of judgment by the expert respondents. The time indicators reported in *Doing Business* represent median perceived values of several respondents under the assumptions of the case study.

Updated indicators, as well as any revisions of or corrections to the printed data, are available on the *Doing Business* website: http://rru.worldbank.org/doingbusiness/.

Economy Characteristics

Region and Income group

Doing Business reports the World Bank regional and income groupings, available at http://www.worldbank.org/data/countryclass/countryclass.html

Gross National Income (GNI) per Capita

Doing Business reports income per capita as at end 2002, calculated using the Atlas method (current US$), as published in the World Development Indicators.

Population

Doing Business reports population statistics as published in the World Development Indicators 2002.

Legal origin

Legal origin identifies the origin of the Company Law or Commercial Code in each country. It is compiled by the *Doing Business* team using several sources, including La Porta et al. (1999) and the CIA Factbook (2002). There are five possible origins: English, French, German, Nordic, and Socialist. The English origin comprises jurisdictions in the common law tradition. French legal origin includes the majority of countries in the civil law tradition. Laws have been transplanted through voluntary adoption or colonization. Jurisdictions in the Socialist law tradition include only the countries which did not have well-developed commercial laws prior to the founding of the Soviet Union and the Socialist Bloc after World War II. The countries of Central and Eastern Europe and the Baltics thus belong to the German or French legal tradition.

Informal economy
Measures the output in the informal economy as a share of gross national income. Source: Schneider (2002).

Starting a Business

Doing Business compiles a comprehensive list of entry regulations by recording the procedures that are officially required for an entrepreneur to obtain all necessary permits, and to notify and file with all requisite authorities, in order to legally operate a business. The current mark of the data refers to January 2003.

The survey divides the process of starting up a business into distinct procedures, and then calculates the costs and time necessary for the accomplishment of each procedure under normal circumstances. The assumption is that the required information is readily available and that all government and nongovernment entities involved in the process function efficiently and without corruption.

There are a number of procedures necessary to legally operate industrial or commercial businesses. These include (1) obtaining the necessary permits and licenses, and (2) completing all of the required inscriptions, verifications, and notifications to enable the company to start operation. A "procedure" is defined as any interaction of the company founder with external parties (government agencies, lawyers, auditors, notaries, etc). Interactions between company founders or company officers and employees are not considered as separate procedures. For example, an inauguration meeting where shareholders elect the directors and secretary of the company is not considered a procedure, as there are no outside parties involved.

All procedures that are required for establishing a business are recorded, even if they may be avoided in exceptional cases or for exceptional types of business. In general, there are four types of procedures: (1) procedures that are always required; (2) procedures that are generally required but that can be avoided in exceptional cases or for exceptional types of businesses; (3) mandatory procedures that are not generally required (industry-specific and procedures specific to large businesses); and (4) voluntary procedures. The data cover only procedures in the first two categories.

Assumptions about the Business
To make the business comparable across countries, several assumptions are employed:

- The business is a limited liability company. If there is more than one type of limited liability company in the country, the most popular limited liability form among domestic firms is chosen. Information on the most popular form is obtained from incorporation lawyers or the statistical office.
- It operates in the country's most populous city.
- The business is 100 percent domestically owned, and has five owners, none of whom is a legal entity.
- The business has a start-up capital of 10 times income per capita in 2002. The company's start-up capital cannot include contributions in kind, i.e., it is composed of 100 percent cash.
- It performs general industrial or commercial activities, such as the production or sale of products or services to the public. It does not perform activities of foreign trade and does not handle products subject to a special tax regime, for example, liquor or tobacco. The company is not using heavily polluting production processes.
- The business leases the commercial plant and offices and is not a proprietor of real estate.
- It does not qualify for investment incentives or any special benefits.
- The business has up to 50 employees one month after the commencement of operations. All employees are nationals.
- It has a turnover up to 10 times its start-up capital.
- The company deed is 10 pages long.

Assumptions about Procedures

To make the procedures comparable across countries, several assumptions are employed:

- On facilitators: It is assumed that the founders complete all procedures themselves, without middlemen, facilitators, accountants, lawyers, etc., unless the use of such third party is mandated by law. In all countries, it is possible to hire a consultant or middleman to perform most or all of the entry procedures, but this tends to be expensive.
- On voluntary procedures: Procedures that are not required by law for getting the business started are ignored. For example, the procedure of obtaining exclusive rights over the company name is not counted in a country where businesses can use a "number" as identification.
- On nonmandatory lawful shortcuts: Lawful shortcuts are counted as "required" procedures. These are procedures fulfilling the following four requirements: (1) they are not mandatory by law; (2) they are not illegal; (3) they are available for the general public (i.e. they are not specifically designed for special people); and (4) avoiding them causes substantial delays.
- On industry-specific requirements: Only procedures that are required of all businesses are covered. The study does not track procedures that only businesses in specific industries undergo. For example, procedures to comply with environmental regulations are included only when they apply to all businesses.
- On utilities: Procedures that the business undergoes in order to hook up for electricity, water, gas, and waste-disposal services are not included, unless these constitute required inspections for the business to legally start operations.

Cost Measure

The text of the Company Law, the Commercial Code, or specific regulations are used as a source for the costs associated with starting up a business. If there are conflicting sources and the laws are not clear, the most authoritative source is used. If the sources have the same rank, the source indicating the most costly procedure is used, since an entrepreneur never second-guesses a government official. In the absence of express legal fee schedules, a governmental officer's estimate is taken as an official source. If several sources have different estimates, the median reported value is used. In the absence of government officers' estimates, estimates of incorporation lawyers are used instead. If these differ, the median reported value is computed. In all cases, the cost estimate excludes bribes.

Time Measure

Time is recorded in calendar days. For the sake of uniformity, for all countries it is assumed that the minimum time required to fulfill a procedural requirement is one day. Therefore, the shortest procedure lasts one calendar day. The time variable captures the average duration that incorporation lawyers estimate is necessary to complete a procedure. If a procedure can be accelerated at additional cost, the fastest procedure, independent of cost, is chosen. It is assumed that the entrepreneur does not waste time and commits to the completion of each remaining procedure from the previous day, unless the law stipulates the contrary. When estimating the time needed for complying with entry regulations, the time that the entrepreneur spends in information gathering is ignored. The entrepreneur is aware of all entry regulations and their sequence from the very beginning. Information is collected on the sequence in which the procedures are to be completed, as well as any procedures that lend themselves to being carried out simultaneously.

Minimum Capital Requirement

The minimum capital requirement reflects the amount that the entrepreneur needs to deposit in a bank account in order to obtain a business registration number. This amount is typically specified in the Commercial Code or the Company Law.

This methodology is originally developed in Djankov, Simeon, Rafael La Porta, Florencio Lopez-de-Silanes, and Andrei Shleifer, "The Regulation of Entry," Quarterly Journal of Economics, *117, 1–37, Feb. 2002 and adopted with minor changes here.*

Hiring and Firing Workers

The data on hiring and firing workers are based on an assessment of employment laws and regulations as well as specific constitutional provisions governing this area. The employment laws of most countries are available online in the NATLEX database, published by the International Labour Organization. Constitutions are available online on the U.S. Law Library of Congress website. The main secondary sources include the *International Encyclopaedia for Labour Law and Industrial Relations*, and *Social Security Programs Throughout the World*. Data were confirmed with more than one source. In most cases both the actual laws and a secondary source were used to ensure accuracy. All conflicting answers were checked with two additional sources, including a local legal treatise on labor and social security laws. Legal advice from leading local law firms was solicited to confirm accuracy in all cases. The current mark of the data refers to January 2003.

Following the OECD Job Study and the *International Encyclopaedia for Labour Law and Industrial Relations*, the areas subject to statutory regulation in all countries were identified. Those include hiring of workers, conditions of employment, and firing of workers.

Assumptions about the Worker

To make the worker comparable across countries, several assumptions are employed.

- The worker is a nonexecutive full-time male employee who has worked in the same company for 20 years.
- His salary plus benefits equal the country's average wage during the entire period of his employment.
- The worker has a nonworking wife and two children. The family has always resided in the country's most populous city.
- The worker is a lawful citizen who belongs to the same race and religion as the majority of the country's population.
- He is not a member of the labor union (unless membership is mandatory).

Assumptions about the Business

To make the business comparable across countries, several assumptions are employed.

- The business is a limited liability corporation.
- It operates in the country's most populous city.
- The business is 100 percent domestically owned.
- It operates in the manufacturing sector.
- The business has 201 employees.
- It abides by every law and regulation, but does not grant workers more benefits than what is legally mandated.

Construction of Indices

Four indices are constructed: a flexibility-of-hiring index, the conditions-of-employment index, a flexibility-of-firing index and an overall employment-regulation index. Each index may take values between 0 and 100, with higher values indicating more rigid regulation.

The flexibility of hiring index covers the availability of part-time and fixed-term contracts. Conditions of employment cover working time requirements, including mandatory minimum daily rest, maximum number of hours in a normal workweek, premium for overtime work, restrictions on weekly holiday, mandatory payment for nonworking days, (which includes days of annual leave with pay and paid time off for holidays), and minimum wage legislation. The constitutional principles dealing with the minimum conditions of employment are also coded. Flexibility of firing covers workers' legal protections against dismissal, including grounds for dismissal, procedures for dismissal (individual and collective), notice period, and severance payment. The constitutional principles dealing with protection against dismissal are also coded.

The index of employment regulation is a simple average of the flexibility-of-hiring index, the conditions-of-employment index, and the flexibility-of-firing index.

This methodology is developed in Botero, Juan, Simeon Djankov, Rafael La Porta, Florencio Lopez-de-Silanes, and Andrei Shleifer, "The Regulation of Labor," Working Paper 9756, National Bureau of Economic Research, June 2003, and adopted with minor changes here.

Enforcing a Contract

The data on enforcing a contract are derived from questionnaires answered by attorneys at private law firms. The current mark of the data refers to January 2003. The questionnaire covers the step-by-step evolution of a debt recovery case before local courts in the country's most populous city. The respondent firms were provided with significant detail, including the amount of the claim, the location and main characteristics of the litigants, the presence of city regulations, the nature of the remedy requested by the plaintiff, the merit of the plaintiff's and the defendant's claims, and the social implications of the judicial outcomes. These standardized details enabled the respondent law firms to describe the procedures explicitly and in full detail.

Assumptions about the case

- The debt value equals 50 percent of the country's income per capita.
- The plaintiff has fully complied with the contract (the plaintiff is 100 percent right).
- The case presents a lawful transaction between businesses residing in the country's most populous city.
- The bank refuses payment for lack of funds in the borrower's account.
- The plaintiff files a lawsuit to collect the debt.
- The debtor attempts to delay service of process but it is finally accomplished.
- The debtor opposes the complaint (default judgment is not an option).
- The judge decides every motion for the plaintiff.
- The plaintiff attempts to introduce documentary evidence and to call one witness. The debtor attempts to call one witness. Neither party presents objections.
- The judgment is in favor of the plaintiff.
- No appeals or post-judgment motions are filed by either party to the case.
- The debt is successfully collected.

The study develops three main indicators of the efficiency of the judicial system on the enforcement of commercial contracts. The first indicator is the number of procedures mandated by law or court regulation that demand interaction between the parties or between them and the judge or court officer.

The second indicator of efficiency is an estimate—in calendar days—of the duration of the dispute resolution process. Time is measured as the number of days from the moment the plaintiff files the lawsuit in court, until the moment of actual payment. This measure includes both the days where actions take place and waiting periods between actions. The respondents make separate estimates of the average duration until the completion of service of process, the issuance of judgment (duration of trial), and the moment of payment or repossession (duration of enforcement).

The third indicator is cost, including court costs and attorney fees, as well as payments to other professionals like accountants and bailiffs.

The study also develops an index of the procedural complexity of contract enforcement. This index measures substantive and procedural statutory intervention in civil cases in the courts, and is formed by averaging the following subindices:

1. Use of professionals: This subindex measures whether the resolution of the case provided would rely mostly in the intervention of professional judges and attorneys, as opposed to the intervention of other types of adjudicators and lay people.
2. Nature of actions: This subindex measures the written or oral nature of the actions involved in the procedure, from the filing of the complaint to enforcement.
3. Legal justification: This subindex measures the level of legal justification required in the process of dispute resolution.
4. Statutory regulation of evidence: This subindex measures the level of statutory control or intervention of the administration, admissibility, evaluation, and recording of evidence.
5. Control of superior review: This subindex measures the level of control or intervention of the appellate court's review of the first instance judgment.

6. Other statutory interventions: This subindex measures the formalities required to engage someone into the procedure or to hold him/her accountable for the judgment.

The procedural-complexity index varies from 0 to 100, with higher values indicating more procedural complexity in enforcing a contract.

The methodology is developed in Djankov, Simeon, Rafael La Porta, Florencio Lopez-de-Silanes, and Andrei Shleifer, "Courts," Quarterly Journal of Economics, 118, 453–517, May 2003.

Getting Credit

Two sets of measures on getting credit are constructed: indicators on credit information sharing and an indicator of the legal protection of creditor rights.

The data on credit information sharing institutions were built starting with a survey of banking supervisors, designed to:

- confirm the presence/absence of public credit registries and private credit information bureaus,
- collect descriptive data on credit market outcomes (banking concentration rates, loan default rates), and
- collect information on related rules in credit markets (interest rate controls, collateral, laws on credit information sharing).

For countries that confirmed the presence of a public credit registry, a detailed survey on the registry's structure, laws, and associated rules followed. Similar surveys were sent to major private credit bureaus.

These surveys were designed as a joint cooperative effort with the "Credit Reporting Systems Project" in the World Bank Group, adapting previous surveys conducted by this project. Input was also received from Professor Marco Pagano of the University of Salerno. Variables assessed include:

- coverage of the market
- scope of information collected
- scope of information distributed
- accessibility of the data available
- quality of information available
- legal framework for information sharing and quality of data.

Public Credit Registry Coverage

A public credit registry is defined as a database managed by the public sector, usually by the Central Bank or Superintendent of Banks, that collects information on the standing of borrowers (persons and/or businesses) in the financial system and makes it available to financial institutions.

The coverage indicator reports the number of individuals and/or firms listed in the public credit registry as of January 2003 with current information on repayment history, unpaid debts, or credit outstanding. The number is scaled to country's population (per 1,000 capita). A coverage value of zero indicates that no public registry operates.

Extensiveness-of-Public-Credit-Registries Index

Scores can range from 0 to 100, where higher values indicate that the rules of the public credit registry are better designed to support credit transactions. The overall index of the extensiveness of public credit registries is a simple average of the collection, distribution, access, and quality indices, described below.

- Collection index

Assigns a positive score if the registry: lists both firms and individuals; shares information with other registries; collects information that is submitted voluntarily; has laws/regulations that require lenders to consult the registry when making loans; requires that participating institutions report data at least weekly; maintains historical records of more than seven years; maintains records of defaults even after they have been repaid; collects data from nonbank financial institutions; reports all loans, regardless of the amount; and if there is a minimum loan size for inclusion, the minimum loan size is lower than the sample median.

Higher values indicate broader rules on collection of information.

- Distribution index

Assigns a positive score if the registry: allows distribution of historical data (more than current month); distributes disaggregated loan information; distributes both positive and negative data on individuals; distributes both positive and negative data on firms; distributes an extensive number of types of information on individuals; distributes an extensive number of types of information on firms.

Higher values indicate broader rules on distribution of information.

- Access index

Assigns a positive score if: the registry allows access for other parties beyond banking supervisors, information submitters, and consumers; the registry does not require financial institutions to submit data in order to access the registry; borrower's authorization is not required for access; information can be accessed not only on certain types of borrower; information can be accessed within a day; access is electronic; time to distribute data is regulated.

Higher values indicate broader rules on access to information in the registry.

- Quality index

Assigns a positive score if: legal penalties for reporting inaccurate data are imposed; by law consumers may inspect data; there is a legal requirement to respond to borrower complaints; routine checks with other data, borrower complaints, statistical checks and software programs are used as quality checks; by law data needs to be submitted within two weeks of reporting period; more than 95 percent of financial institutions submit data on time; time to correct reported errors is less than two weeks; data is available for distribution within one week of submission; registry has been in existence for an extended period of time.

Higher values indicate more extensive rules on the quality of information in the registry.

Private Credit Bureau Coverage

A private credit bureau is defined as a private firm or a non-profit organization that maintains a database on the standing of borrowers (persons or businesses) in the financial system, and its primary role is to facilitate exchange of credit information amongst banks and financial institutions.

Credit investigative bureaus and credit reporting firms that do not directly facilitate exchange of information between financial institutions exist in many countries, but are not considered here.

The coverage indicator reports the number of individuals and/or firms listed in the private credit bureau as of January 2003 with current information on repayment history, unpaid debts, or credit outstanding. The number is scaled to country's population (per 1,000 capita). A coverage value of 0 indicates that no private credit bureau operates.

Creditor-Rights Index

Doing Business reports an indicator of creditor rights in insolvency, based on the methodology of La Porta and others (1998). The indicator measures four powers of secured lenders in liquidation and reorganization:

- Restrictions on entering reorganization: whether there are restrictions, such as creditor consent, when a debtor files for reorganization—as opposed to cases where debtors can seek unilateral protection from creditors' claims by filing for reorganization.
- No automatic stay: whether secured creditors are able to seize their collateral after the decision for reorganization is approved, in other words whether there is no "automatic stay" or "asset freeze" imposed by the court.
- Secured creditors are paid first: whether secured creditors are paid first out of the proceeds from liquidating a bankrupt firm, as opposed to other parties such as government (e.g., for taxes) or workers.
- Management does not stay in reorganization: Whether an administrator is responsible for management of the business during the resolution of reorganization, instead of having the management of the bankrupt debtor continue to run the business.

A value of one is assigned for each variable when a country's laws and regulations provide these powers for secured creditors. The aggregate creditor rights index sums the total score across all four variables. A minimum score of zero represents weak creditor rights and the maximum score of four represents strong creditor rights.

This methodology is developed in Djankov, Simeon, Caralee McLiesh, and Andrei Shleifer, "Remedies in Credit Markets," working paper, Department of Economics, Harvard University, July 2003; and La Porta, Rafael, Florencio Lopez-de-Silanes, Andrei Shleifer, and Robert Vishny, "Law and Finance," Journal of Political Economy, *106, 1113–55, 1998.*

Closing a Business

Members of the International Bar Association's Committee on Insolvency were asked to fill out a questionnaire relating to a hypothetical corporate bankruptcy. A first draft of the survey was prepared with scholars from Harvard University, and with advice from practicing attorneys in Argentina, Bulgaria, Germany, Italy, the Netherlands, Nigeria, the United Kingdom, and the United States. This survey was then piloted in the Czech Republic, Italy, Latvia, the Russian Federation, Spain, and Uzbekistan. Responses from these countries were used to revise the initial questionnaire. Next, participating law firms or bankruptcy judges from around the world were sent a final questionnaire to fill out. Answers were provided by a senior partner at each firm, in cooperation with one or two junior associates. In all cases, respondents were contacted for additional information following focus group presentations at the International Bar Association's Committee on Insolvency meetings in Dublin, Ireland, Durban, South Africa, and Rome, Italy. This helped the accurate interpretation of answers, to complete missing information, and to clarify possible inconsistencies. After this second round, a file was completed for each country and sent back to the respondents for final clearance.

Participants were asked to base their responses on the following scenario:

Assumptions about the Business

- The business is a limited liability corporation.
- It operates in the country's most populous city.
- The business is 100 percent domestically owned, of which 51 percent is owned by its founder, who is also the chairman of the supervisory board. Aside from the founder, there is no other shareholder who has above 1 percent of shares.
- Its only asset is downtown real estate.
- The business runs a hotel in the real estate it owns.
- There is a professional general manager.
- The business has average annual revenue of 1,000 times income per capita over the last three years.
- The business has 201 employees, and 50 suppliers, each of whom is owed money for the last delivery.
- Five years ago the business borrowed from a domestic bank, and bought real estate (the hotel building), using it as a security for the bank loan.
- The loan has 10 years to full repayment.
- The business has observed the payment schedule and all other conditions of the loan up to now.
- The current value of the mortgage principal is exactly equal to the market value of the hotel.
- The entire case involves domestic entities (i.e., there are no cross-border issues).

Assumptions about Procedures

In January 2003, the hypothetical business is experiencing liquidity problems. The company's loss in 2002 brought its net worth to a negative figure. The cash flow available in 2003 will cover all operating expenses: supplier payments, salaries, hotel upkeep costs, and outstanding taxes. However, there will be no cash left to pay the bank either interest or principal in full, due on January 2, 2003. Therefore, the business will default on its loan. Management believes that losses will be incurred in 2003 and 2004 as well.

In countries where floating charges are possible, it is assumed that the bank holds a floating charge against the hotel. If the law does not permit a floating charge, but contracts nevertheless commonly use some other provision to that effect, such as allowing the lender rights to the future stream of profits or other proceeds of the collateral, the assumption is that this provision is specified in the lending contract.

If the bank were to have the new projections, it might try to salvage as much of its loan as possible, for example by seizing and selling the security (the hotel building) or by filing for formal liquidation. The bank prefers to act fast.

The only argument management can make in favor of keeping the company in operation is that the value of the firm is higher as a going concern than if it is liquidated. In contrast, the market price of the bank's security is decreasing every day. Further, it seems unlikely that, without a strong boom in the industry, the business will be able to catch up with back payments.

Assumptions about Information Availability

- Management has the entire information.
- The bank will observe the payment default by the company on January 2, 2003, when its interest and principal payments are due. However, the bank does not know the new projections for the future. The latter information will only be available in the 2002 Annual Report published on March 31, 2003.
- The shareholders will have access to the Annual Report on March 31, 2003, will be present at the general shareholders meeting, and know any public information.
- Suppliers do not have access to the new projections, and are therefore not aware of the hotel's financial problems. Unless any developments are publicly announced, or unless suppliers are contacted as creditors as part of a legal proceeding, they will not foresee problems before October 1, 2003, when their payment and the new inventory delivery is due. Suppliers can consult the Annual Report on March 31, 2003.
- The tax agency and any other institutions that supervise firms do not have access to the new projections, but will have the financial statements for 2002 once the company publishes its Annual Report on March 31, 2003.
- Employees will have access to the Annual Report on March 31, 2003.

The claims as of January 1, 2003, and a set of financial statements for 2001–2002, as well as 2003–2005 projections, are attached as an appendix to the case sent to the respondents.

Assumptions about Legal Options

The business has too many creditors to renegotiate out of court. Its options are:

- A procedure aimed at rehabilitation or any procedure that will reorganize the business to permit further operation,
- A procedure aimed at liquidation,
- A procedure aimed at selling the hotel, either as a going concern or piecemeal, either enforced through court (or a government authority like a debt collection agency) or out of court (receivership).

Cost Measure

The answers of practicing insolvency lawyers are used as a source for the costs associated with resolving insolvency in the courts. If several respondents report different estimates, the median reported value is used. Cost is defined as the cost of the entire bankruptcy process, including court costs, insolvency practitioners' costs, the cost of independent assessors, lawyers, accountants, etc. In all cases, the cost estimate excludes bribes. The cost figures are averages of the estimates in a multiple-choice question, where the respondents choose among the following options: 0–2 percent, 3–5 percent, 6–10 percent, 11–25 percent, 26–50 percent, and more than 50 percent of the insolvency estate value.

Time Measure

Time is recorded in calendar years. The time measure captures the average duration that insolvency lawyers estimate is necessary to complete a procedure. If a procedure can be accelerated at additional cost, the fastest procedure, independent of cost, is chosen. The legal team of the party filing for insolvency is aware of all procedures and their sequence from the very beginning. The study collects information on the sequence in which the insolvency procedures are to be completed, as well as any procedures that can be carried out simultaneously. The time measure includes all delays due to legal derailment tactics that parties to the insolvency may use. In particular, it includes delays due to extension of response periods or to appeals, if these are allowed under the law. As such, the measure represents the actual time of the insolvency proceedings, not the time that the law may mandate.

Absolute Priority Preserved

The measure documents the order in which claims are paid in the insolvency process, including payment of post-petition claims. The measure is scaled so that higher values imply stricter observance of priority. A 100 on Absolute Priority Preserved means that secured creditors are paid before labor claims, tax claims and shareholders. A 67 means that secured creditors get paid second, and 33 means they get paid third. A zero on Absolute Priority Preserved means that secured creditors get paid after all labor claims and tax claims are satisfied, and after shareholders have received payments as well.

Efficient Outcome Achieved

The measure documents the success of the insolvency regime in reaching the economically efficient outcome. A one on Efficient Outcome Achieved means that the insolvency process results in either foreclosure or liquidation with a going-concern sale or in a successful rehabilitation maintaining the business but hiring new management. A zero indicates that the efficient outcome is not achieved.

Goals-of-Insolvency Index

The measure documents the success in reaching the three goals of insolvency, as stated in Hart (1999). It is calculated as the simple average of the cost of insolvency (rescaled from 0 to 100, where higher scores indicate less cost), time of insolvency (rescaled from 0 to 100, where higher scores indicate less time), the observance of absolute priority of claims, and the efficient outcome achieved. The total Goals-of-Insolvency Index ranges from 0 to 100: a score 100 on the index means perfect efficiency (Finland, Norway, and Singapore have 99), a 0 means that the insolvency system does not function at all.

Court-Powers Index

The measure documents the degree to which the court drives insolvency proceedings. It is an average of three indicators: whether the court appoints and replaces the insolvency administrator with no restrictions imposed by law, whether the reports of the administrator are accessible only to the court and not creditors, and whether the court decides on the adoption of the rehabilitation plan. The index is scaled from 0 to 100, where higher values indicate more court involvement in the insolvency process.

This methodology is developed in Djankov, Simeon, Oliver Hart, Tatiana Nenova, and Andrei Shleifer, "Efficiency in Bankruptcy," working paper, Department of Economics, Harvard University, July 2003.

ECONOMY CHARACTERISTICS

Country	Region	Income group	Legal origin
Albania	Europe & Central Asia	Lower-middle	French
Algeria	Middle East & North Africa	Lower-middle	French
Angola	Sub-Saharan Africa	Low	French
Argentina	Latin America & Caribbean	Upper-middle	French
Armenia	Europe & Central Asia	Low	Socialist
Australia	OECD: High Income	High	English
Austria	OECD: High Income	High	German
Azerbaijan	Europe & Central Asia	Low	Socialist
Bangladesh	South Asia	Low	English
Belarus	Europe & Central Asia	Lower-middle	Socialist
Belgium	OECD: High Income	High	French
Benin	Sub-Saharan Africa	Low	French
Bolivia	Latin America & Caribbean	Lower-middle	French
Bosnia and Herzegovina	Europe & Central Asia	Lower-middle	German
Botswana	Sub-Saharan Africa	Upper-middle	English
Brazil	Latin America & Caribbean	Lower-middle	French
Bulgaria	Europe & Central Asia	Lower-middle	German
Burkina Faso	Sub-Saharan Africa	Low	French
Burundi	Sub-Saharan Africa	Low	French
Cambodia	East Asia & Pacific	Low	French
Cameroon	Sub-Saharan Africa	Low	French
Canada	OECD: High Income	High	English
Central African Republic	Sub-Saharan Africa	Low	French
Chad	Sub-Saharan Africa	Low	French
Chile	Latin America & Caribbean	Upper-middle	French
China	East Asia & Pacific	Lower-middle	German
Colombia	Latin America & Caribbean	Lower-middle	French
Congo, Dem. Rep. of	Sub-Saharan Africa	Low	French
Congo, Rep. of	Sub-Saharan Africa	Low	French
Costa Rica	Latin America & Caribbean	Upper-middle	French
Côte d'Ivoire	Sub-Saharan Africa	Low	French
Croatia	Europe & Central Asia	Upper-middle	German
Czech Republic	Europe & Central Asia	Upper-middle	German
Denmark	OECD: High Income	High	Nordic
Dominican Republic	Latin America & Caribbean	Lower-middle	French
Ecuador	Latin America & Caribbean	Lower-middle	French

Country	Region	Income group	Legal origin
Egypt, Arab Rep. of	Middle East & North Africa	Lower-middle	French
El Salvador	Latin America & Caribbean	Lower-middle	French
Ethiopia	Sub-Saharan Africa	Low	English
Finland	OECD: High Income	High	Nordic
France	OECD: High Income	High	French
Georgia	Europe & Central Asia	Low	Socialist
Germany	OECD: High Income	High	German
Ghana	Sub-Saharan Africa	Low	English
Greece	OECD: High Income	High	French
Guatemala	Latin America & Caribbean	Lower-middle	French
Guinea	Sub-Saharan Africa	Low	French
Haiti	Latin America & Caribbean	Low	French
Honduras	Latin America & Caribbean	Lower-middle	French
Hong Kong, China	East Asia & Pacific	High	English
Hungary	Europe & Central Asia	Upper-middle	German
India	South Asia	Low	English
Indonesia	East Asia & Pacific	Low	French
Iran, Islamic Rep. of	Middle East & North Africa	Lower-middle	English
Ireland	OECD: High Income	High	English
Israel	Middle East & North Africa	High	English
Italy	OECD: High Income	High	French
Jamaica	Latin America & Caribbean	Lower-middle	English
Japan	OECD: High Income	High	German
Jordan	Middle East & North Africa	Lower-middle	French
Kazakhstan	Europe & Central Asia	Lower-middle	Socialist
Kenya	Sub-Saharan Africa	Low	English
Korea, Rep. of	OECD: High Income	High	German
Kuwait	Middle East & North Africa	High	French
Kyrgyz Republic	Europe & Central Asia	Low	Socialist
Lao PDR	East Asia & Pacific	Low	French
Latvia	Europe & Central Asia	Upper-middle	German
Lebanon	Middle East & North Africa	Upper-middle	French
Lesotho	Sub-Saharan Africa	Low	English
Lithuania	Europe & Central Asia	Upper-middle	French
Macedonia, FYR	Europe & Central Asia	Lower-middle	German
Madagascar	Sub-Saharan Africa	Low	French
Malawi	Sub-Saharan Africa	Low	English
Malaysia	East Asia & Pacific	Upper-middle	English
Mali	Sub-Saharan Africa	Low	French
Mauritania	Sub-Saharan Africa	Low	French
Mexico	Latin America & Caribbean	Upper-middle	French
Moldova	Europe & Central Asia	Low	Socialist
Mongolia	East Asia & Pacific	Low	Socialist
Morocco	Middle East & North Africa	Lower-middle	French
Mozambique	Sub-Saharan Africa	Low	French
Namibia	Sub-Saharan Africa	Lower-middle	English
Nepal	South Asia	Low	English
Netherlands	OECD: High Income	High	French
New Zealand	OECD: High Income	High	English
Nicaragua	Latin America & Caribbean	Low	French

Country	Region	Income group	Legal origin
Niger	Sub-Saharan Africa	Low	French
Nigeria	Sub-Saharan Africa	Low	English
Norway	OECD: High Income	High	Nordic
Oman	Middle East & North Africa	Upper-middle	French
Pakistan	South Asia	Low	English
Panama	Latin America & Caribbean	Upper-middle	French
Papua New Guinea	East Asia & Pacific	Low	English
Paraguay	Latin America & Caribbean	Lower-middle	French
Peru	Latin America & Caribbean	Lower-middle	French
Philippines	East Asia & Pacific	Lower-middle	French
Poland	Europe & Central Asia	Upper-middle	German
Portugal	OECD: High Income	High	French
Puerto Rico	Latin America & Caribbean	High	French
Romania	Europe & Central Asia	Lower-middle	French
Russian Federation	Europe & Central Asia	Lower-middle	Socialist
Rwanda	Sub-Saharan Africa	Low	French
Saudi Arabia	Middle East & North Africa	Upper-middle	English
Senegal	Sub-Saharan Africa	Low	French
Serbia and Montenegro	Europe & Central Asia	Lower-middle	German
Sierra Leone	Sub-Saharan Africa	Low	English
Singapore	East Asia & Pacific	High	English
Slovak Republic	Europe & Central Asia	Upper-middle	German
Slovenia	Europe & Central Asia	High	German
South Africa	Sub-Saharan Africa	Lower-middle	English
Spain	OECD: High Income	High	French
Sri Lanka	South Asia	Lower-middle	English
Sweden	OECD: High Income	High	Nordic
Switzerland	OECD: High Income	High	German
Syrian Arab Republic	Middle East & North Africa	Lower-middle	French
Taiwan, China	East Asia & Pacific	High	German
Tanzania	Sub-Saharan Africa	Low	English
Thailand	East Asia & Pacific	Lower-middle	English
Togo	Sub-Saharan Africa	Low	French
Tunisia	Middle East & North Africa	Lower-middle	French
Turkey	Europe & Central Asia	Lower-middle	French
Uganda	Sub-Saharan Africa	Low	English
Ukraine	Europe & Central Asia	Lower-middle	Socialist
United Arab Emirates	Middle East & North Africa	High	English
United Kingdom	OECD: High Income	High	English
United States	OECD: High Income	High	English
Uruguay	Latin America & Caribbean	Upper-middle	French
Uzbekistan	Europe & Central Asia	Low	Socialist
Venezuela, RB	Latin America & Caribbean	Upper-middle	French
Vietnam	East Asia & Pacific	Low	French
Yemen, Rep. of	Middle East & North Africa	Low	English
Zambia	Sub-Saharan Africa	Low	English
Zimbabwe	Sub-Saharan Africa	Low	English

Starting-a-Business Indicators—measure the procedures, time, cost, and minimum capital requirements to register a business formally

Country	Number of procedures	Time (days)	Cost (US$)	Cost (% of income per capita)	Min. capital (% of income per capita)
Albania	11	47	897	65.0	51.7
Algeria	18	29	548	31.9	73.0
Angola	14	146	5531	838.0	174.0
Argentina	15	68	324	8.0	0.0
Armenia	10	25	68	8.7	11.0
Australia	2	2	402	2.0	0.0
Austria	9	29	1534	6.6	140.8
Azerbaijan	14	106	119	16.8	0.0
Bangladesh	7	30	272	75.5	0.0
Belarus	19	118	369	27.1	110.7
Belgium	7	56	2633	11.3	75.1
Benin	9	63	719	189.2	377.6
Bolivia	18	67	1499	166.6	0.0
Bosnia and Herzegovina	12	59	657	51.8	379.1
Botswana	10	97	1076	36.1	0.0
Brazil	15	152	331	11.6	0.0
Bulgaria	10	30	148	8.3	134.4
Burkina Faso	15	136	716	325.2	652.2
Burundi	..	..	..	..	..
Cambodia	11	94	1551	553.8	1825.8
Cameroon	12	37	1068	190.7	243.6
Canada	2	3	127	0.6	0.0
Central African Republic	..	..	..	..	..
Chad	19	73	870	395.3	652.2
Chile	10	28	493	11.6	0.0
China	12	46	135	14.3	3855.9
Colombia	19	60	498	27.2	0.0
Congo, Dem. Rep. of	13	215	785	871.9	320.7
Congo, Rep. of	8	67	1897	271.0	205.0
Costa Rica	11	80	879	21.4	0.0
Côte d'Ivoire	10	77	873	143.1	235.2
Croatia	13	50	843	18.2	50.7
Czech Republic	10	88	648	11.7	110.0
Denmark	4	4	0	0.0	52.3
Dominican Republic	12	78	1115	48.1	23.2
Ecuador	14	90	914	63.0	27.6
Egypt, Arab Rep. of	13	43	900	61.2	788.6
El Salvador	12	115	2690	129.3	549.5
Ethiopia	8	44	422	421.6	1756.1
Finland	4	33	739	3.1	32.0
France	10	53	663	3.0	32.1
Georgia	9	30	171	26.3	140.1
Germany	9	45	1341	5.9	103.8
Ghana	10	84	302	111.7	1.2
Greece	16	45	8115	69.6	145.3

Note: .. means no data available.

Country	Number of procedures	Time (days)	Cost (US$)	Cost (% of income per capita)	Min. capital (% of income per capita)
Guatemala	13	39	1167	66.7	36.5
Guinea	13	71	941	229.9	396.6
Haiti	12	203	875	198.9	209.8
Honduras	14	80	670	72.8	165.4
Hong Kong, China	5	11	581	2.3	0.0
Hungary	5	65	3396	64.3	220.3
India	10	88	239	49.8	430.4
Indonesia	11	168	103	14.5	302.5
Iran, Islamic Rep. of	9	48	113	6.6	7.4
Ireland	3	12	2473	10.4	0.0
Israel	5	34	784	4.7	0.0
Italy	9	23	4565	24.1	49.6
Jamaica	7	31	458	16.2	0.0
Japan	11	31	3518	10.5	71.3
Jordan	14	98	876	49.8	2404.2
Kazakhstan	10	25	153	10.1	35.2
Kenya	11	61	194	54.0	0.0
Korea, Rep. of	12	33	1776	17.9	402.5
Kuwait	13	34	329	1.8	910.6
Kyrgyz Republic	9	26	39	13.4	74.8
Lao PDR	9	198	60	19.5	150.7
Latvia	7	11	513	14.7	93.0
Lebanon	6	46	5185	129.9	83.1
Lesotho	9	92	317	67.4	20.2
Lithuania	9	26	231	6.3	74.4
Macedonia, FYR	13	48	223	13.1	138.4
Madagascar	15	67	151	62.8	30.5
Malawi	11	45	201	125.4	0.0
Malaysia	8	31	961	27.1	0.0
Mali	13	61	557	232.2	597.8
Mauritania	11	73	452	110.2	896.7
Mexico	7	51	1110	18.8	87.6
Moldova	11	42	121	26.2	86.3
Mongolia	8	31	53	12.0	2046.9
Morocco	11	36	227	19.1	762.5
Mozambique	15	153	209	99.6	30.2
Namibia	10	85	332	18.7	0.0
Nepal	8	25	439	191.0	0.0
Netherlands	7	11	3276	13.7	70.7
New Zealand	3	3	28	0.2	0.0
Nicaragua	12	71	1335	337.8	0.0
Niger	11	27	759	446.6	844.0
Nigeria	10	44	268	92.3	28.6
Norway	4	24	1460	3.9	33.1
Oman	9	34	385	5.3	720.9
Pakistan	10	22	192	46.8	0.0
Panama	7	19	1057	26.3	0.0
Papua New Guinea	7	69	140	26.4	0.0

Note: .. means no data available.

Country	Number of procedures	Time (days)	Cost (US$)	Cost (% of income per capita)	Min. capital (% of income per capita)
Paraguay	18	73	1883	160.9	0.0
Peru	9	100	510	24.9	0.0
Philippines	11	59	249	24.4	9.5
Poland	12	31	925	20.3	21.4
Portugal	11	95	1360	12.5	43.4
Puerto Rico	6	6	300	2.8	0.0
Romania	6	27	217	11.7	3.3
Russian Federation	12	29	200	9.3	29.8
Rwanda	9	43	534	232.3	457.3
Saudi Arabia	14	95	10814	130.5	1610.5
Senegal	9	58	581	123.6	296.1
Serbia and Montenegro	10	44	186	13.3	5.5
Sierra Leone	9	26	1817	1297.6	0.0
Singapore	7	8	249	1.2	0.0
Slovak Republic	10	98	401	10.2	111.8
Slovenia	10	61	1518	15.5	89.1
South Africa	9	38	227	8.7	0.0
Spain	11	115	2366	16.4	19.6
Sri Lanka	8	58	154	18.3	0.0
Sweden	3	16	190	0.8	41.4
Switzerland	6	20	3228	8.5	33.8
Syrian Arab Republic	10	42	189	16.7	5627.2
Taiwan, China	8	48	807	6.1	217.4
Tanzania	13	35	557	199.0	0.0
Thailand	9	42	144	7.3	0.0
Togo	14	63	760	281.4	531.4
Tunisia	10	46	327	16.4	351.7
Turkey	13	38	927	37.1	13.2
Uganda	17	36	338	135.1	0.0
Ukraine	14	40	210	27.3	450.8
United Arab Emirates	10	29	4944	24.5	404.0
United Kingdom	6	18	264	1.0	0.0
United States	5	4	210	0.6	0.0
Uruguay	10	27	2043	46.7	699.0
Uzbekistan	9	33	72	16.0	64.3
Venezuela, RB	14	119	788	19.3	0.0
Vietnam	11	63	129	29.9	0.0
Yemen, Rep. of	13	96	1294	264.1	1716.9
Zambia	6	40	80	24.1	137.8
Zimbabwe	10	122	1322	285.3	0.0

Note: .. means no data available.

Hiring-and-Firing Indicators—measure the degree of rigidity in employment laws

Country	Flexibility-of-hiring index	Conditions-of-employment index	Flexibility-of-firing index	Employment-laws index
Albania	33	76	15	41
Algeria	58	60	19	46
Angola	71	89	74	78
Argentina	71	81	46	66
Armenia	51	84	37	57
Australia	33	61	13	36
Austria	33	41	14	30
Azerbaijan	71	90	27	63
Bangladesh	33	85	32	50
Belarus	71	89	71	77
Belgium	33	90	22	48
Benin	48	86	20	52
Bolivia	58	95	45	66
Bosnia and Herzegovina	53	63	31	49
Botswana	33	55	17	35
Brazil	78	89	68	78
Bulgaria	43	90	26	53
Burkina Faso	53	79	27	53
Burundi	58	76	51	62
Cambodia	33	81	49	54
Cameroon	48	43	39	44
Canada	33	52	16	34
Central African Republic	53	84	50	62
Chad	78	93	27	66
Chile	56	65	29	50
China	17	67	57	47
Colombia	33	85	60	59
Congo, Dem. Rep. of	73	63	43	60
Congo, Rep. of	53	78	49	60
Costa Rica	58	83	46	63
Côte d'Ivoire	53	61	45	53
Croatia	76	89	31	65
Czech Republic	17	63	27	36
Denmark	33	25	17	25
Dominican Republic	33	79	35	49
Ecuador	37	63	65	55
Egypt, Arab Rep. of	33	83	61	59
El Salvador	81	75	52	69
Ethiopia	58	67	29	51
Finland	71	43	52	55
France	63	61	26	50
Georgia	51	66	49	55
Germany	63	46	45	51
Ghana	33	56	17	35
Greece	78	81	43	67
Guatemala	58	85	51	65
Guinea	78	44	57	60

Country	Flexibility-of-hiring index	Conditions-of-employment index	Flexibility-of-firing index	Employment-laws index
Haiti	58	85	35	60
Honduras	33	87	47	56
Hong Kong, China	58	22	1	27
Hungary	46	92	23	54
India	33	75	45	51
Indonesia	76	53	43	57
Iran, Islamic Rep. of	33	77	47	52
Ireland	48	68	30	49
Israel	33	64	16	38
Italy	76	62	40	59
Jamaica	33	52	18	34
Japan	39	64	9	37
Jordan	33	82	64	60
Kazakhstan	33	89	42	55
Kenya	33	53	16	34
Korea, Rep. of	33	88	32	51
Kuwait	33	40	50	41
Kyrgyz Republic	71	90	33	64
Lao PDR	33	87	44	54
Latvia	58	87	42	62
Lebanon	53	50	35	46
Lesotho	58	51	25	45
Lithuania	71	90	31	64
Macedonia, FYR	65	53	32	50
Madagascar	48	86	49	61
Malawi	33	68	54	52
Malaysia	33	26	15	25
Mali	53	86	23	54
Mauritania	62	47	66	59
Mexico	81	81	70	77
Moldova	71	75	54	67
Mongolia	33	90	25	50
Morocco	56	63	33	51
Mozambique	73	85	64	74
Namibia	17	57	54	43
Nepal	33	54	47	45
Netherlands	51	79	33	54
New Zealand	33	43	20	32
Nicaragua	33	90	58	61
Niger	53	89	34	59
Nigeria	17	76	36	43
Norway	58	39	25	41
Oman	58	78	25	54
Pakistan	65	75	33	58
Panama	81	87	68	79
Papua New Guinea	17	57	4	26
Paraguay	58	90	71	73
Peru	71	81	69	73
Philippines	58	73	50	60

Country	Flexibility-of-hiring index	Conditions-of-employment index	Flexibility-of-firing index	Employment-laws index
Poland	33	92	39	55
Portugal	76	88	73	79
Puerto Rico	33	67	24	41
Romania	48	85	29	54
Russian Federation	33	77	71	61
Rwanda	53	94	32	60
Saudi Arabia	33	58	16	36
Senegal	48	83	30	54
Serbia and Montenegro	51	88	29	56
Sierra Leone	56	84	62	67
Singapore	33	26	1	20
Slovak Republic	34	89	60	61
Slovenia	53	84	41	59
South Africa	42	36	30	36
Spain	76	88	45	70
Sri Lanka	33	52	40	42
Sweden	56	39	31	42
Switzerland	33	53	23	36
Syrian Arab Republic	33	79	22	45
Taiwan, China	81	59	32	57
Tanzania	57	77	49	61
Thailand	78	73	30	61
Togo	53	80	36	57
Tunisia	73	53	44	57
Turkey	58	91	17	55
Uganda	33	44	50	42
Ukraine	58	93	69	73
United Arab Emirates	33	66	37	45
United Kingdom	33	42	9	28
United States	33	29	5	22
Uruguay	58	56	3	39
Uzbekistan	46	69	50	55
Venezuela, RB	78	88	60	75
Vietnam	43	77	48	56
Yemen, Rep. of	33	66	28	43
Zambia	33	64	40	46
Zimbabwe	33	22	26	27

Enforcing-a-Contract Indicators—cover the procedures, time, cost, and degree of complexity in the procedures to resolve a payment dispute

Country	Number of procedures	Time (days)	Cost (US$)	Cost (% income per capita)	Procedural-complexity index
Albania	37	220	794	72.6	76
Algeria	20	387	..	..	72
Angola	46	865	83	15.7	65
Argentina	32	300	621	8.5	80
Armenia	22	65	80	15.3	46
Australia	11	320	1623	8.0	29
Austria	20	434	240	1.0	54
Azerbaijan	25	115	20	3.3	53
Bangladesh	15	270	1019	48.2	51
Belarus	19	135	564	43.6	56
Belgium	22	365	2205	9.1	53
Benin	44	248	114	31.0	53
Bolivia	44	464	52	5.3	78
Bosnia and Herzegovina	31	630	260	21.3	63
Botswana	22	56	..	..	52
Brazil	16	380	83	2.4	48
Bulgaria	26	410	95	6.4	69
Burkina Faso	24	376	375	172.8	71
Burundi	62	367	29	27.6	58
Cambodia	20	210	752	268.5	78
Cameroon	46	548	367	62.9	63
Canada	17	425	6065	28.0	29
Central African Republic	..	..	..	..	..
Chad	50	604	121	58.4	72
Chile	21	200	663	14.7	73
China	20	180	268	32.0	52
Colombia	37	527	119	5.9	56
Congo, Dem. Rep. of	55	414	800	92.3	54
Congo, Rep. of	44	500	330	51.0	67
Costa Rica	21	370	857	22.6	86
Côte d'Ivoire	18	150	572	83.3	57
Croatia	20	330	305	6.6	50
Czech Republic	16	270	967	18.5	65
Denmark	14	83	1210	3.8	40
Dominican Republic	19	495	9250	440.5	69
Ecuador	33	333	132	10.5	72
Egypt, Arab Rep. of	19	202	450	30.7	50
El Salvador	42	240	149	7.3	81
Ethiopia	24	895	35	34.6	52
Finland	19	240	3886	15.8	48
France	21	210	896	3.8	79
Georgia	17	180	408	63.1	48
Germany	22	154	1483	6.0	61
Ghana	21	90	80	23.8	33
Greece	15	315	980	8.2	64
Guatemala	19	1460	338	20.0	90

Note: .. means no data available.

Country	Number of procedures	Time (days)	Cost (US$)	Cost (% income per capita)	Procedural-complexity index
Guinea	41	150	171	40.0	77
Haiti	41	76	87	18.4	69
Honduras	32	225	57	6.7	72
Hong Kong, China	17	180	1737	6.9	50
Hungary	17	365	256	5.4	57
India	22	365	444	95.0	50
Indonesia	29	225	1754	269.0	67
Iran, Islamic Rep. of	23	150	96	5.8	67
Ireland	16	183	1604	7.2	42
Israel	19	315	5635	34.1	51
Italy	16	645	780	3.9	64
Jamaica	14	202	1138	42.1	38
Japan	16	60	2223	6.4	39
Jordan	32	147	5	0.3	49
Kazakhstan	41	120	103	7.9	65
Kenya	25	255	173	49.5	44
Korea, Rep. of	23	75	402	4.5	50
Kuwait	17	195	788	4.4	76
Kyrgyz Republic	44	365	730	254.7	48
Lao PDR	..	..	..	..	..
Latvia	19	189	218	7.5	56
Lebanon	27	721	2160	54.3	67
Lesotho	..	..	..	..	..
Lithuania	17	74	580	13.0	58
Macedonia, FYR	27	509	750	43.0	67
Madagascar	29	166	304	120.2	63
Malawi	16	108	920	520.6	48
Malaysia	22	270	671	19.4	41
Mali	27	150	16	7.0	71
Mauritania	..	..	..	..	..
Mexico	47	325	504	10.0	62
Moldova	36	210	56	14.2	48
Mongolia	26	224	7	1.8	71
Morocco	17	192	108	9.1	69
Mozambique	18	540	20	9.1	71
Namibia	..	..	..	..	..
Nepal	24	350	106	44.2	63
Netherlands	21	39	120	0.5	46
New Zealand	19	50	1526	11.6	31
Nicaragua	17	125	70	17.7	79
Niger	29	365	103	57.1	63
Nigeria	23	730	18	6.6	52
Norway	12	87	3606	10.4	48
Oman	54	250	346	4.8	51
Pakistan	30	365	200	45.8	53
Panama	44	197	642	20.0	82
Papua New Guinea	22	270	244	41.1	45
Paraguay	46	188	461	34.0	67

Note: .. means no data available.

Country	Number of procedures	Time (days)	Cost (US$)	Cost (% income per capita)	Procedural-complexity index
Peru	35	441	613	29.7	82
Philippines	28	164	1086	103.7	75
Poland	18	1000	466	11.2	65
Portugal	22	420	534	4.9	54
Puerto Rico	55	365	2250	20.9	52
Romania	28	225	217	13.1	60
Russian Federation	16	160	350	20.2	48
Rwanda	..	..	..	..	..
Saudi Arabia	19	195	..	..	50
Senegal	30	335	238	48.6	75
Serbia and Montenegro	40	1028	200	20.0	61
Sierra Leone	48	114	11	8.3	29
Singapore	23	50	3521	14.4	49
Slovak Republic	26	420	494	13.3	40
Slovenia	22	1003	360	3.6	65
South Africa	26	207	510	16.7	56
Spain	20	147	1600	10.7	83
Sri Lanka	17	440	64	7.6	59
Sweden	21	190	4590	7.6	44
Switzerland	14	224	1490	3.9	44
Syrian Arab Republic	36	596	300	31.3	69
Taiwan, China	15	210	68	0.5	37
Tanzania	14	127	10	3.8	62
Thailand	19	210	589	29.6	53
Togo	43	503	59	21.4	63
Tunisia	14	7	86	4.1	60
Turkey	18	105	154	5.4	38
Uganda	16	99	30	10.0	40
Ukraine	20	224	80	11.0	51
United Arab Emirates	27	559	2148	10.6	56
United Kingdom	12	101	120	0.5	36
United States	17	365	120	0.4	46
Uruguay	38	360	822	13.7	55
Uzbekistan	34	258	13	2.1	57
Venezuela, RB	41	360	2000	46.9	81
Vietnam	28	120	33	8.5	46
Yemen, Rep. of	27	240	2	0.5	60
Zambia	16	188	50	15.8	32
Zimbabwe	13	197	183	39.5	50

Note: .. means no data available.

Getting-Credit Indicators—measure institutions for credit information and legal protection of creditors

Country	Public registry coverage (borrowers/1,000 cap.)	Extensiveness-of public-credit-registries index	Private bureau coverage (borrowers/1,000 cap.)	Creditor-rights index
Albania	0	0	0	3
Algeria	0	0	0	1
Angola	10	60	0	3
Argentina	149	61	475	1
Armenia	0	0	0	2
Australia	0	0	722	3
Austria	9	66	308	3
Azerbaijan	0	0	0	3
Bangladesh	1	51	0	2
Belarus	..	42	0	2
Belgium	68	63	42	2
Benin	1	22	0	1
Bolivia	55	58	134	2
Bosnia and Herzegovina	0	0	67	3
Botswana	0	0	382	3
Brazil	44	50	439	1
Bulgaria	5	47	0	3
Burkina Faso	1	22	0	1
Burundi	1	49	0	1
Cambodia	0	0	0	2
Cameroon	<1	49	0	1
Canada	0	0	806	1
Central African Republic	<1	49	0	2
Chad	<1	49	0	1
Chile	209	45	227	2
China	3	56	0	2
Colombia	0	0	187	0
Congo, Dem. Rep. of	0	0	0	2
Congo, Rep. of	<1	49	0	0
Costa Rica	7	44	55	1
Côte d'Ivoire	1	22	0	1
Croatia	0	0	0	3
Czech Republic	10	60	136	3
Denmark	0	0	58	3
Dominican Republic	..	42	423	2
Ecuador	82	55	0	1
Egypt, Arab Rep. of	..	48	0	1
El Salvador	130	50	128	3
Ethiopia	0	0	0	3
Finland	0	0	96	1
France	12	53	0	0
Georgia	0	0	0	2
Germany	5	44	693	3
Ghana	0	0	<1	1
Greece	0	0	86	1

Notes: .. means no data available.

A zero for public registry coverage or private bureau coverage means no public credit registry or private credit bureau operates in the country.

Country	Public registry coverage (borrowers/1,000 cap.)	Extensiveness-of public-credit-registries index	Private bureau coverage (borrowers/1,000 cap.)	Creditor-rights index
Guatemala	0	0	35	1
Guinea	..	..	0	1
Haiti	1	59	0	2
Honduras	45	42	0	2
Hong Kong, China	0	0	200	4
Hungary	0	0	15	2
India	0	0	0	3
Indonesia	3	61	0	2
Iran, Islamic Rep. of	..	45	0	2
Ireland	0	0	730	1
Israel	0	0	47	3
Italy	55	61	416	1
Jamaica	0	0	0	2
Japan	0	0	777	2
Jordan	19	47	0	1
Kazakhstan	0	0	0	2
Kenya	0	0	309	4
Korea, Rep. of	0	0	530	3
Kuwait	0	0	147	2
Kyrgyz Republic	0	0	0	3
Lao PDR	..	..	0	0
Latvia	0	0	0	3
Lebanon	0	0	0	4
Lesotho	0	0	0	2
Lithuania	7	63	0	2
Macedonia, FYR	2	42	0	3
Madagascar	2	46	0	2
Malawi	0	0	0	2
Malaysia	105	59	461	2
Mali	1	22	0	1
Mauritania	..	..	0	3
Mexico	0	0	382	0
Moldova	0	0	0	2
Mongolia	15	68	0	1
Morocco	..	33	0	1
Mozambique	1	52	0	2
Namibia	0	0	..	..
Nepal	0	0	0	2
Netherlands	0	0	530	3
New Zealand	0	0	818	4
Nicaragua	50	45	0	4
Niger	1	22	0	1
Nigeria	<1	55	0	4
Norway	0	0	945	2
Oman	0	0	0	0
Pakistan	1	42	<1	1
Panama	0	0	302	4

Notes: .. means no data available.

A zero for public registry coverage or private bureau coverage means no public credit registry or private credit bureau operates in the country.

Country	Public registry coverage (borrowers/1,000 cap.)	Extensiveness-of public-credit-registries index	Private bureau coverage (borrowers/1,000 cap.)	Creditor-rights index
Papua New Guinea	0	0	0	2
Paraguay	..	..	..	2
Peru	92	54	185	0
Philippines	0	0	22	1
Poland	0	0	543	2
Portugal	496	61	24	1
Puerto Rico	0	0	..	1
Romania	1	59	0	0
Russian Federation	0	0	0	2
Rwanda	<1	57	0	1
Saudi Arabia	<1	42	0	2
Senegal	2	22	0	1
Serbia and Montenegro	<1	33	0	2
Sierra Leone	0	0	0	2
Singapore	0	0	0	3
Slovak Republic	2	48	0	2
Slovenia	14	60	0	3
South Africa	0	0	469	3
Spain	305	64	48	2
Sri Lanka	0	0	9	2
Sweden	0	0	489	1
Switzerland	0	0	178	1
Syrian Arab Republic	0	0	0	3
Taiwan, China	27	70	..	1
Tanzania	0	0	0	2
Thailand	0	0	98	3
Togo	1	22	0	2
Tunisia	4	48	0	0
Turkey	7	44	204	2
Uganda	0	0	0	2
Ukraine	0	0	0	2
United Arab Emirates	12	44	0	2
United Kingdom	0	0	652	4
United States	0	0	810	1
Uruguay	49	57	479	3
Uzbekistan	0	0	0	2
Venezuela, RB	97	46	0	2
Vietnam	2	67	0	0
Yemen, Rep. of	7	38	0	0
Zambia	0	0	0	1
Zimbabwe	0	0	0	4

Notes: .. means no data available.

A zero for public registry coverage or private bureau coverage means no public credit registry or private credit bureau operates in the country.

Closing-a-Business Indicators—measure the procedures, time, and cost to go through insolvency proceedings as well as court powers in insolvency proceedings

Country	Time (years)	Cost (% of estate)	Absolute priority preserved	Efficient outcome achieved	Goals-of-insolvency index	Court-powers index
Albania	no practice	no practice	67	1	42	67
Algeria	3.5	4	33	0	45	33
Angola	no practice	no practice	33	0	8	67
Argentina	2.8	18	67	0	43	67
Armenia	1.9	4	100	0	65	33
Australia	1.0	18	100	1	80	0
Austria	1.3	18	67	1	71	33
Azerbaijan	2.7	8	67	0	49	100
Bangladesh	no practice	no practice	100	0	25	67
Belarus	2.2	4	0	0	40	67
Belgium	0.9	4	100	1	93	67
Benin	3.2	18	33	0	33	100
Bolivia	2.0	18	100	0	53	100
Bosnia and Herzegovina	1.9	8	67	0	51	67
Botswana	2.2	18	100	1	77	33
Brazil	10.0	8	33	0	24	67
Bulgaria	3.8	18	100	0	48	67
Burkina Faso	4.0	8	0	0	29	100
Burundi	no practice	no practice	33	0	8	67
Cambodia	no practice	no practice	100	0	25	67
Cameroon	2.0	18	67	0	44	100
Canada	0.8	4	100	1	93	33
Central African Republic	..	..	..	..	..	..
Chad	10.0	38	33	0	11	100
Chile	5.8	18	0	0	19	67
China	2.6	18	100	0	51	67
Colombia	3.0	1	33	1	77	33
Congo, Dem. Rep. of	no practice	no practice	33	0	8	33
Congo, Rep. of	3.0	18	67	0	42	100
Costa Rica	2.5	18	67	0	43	100
Côte d'Ivoire	2.2	18	67	0	44	100
Croatia	3.1	18	100	0	50	67
Czech Republic	9.2	38	67	0	22	0
Denmark	4.2	8	100	1	79	33
Dominican Republic	3.5	4	0	0	37	67
Ecuador	3.5	18	0	0	24	67
Egypt, Arab Rep. of	4.3	18	67	0	39	67
El Salvador	no practice	no practice	67	1	42	67
Ethiopia	2.2	8	67	1	75	33
Finland	0.9	1	100	1	99	0
France	2.4	18	67	0	43	100
Georgia	3.2	1	100	0	69	33
Germany	1.2	8	100	0	61	33
Ghana	no practice	no practice	67	0	17	33
Greece	2.2	8	33	0	42	33

Notes: .. means no data available.

Country	Time (years)	Cost (% of estate)	Absolute priority preserved	Efficient outcome achieved	Goals-of-insolvency index	Court-powers index
Guatemala	4.0	18	67	0	40	67
Guinea	no practice	no practice	33	0	8	100
Haiti	no practice	no practice	67	1	42	67
Honduras	no practice	no practice	67	0	17	67
Hong Kong, China	1.0	18	33	1	63	67
Hungary	2.0	38	67	0	38	33
India	11.3	8	33	0	21	33
Indonesia	6.0	18	67	0	35	100
Iran, Islamic Rep. of	1.8	8	100	1	84	67
Ireland	0.4	8	100	1	88	33
Israel	4.0	38	100	1	67	67
Italy	1.3	18	67	0	46	33
Jamaica	1.1	18	33	1	63	67
Japan	0.6	4	100	1	93	33
Jordan	4.3	8	33	0	37	33
Kazakhstan	3.3	18	67	1	66	67
Kenya	4.6	18	100	0	47	33
Korea, Rep. of	1.5	4	100	1	91	67
Kuwait	4.2	1	67	1	83	67
Kyrgyz Republic	4.0	4	100	0	61	33
Lao PDR	no practice	no practice	0	0	14	67
Latvia	1.2	4	100	1	92	67
Lebanon	4.0	18	33	0	31	67
Lesotho	..	..	..	..	..	..
Lithuania	1.2	18	100	0	54	67
Macedonia, FYR	3.6	38	67	0	34	67
Madagascar	no practice	no practice	100	0	25	67
Malawi	2.8	8	33	0	40	67
Malaysia	2.2	18	100	0	52	33
Mali	3.5	18	33	0	32	100
Mauritania	8.0	8	33	0	28	67
Mexico	2.0	18	33	1	61	67
Moldova	2.8	8	67	0	49	67
Mongolia	4.0	8	100	0	54	67
Morocco	1.9	18	33	0	36	100
Mozambique	no practice	no practice	100	0	25	67
Namibia	..	..	..	..	..	..
Nepal	5.0	8	33	0	35	33
Netherlands	2.6	1	100	1	95	33
New Zealand	2.0	4	100	1	90	0
Nicaragua	2.3	8	100	0	58	67
Niger	5.0	18	67	0	37	100
Nigeria	1.6	18	67	0	45	67
Norway	0.9	1	100	1	99	67
Oman	7.0	4	0	0	29	67
Pakistan	2.8	4	100	0	63	33
Panama	6.5	38	100	0	36	33

Notes: .. means no data available.

Country	Time (years)	Cost (% of estate)	Absolute priority preserved	Efficient outcome achieved	Goals-of-insolvency index	Court-powers index
Papua New Guinea	..	..	..	..	..	..
Paraguay	3.9	8	67	0	46	67
Peru	2.1	8	33	1	67	33
Philippines	5.7	38	100	0	38	100
Poland	1.5	18	67	1	70	67
Portugal	2.6	8	33	1	66	33
Puerto Rico	3.8	8	67	1	71	33
Romania	3.2	8	33	0	39	33
Russian Federation	1.5	4	67	0	58	67
Rwanda	no practice	no practice	33	0	8	33
Saudi Arabia	3.0	18	100	0	50	33
Senegal	3.0	8	67	1	73	100
Serbia and Montenegro	7.3	38	33	1	42	67
Sierra Leone	2.5	38	0	0	20	33
Singapore	0.7	1	100	1	99	33
Slovak Republic	4.8	18	100	1	71	67
Slovenia	3.7	18	67	0	41	67
South Africa	2.0	18	100	0	53	67
Spain	1.5	8	33	1	68	33
Sri Lanka	2.3	18	33	0	35	67
Sweden	2.0	8	100	1	84	33
Switzerland	4.6	4	100	0	59	67
Syrian Arab Republic	4.1	8	33	0	37	67
Taiwan, China	0.8	4	0	1	68	100
Tanzania	3.0	8	33	1	65	67
Thailand	2.6	38	67	1	62	33
Togo	no practice	no practice	33	0	8	100
Tunisia	2.5	8	67	0	50	67
Turkey	1.8	8	67	0	51	67
Uganda	2.0	38	33	1	55	67
Ukraine	3.0	18	67	0	42	33
United Arab Emirates	5.0	38	33	0	23	33
United Kingdom	1.0	8	100	1	86	0
United States	3.0	4	100	1	88	33
Uruguay	4.0	8	100	0	54	67
Uzbekistan	3.3	4	33	0	46	67
Venezuela, RB	4.0	38	100	1	67	67
Vietnam	no practice	no practice	33	1	33	67
Yemen, Rep. of	2.4	4	33	0	47	33
Zambia	3.7	8	100	0	55	33
Zimbabwe	2.3	18	100	0	52	67

Notes: .. means no data available.

ALBANIA
Europe and Central Asia

Economy Characteristics		Starting a Business	
GNI per capita (US$)	1,380	Number of procedures	11
Population	3,164,400	Time (days)	47
Informal economy (% of income)	33.40	Cost (% of income per capita)	65.0
Legal origin	French	Minimum capital (% of income per capita)	51.7
Hiring and Firing Workers		**Enforcing a Contract**	
Flexibility-of-hiring index	33	Number of procedures	37
Conditions-of-employment index	76	Time (days)	220
Flexibility-of-firing index	15	Cost (% of income per capita)	72.6
Employment-law index	41	Procedural-complexity index	76
Getting Credit		**Closing a Business**	
Public credit registry operates?	No	Time to go through insolvency (years)	No practice
Public registry coverage (borrowers/1,000 capita)	0	Cost to go through insolvency (% estate)	No practice
Public-registry index	0	Absolute priority preserved	67
Private credit-information bureau operates?	No	Efficient outcome achieved	1
Private bureau coverage (borrowers/1,000 capita)	0	Goals-of-insolvency index	42
Creditor-rights index	3	Court-powers index	67

ALGERIA
Middle East and North Africa

Economy Characteristics		Starting a Business	
GNI per capita (US$)	1,720	Number of procedures	18
Population	30,835,000	Time (days)	29
Informal economy (% of income)	34.1	Cost (% of income per capita)	31.9
Legal origin	French	Minimum capital (% of income per capita)	73.0
Hiring and Firing Workers		**Enforcing a Contract**	
Flexibility-of-hiring index	58	Number of procedures	20
Conditions-of-employment index	60	Time (days)	387
Flexibility-of-firing index	19	Cost (% of income per capita)	..
Employment-law index	46	Procedural-complexity index	72
Getting Credit		**Closing a Business**	
Public credit registry operates?	No	Time to go through insolvency (years)	3.5
Public registry coverage (borrowers/1,000 capita)	0	Cost to go through insolvency (% estate)	4
Public-registry index	0	Absolute priority preserved	33
Private credit-information bureau operates?	No	Efficient outcome achieved	0
Private bureau coverage (borrowers/1,000 capita)	0	Goals-of-insolvency index	45
Creditor-rights index	1	Court-powers index	33

Note: .. means no data available.

ANGOLA
Sub-Saharan Africa

Economy Characteristics		Starting a Business	
GNI per capita (US$)	660	Number of procedures	14
Population	13,512,450	Time (days)	146
Informal economy (% of income)	..	Cost (% of income per capita)	838.0
Legal origin	French	Minimum capital (% of income per capita)	174.0
Hiring and Firing Workers		**Enforcing a Contract**	
Flexibility-of-hiring index	71	Number of procedures	46
Conditions-of-employment index	89	Time (days)	865
Flexibility-of-firing index	74	Cost (% of income per capita)	15.7
Employment-law index	78	Procedural-complexity index	65
Getting Credit		**Closing a Business**	
Public credit registry operates?	Yes	Time to go through insolvency (years)	No practice
Public registry coverage (borrowers/1,000 capita)	10	Cost to go through insolvency (% estate)	No practice
Public-registry index	60	Absolute priority preserved	33
Private credit-information bureau operates?	No	Efficient outcome achieved	0
Private bureau coverage (borrowers/1,000 capita)	0	Goals-of-insolvency index	8
Creditor-rights index	3	Court-powers index	67

ARGENTINA
Latin America and Caribbean

Economy Characteristics		Starting a Business	
GNI per capita (US$)	4,060	Number of procedures	15
Population	37,488,000	Time (days)	68
Informal economy (% of income)	25.4	Cost (% of income per capita)	8.0
Legal origin	French	Minimum capital (% of income per capita)	0.0
Hiring and Firing Workers		**Enforcing a Contract**	
Flexibility-of-hiring index	71	Number of procedures	32
Conditions-of-employment index	81	Time (days)	300
Flexibility-of-firing index	46	Cost (% of income per capita)	8.5
Employment-law index	66	Procedural-complexity index	80
Getting Credit		**Closing a Business**	
Public credit registry operates?	Yes	Time to go through insolvency (years)	2.8
Public registry coverage (borrowers/1,000 capita)	149	Cost to go through insolvency (% estate)	18
Public-registry index	61	Absolute priority preserved	67
Private credit-information bureau operates?	Yes	Efficient outcome achieved	0
Private bureau coverage (borrowers/1,000 capita)	475	Goals-of-insolvency index	43
Creditor-rights index	1	Court-powers index	67

ARMENIA
Europe and Central Asia

Economy Characteristics		Starting a Business	
GNI per capita (US$)	790	Number of procedures	10
Population	3,088,000	Time (days)	25
Informal economy (% of income)	46.3	Cost (% of income per capita)	8.7
Legal origin	Socialist	Minimum capital (% of income per capita)	11.0
Hiring and Firing Workers		**Enforcing a Contract**	
Flexibility-of-hiring index	51	Number of procedures	22
Conditions-of-employment index	84	Time (days)	65
Flexibility-of-firing index	37	Cost (% of income per capita)	15.3
Employment-law index	57	Procedural-complexity index	46
Getting Credit		**Closing a Business**	
Public credit registry operates?	No	Time to go through insolvency (years)	1.9
Public registry coverage (borrowers/1,000 capita)	0	Cost to go through insolvency (% estate)	4
Public-registry index	0	Absolute priority preserved	100
Private credit-information bureau operates?	No	Efficient outcome achieved	0
Private bureau coverage (borrowers/1,000 capita)	0	Goals-of-insolvency index	65
Creditor-rights index	2	Court-powers index	33

Note: .. means no data available.

AUSTRALIA
OECD: High Income

Economy Characteristics		Starting a Business	
GNI per capita (US$)	19,740	Number of procedures	2
Population	19,386,820	Time (days)	2
Informal economy (% of income)	15.3	Cost (% of income per capita)	2.0
Legal origin	English	Minimum capital (% of income per capita)	0.0
Hiring and Firing Workers		**Enforcing a Contract**	
Flexibility-of-hiring index	33	Number of procedures	11
Conditions-of-employment index	61	Time (days)	320
Flexibility-of-firing index	13	Cost (% of income per capita)	8.0
Employment-law index	36	Procedural-complexity index	29
Getting Credit		**Closing a Business**	
Public credit registry operates?	No	Time to go through insolvency (years)	1.0
Public registry coverage (borrowers/1,000 capita)	0	Cost to go through insolvency (% estate)	18
Public-registry index	0	Absolute priority preserved	100
Private credit-information bureau operates?	Yes	Efficient outcome achieved	1
Private bureau coverage (borrowers/1,000 capita)	722	Goals-of-insolvency index	80
Creditor-rights index	3	Court-powers index	0

AUSTRIA
OECD: High Income

Economy Characteristics		Starting a Business	
GNI per capita (US$)	23,390	Number of procedures	9
Population	8,132,000	Time (days)	29
Informal economy (% of income)	10.2	Cost (% of income per capita)	6.6
Legal origin	German	Minimum capital (% of income per capita)	140.8
Hiring and Firing Workers		**Enforcing a Contract**	
Flexibility-of-hiring index	33	Number of procedures	20
Conditions-of-employment index	41	Time (days)	434
Flexibility-of-firing index	14	Cost (% of income per capita)	1.0
Employment-law index	30	Procedural-complexity index	54
Getting Credit		**Closing a Business**	
Public credit registry operates?	Yes	Time to go through insolvency (years)	1.3
Public registry coverage (borrowers/1,000 capita)	9	Cost to go through insolvency (% estate)	18
Public-registry index	66	Absolute priority preserved	67
Private credit-information bureau operates?	Yes	Efficient outcome achieved	1
Private bureau coverage (borrowers/1,000 capita)	308	Goals-of-insolvency index	71
Creditor-rights index	3	Court-powers index	33

AZERBAIJAN
Europe and Central Asia

Economy Characteristics		Starting a Business	
GNI per capita (US$)	710	Number of procedures	14
Population	8,116,110	Time (days)	106
Informal economy (% of income)	60.6	Cost (% of income per capita)	16.8
Legal origin	Socialist	Minimum capital (% of income per capita)	0.0
Hiring and Firing Workers		**Enforcing a Contract**	
Flexibility-of-hiring index	71	Number of procedures	25
Conditions-of-employment index	90	Time (days)	115
Flexibility-of-firing index	27	Cost (% of income per capita)	3.3
Employment-law index	63	Procedural-complexity index	53
Getting Credit		**Closing a Business**	
Public credit registry operates?	No	Time to go through insolvency (years)	2.6
Public registry coverage (borrowers/1,000 capita)	0	Cost to go through insolvency (% estate)	8
Public-registry index	0	Absolute priority preserved	67
Private credit-information bureau operates?	No	Efficient outcome achieved	0
Private bureau coverage (borrowers/1,000 capita)	0	Goals-of-insolvency index	49
Creditor-rights index	3	Court-powers index	100

Note: .. means no data available.

BANGLADESH
South Asia

Economy Characteristics	
GNI per capita (US$)	360
Population	133,345,160
Informal economy (% of income)	35.6
Legal origin	English

Hiring and Firing Workers	
Flexibility-of-hiring index	33
Conditions-of-employment index	85
Flexibility-of-firing index	32
Employment-law index	0

Getting Credit	
Public credit registry operates?	Yes
Public registry coverage (borrowers/1,000 capita)	1
Public-registry index	51
Private credit-information bureau operates?	No
Private bureau coverage (borrowers/1,000 capita)	0
Creditor-rights index	2

Starting a Business	
Number of procedures	7
Time (days)	30
Cost (% of income per capita)	75.5
Minimum capital (% of income per capita)	0.0

Enforcing a Contract	
Number of procedures	15
Time (days)	270
Cost (% of income per capita)	48.2
Procedural-complexity index	51

Closing a Business	
Time to go through insolvency (years)	No practice
Cost to go through insolvency (% estate)	No practice
Absolute priority preserved	100
Efficient outcome achieved	0
Goals-of-insolvency index	25
Court-powers index	67

BELARUS
Europe and Central Asia

Economy Characteristics	
GNI per capita (US$)	1,360
Population	9,970,260
Informal economy (% of income)	48.1
Legal origin	Socialist

Hiring and Firing Workers	
Flexibility-of-hiring index	71
Conditions-of-employment index	89
Flexibility-of-firing index	71
Employment-law index	77

Getting Credit	
Public credit registry operates?	Yes
Public registry coverage (borrowers/1,000 capita)	..
Public-registry index	42
Private credit-information bureau operates?	No
Private bureau coverage (borrowers/1,000 capita)	0
Creditor-rights index	2

Starting a Business	
Number of procedures	19
Time (days)	118
Cost (% of income per capita)	27.1
Minimum capital (% of income per capita)	110.7

Enforcing a Contract	
Number of procedures	19
Time (days)	135
Cost (% of income per capita)	43.6
Procedural-complexity index	56

Closing a Business	
Time to go through insolvency (years)	2.1
Cost to go through insolvency (% estate)	4
Absolute priority preserved	0
Efficient outcome achieved	0
Goals-of-insolvency index	40
Court-powers index	67

BELGIUM
OECD: High Income

Economy Characteristics	
GNI per capita (US$)	23,250
Population	10,286,000
Informal economy (% of income)	23.2
Legal origin	French

Hiring and Firing Workers	
Flexibility-of-hiring index	33
Conditions-of-employment index	90
Flexibility-of-firing index	22
Employment-law index	48

Getting Credit	
Public credit registry operates?	Yes
Public registry coverage (borrowers/1,000 capita)	68
Public-registry index	63
Private credit-information bureau operates?	Yes
Private bureau coverage (borrowers/1,000 capita)	42
Creditor-rights index	2

Starting a Business	
Number of procedures	7
Time (days)	56
Cost (% of income per capita)	11.3
Minimum capital (% of income per capita)	75.1

Enforcing a Contract	
Number of procedures	22
Time (days)	365
Cost (% of income per capita)	9.1
Procedural-complexity index	53

Closing a Business	
Time to go through insolvency (years)	0.8
Cost to go through insolvency (% estate)	4
Absolute priority preserved	100
Efficient outcome achieved	1
Goals-of-insolvency index	93
Court-powers index	67

Note: .. means no data available.

BENIN
Sub-Saharan Africa

Economy Characteristics		Starting a Business	
GNI per capita (US$)	380	Number of procedures	9
Population	6,436,660	Time (days)	63
Informal economy (% of income)	45.2	Cost (% of income per capita)	189.2
Legal origin	French	Minimum capital (% of income per capita)	377.6
Hiring and Firing Workers		**Enforcing a Contract**	
Flexibility-of-hiring index	48	Number of procedures	44
Conditions-of-employment index	86	Time (days)	248
Flexibility-of-firing index	20	Cost (% of income per capita)	31.0
Employment-law index	52	Procedural-complexity index	53
Getting Credit		**Closing a Business**	
Public credit registry operates?	Yes	Time to go through insolvency (years)	3.2
Public registry coverage (borrowers/1,000 capita)	1	Cost to go through insolvency (% estate)	18
Public-registry index	22	Absolute priority preserved	33
Private credit-information bureau operates?	No	Efficient outcome achieved	0
Private bureau coverage (borrowers/1,000 capita)	0	Goals-of-insolvency index	33
Creditor-rights index	1	Court-powers index	100

BOLIVIA
Latin America and Caribbean

Economy Characteristics		Starting a Business	
GNI per capita (US$)	900	Number of procedures	18
Population	8,515,220	Time (days)	67
Informal economy (% of income)	67.1	Cost (% of income per capita)	166.6
Legal origin	French	Minimum capital (% of income per capita)	0.0
Hiring and Firing Workers		**Enforcing a Contract**	
Flexibility-of-hiring index	58	Number of procedures	44
Conditions-of-employment index	95	Time (days)	464
Flexibility-of-firing index	45	Cost (% of income per capita)	5.3
Employment-law index	66	Procedural-complexity index	78
Getting Credit		**Closing a Business**	
Public credit registry operates?	Yes	Time to go through insolvency (years)	2.0
Public registry coverage (borrowers/1,000 capita)	55	Cost to go through insolvency (% estate)	18
Public-registry index	58	Absolute priority preserved	100
Private credit-information bureau operates?	Yes	Efficient outcome achieved	0
Private bureau coverage (borrowers/1,000 capita)	134	Goals-of-insolvency index	53
Creditor-rights index	2	Court-powers index	100

BOSNIA AND HERZEGOVINA
Europe and Central Asia

Economy Characteristics		Starting a Business	
GNI per capita (US$)	1,270	Number of procedures	12
Population	4,060,000	Time (days)	59
Informal economy (% of income)	34.1	Cost (% of income per capita)	51.8
Legal origin	German	Minimum capital (% of income per capita)	379.1
Hiring and Firing Workers		**Enforcing a Contract**	
Flexibility-of-hiring index	53	Number of procedures	31
Conditions-of-employment index	63	Time (days)	630
Flexibility-of-firing index	31	Cost (% of income per capita)	21.3
Employment-law index	49	Procedural-complexity index	63
Getting Credit		**Closing a Business**	
Public credit registry operates?	No	Time to go through insolvency (years)	1.86
Public registry coverage (borrowers/1,000 capita)	0	Cost to go through insolvency (% estate)	8
Public-registry index	0	Absolute priority preserved	67
Private credit-information bureau operates?	Yes	Efficient outcome achieved	0
Private bureau coverage (borrowers/1,000 capita)	67	Goals-of-insolvency index	51
Creditor-rights index	3	Court-powers index	67

Note: .. means no data available.

BOTSWANA
Sub-Saharan Africa

Economy Characteristics		Starting a Business	
GNI per capita (US$)	2,980	Number of procedures	10
Population	1,695,000	Time (days)	97
Informal economy (% of income)	33.4	Cost (% of income per capita)	36.1
Legal origin	English	Minimum capital (% of income per capita)	0.0
Hiring and Firing Workers		**Enforcing a Contract**	
Flexibility-of-hiring index	33	Number of procedures	22
Conditions-of-employment index	55	Time (days)	56
Flexibility-of-firing index	17	Cost (% of income per capita)	..
Employment-law index	35	Procedural-complexity index	52
Getting Credit		**Closing a Business**	
Public credit registry operates?	No	Time to go through insolvency (years)	2.2
Public registry coverage (borrowers/1,000 capita)	0	Cost to go through insolvency (% estate)	18
Public-registry index	0	Absolute priority preserved	100
Private credit-information bureau operates?	Yes	Efficient outcome achieved	1
Private bureau coverage (borrowers/1,000 capita)	382	Goals-of-insolvency index	77
Creditor-rights index	3	Court-powers index	33

BRAZIL
Latin America and Caribbean

Economy Characteristics		Starting a Business	
GNI per capita (US$)	2,850	Number of procedures	15
Population	172,386,000	Time (days)	152
Informal economy (% of income)	39.8	Cost (% of income per capita)	11.6
Legal origin	French	Minimum capital (% of income per capita)	0.0
Hiring and Firing Workers		**Enforcing a Contract**	
Flexibility-of-hiring index	78	Number of procedures	16
Conditions-of-employment index	89	Time (days)	380
Flexibility-of-firing index	68	Cost (% of income per capita)	2.4
Employment-law index	78	Procedural-complexity index	48
Getting Credit		**Closing a Business**	
Public credit registry operates?	Yes	Time to go through insolvency (years)	10.0
Public registry coverage (borrowers/1,000 capita)	44	Cost to go through insolvency (% estate)	8
Public-registry index	50	Absolute priority preserved	33
Private credit-information bureau operates?	Yes	Efficient outcome achieved	0
Private bureau coverage (borrowers/1,000 capita)	439	Goals-of-insolvency index	24
Creditor-rights index	1	Court-powers index	67

BULGARIA
Europe and Central Asia

Economy Characteristics		Starting a Business	
GNI per capita (US$)	1,790	Number of procedures	10
Population	7,913,000	Time (days)	30
Informal economy (% of income)	36.9	Cost (% of income per capita)	8.3
Legal origin	German	Minimum capital (% of income per capita)	134.4
Hiring and Firing Workers		**Enforcing a Contract**	
Flexibility-of-hiring index	43	Number of procedures	26
Conditions-of-employment index	90	Time (days)	410
Flexibility-of-firing index	26	Cost (% of income per capita)	6.4
Employment-law index	53	Procedural-complexity index	69
Getting Credit		**Closing a Business**	
Public credit registry operates?	Yes	Time to go through insolvency (years)	3.8
Public registry coverage (borrowers/1,000 capita)	5	Cost to go through insolvency (% estate)	18
Public-registry index	47	Absolute priority preserved	100
Private credit-information bureau operates?	No	Efficient outcome achieved	0
Private bureau coverage (borrowers/1,000 capita)	0	Goals-of-insolvency index	48
Creditor-rights index	3	Court-powers index	67

Note: .. means no data available.

BURKINA FASO
Sub-Saharan Africa

Economy Characteristics		Starting a Business	
GNI per capita (US$)	220	Number of procedures	15
Population	11,552,570	Time (days)	136
Informal economy (% of income)	38.4	Cost (% of income per capita)	325.2
Legal origin	French	Minimum capital (% of income per capita)	652.2
Hiring and Firing Workers		**Enforcing a Contract**	
Flexibility-of-hiring index	53	Number of procedures	24
Conditions-of-employment index	79	Time (days)	376
Flexibility-of-firing index	27	Cost (% of income per capita)	172.8
Employment-law index	53	Procedural-complexity index	71
Getting Credit		**Closing a Business**	
Public credit registry operates?	Yes	Time to go through insolvency (years)	4.0
Public registry coverage (borrowers/1,000 capita)	1	Cost to go through insolvency (% estate)	8
Public-registry index	22	Absolute priority preserved	0
Private credit-information bureau operates?	No	Efficient outcome achieved	0
Private bureau coverage (borrowers/1,000 capita)	0	Goals-of-insolvency index	29
Creditor-rights index	1	Court-powers index	100

BURUNDI
Sub-Saharan Africa

Economy Characteristics		Starting a Business	
GNI per capita (US$)	100	Number of procedures	..
Population	6,938,010	Time (days)	..
Informal economy (% of income)	..	Cost (% of income per capita)	..
Legal origin	French	Minimum capital (% of income per capita)	..
Hiring and Firing Workers		**Enforcing a Contract**	
Flexibility-of-hiring index	58	Number of procedures	62
Conditions-of-employment index	76	Time (days)	367
Flexibility-of-firing index	51	Cost (% of income per capita)	27.6
Employment-law index	62	Procedural-complexity index	58
Getting Credit		**Closing a Business**	
Public credit registry operates?	Yes	Time to go through insolvency (years)	No practice
Public registry coverage (borrowers/1,000 capita)	1	Cost to go through insolvency (% estate)	No practice
Public-registry index	49	Absolute priority preserved	33
Private credit-information bureau operates?	No	Efficient outcome achieved	0
Private bureau coverage (borrowers/1,000 capita)	0	Goals-of-insolvency index	8
Creditor-rights index	1	Court-powers index	67

CAMBODIA
East Asia and Pacific

Economy Characteristics		Starting a Business	
GNI per capita (US$)	280	Number of procedures	11
Population	12,265,220	Time (days)	94
Informal economy (% of income)	..	Cost (% of income per capita)	553.8
Legal origin	French	Minimum capital (% of income per capita)	1825.8
Hiring and Firing Workers		**Enforcing a Contract**	
Flexibility-of-hiring index	33	Number of procedures	20
Conditions-of-employment index	81	Time (days)	210
Flexibility-of-firing index	49	Cost (% of income per capita)	268.5
Employment-law index	54	Procedural-complexity index	78
Getting Credit		**Closing a Business**	
Public credit registry operates?	No	Time to go through insolvency (years)	No practice
Public registry coverage (borrowers/1,000 capita)	0	Cost to go through insolvency (% estate)	No practice
Public-registry index	0	Absolute priority preserved	100
Private credit-information bureau operates?	No	Efficient outcome achieved	0
Private bureau coverage (borrowers/1,000 capita)	0	Goals-of-insolvency index	25
Creditor-rights index	2	Court-powers index	67

Note: .. means no data available.

CAMEROON
Sub-Saharan Africa

Economy Characteristics	
GNI per capita (US$)	560
Population	15,197,470
Informal economy (% of income)	32.8
Legal origin	French

Starting a Business	
Number of procedures	12
Time (days)	37
Cost (% of income per capita)	190.7
Minimum capital (% of income per capita)	243.6

Hiring and Firing Workers	
Flexibility-of-hiring index	48
Conditions-of-employment index	43
Flexibility-of-firing index	39
Employment-law index	44

Enforcing a Contract	
Number of procedures	46
Time (days)	548
Cost (% of income per capita)	62.9
Procedural-complexity index	63

Getting Credit	
Public credit registry operates?	Yes
Public registry coverage (borrowers/1,000 capita)	<1
Public-registry index	49
Private credit-information bureau operates?	No
Private bureau coverage (borrowers/1,000 capita)	0
Creditor-rights index	1

Closing a Business	
Time to go through insolvency (years)	2.0
Cost to go through insolvency (% estate)	18
Absolute priority preserved	67
Efficient outcome achieved	0
Goals-of-insolvency index	44
Court-powers index	100

CANADA
OECD: High Income

Economy Characteristics	
GNI per capita (US$)	22,300
Population	31,081,900
Informal economy (% of income)	16.4
Legal origin	English

Starting a Business	
Number of procedures	2
Time (days)	3
Cost (% of income per capita)	0.6
Minimum capital (% of income per capita)	0.0

Hiring and Firing Workers	
Flexibility-of-hiring index	33
Conditions-of-employment index	52
Flexibility-of-firing index	16
Employment-law index	34

Enforcing a Contract	
Number of procedures	17
Time (days)	425
Cost (% of income per capita)	28.0
Procedural-complexity index	29

Getting Credit	
Public credit registry operates?	No
Public registry coverage (borrowers/1,000 capita)	0
Public-registry index	0
Private credit-information bureau operates?	Yes
Private bureau coverage (borrowers/1,000 capita)	806
Creditor-rights index	1

Closing a Business	
Time to go through insolvency (years)	0.8
Cost to go through insolvency (% estate)	4
Absolute priority preserved	100
Efficient outcome achieved	1
Goals-of-insolvency index	93
Court-powers index	33

CENTRAL AFRICAN REPUBLIC
Sub-Saharan Africa

Economy Characteristics	
GNI per capita (US$)	260
Population	3,770,820
Informal economy (% of income)	..
Legal origin	French

Starting a Business	
Number of procedures	..
Time (days)	..
Cost (% of income per capita)	..
Minimum capital (% of income per capita)	..

Hiring and Firing Workers	
Flexibility-of-hiring index	53
Conditions-of-employment index	84
Flexibility-of-firing index	50
Employment-law index	62

Enforcing a Contract	
Number of procedures	..
Time (days)	..
Cost (% of income per capita)	..
Procedural-complexity index	..

Getting Credit	
Public credit registry operates?	Yes
Public registry coverage (borrowers/1,000 capita)	<1
Public-registry index	49
Private credit-information bureau operates?	No
Private bureau coverage (borrowers/1,000 capita)	0
Creditor-rights index	2

Closing a Business	
Time to go through insolvency (years)	..
Cost to go through insolvency (% estate)	..
Absolute priority preserved	..
Efficient outcome achieved	..
Goals-of-insolvency index	..
Court-powers index	..

Note: .. means no data available.

CHAD
Sub-Saharan Africa

Economy Characteristics	
GNI per capita (US$)	220
Population	7,916,010
Informal economy (% of income)	..
Legal origin	French

Starting a Business	
Number of procedures	19
Time (days)	73
Cost (% of income per capita)	395.3
Minimum capital (% of income per capita)	652.2

Hiring and Firing Workers	
Flexibility-of-hiring index	78
Conditions-of-employment index	93
Flexibility-of-firing index	27
Employment-law index	66

Enforcing a Contract	
Number of procedures	50
Time (days)	604
Cost (% of income per capita)	58.4
Procedural-complexity index	72

Getting Credit	
Public credit registry operates?	Yes
Public registry coverage (borrowers/1,000 capita)	<1
Public-registry index	49
Private credit-information bureau operates?	No
Private bureau coverage (borrowers/1,000 capita)	0
Creditor-rights index	1

Closing a Business	
Time to go through insolvency (years)	10.0
Cost to go through insolvency (% estate)	38
Absolute priority preserved	33
Efficient outcome achieved	0
Goals-of-insolvency index	11
Court-powers index	100

CHILE
Latin America and Caribbean

Economy Characteristics	
GNI per capita (US$)	4,260
Population	15,402,000
Informal economy (% of income)	19.8
Legal origin	French

Starting a Business	
Number of procedures	10
Time (days)	28
Cost (% of income per capita)	11.6
Minimum capital (% of income per capita)	0.0

Hiring and Firing Workers	
Flexibility-of-hiring index	56
Conditions-of-employment index	65
Flexibility-of-firing index	29
Employment-law index	50

Enforcing a Contract	
Number of procedures	21
Time (days)	200
Cost (% of income per capita)	14.7
Procedural-complexity index	73

Getting Credit	
Public credit registry operates?	Yes
Public registry coverage (borrowers/1,000 capita)	209
Public-registry index	45
Private credit-information bureau operates?	Yes
Private bureau coverage (borrowers/1,000 capita)	227
Creditor-rights index	2

Closing a Business	
Time to go through insolvency (years)	5.84
Cost to go through insolvency (% estate)	18
Absolute priority preserved	0
Efficient outcome achieved	0
Goals-of-insolvency index	19
Court-powers index	67

CHINA
East Asia and Pacific

Economy Characteristics	
GNI per capita (US$)	940
Population	1,271,849,984
Informal economy (% of income)	13.1
Legal origin	German

Starting a Business	
Number of procedures	12
Time (days)	46
Cost (% of income per capita)	14.3
Minimum capital (% of income per capita)	3855.9

Hiring and Firing Workers	
Flexibility-of-hiring index	17
Conditions-of-employment index	67
Flexibility-of-firing index	57
Employment-law index	47

Enforcing a Contract	
Number of procedures	20
Time (days)	180
Cost (% of income per capita)	32.0
Procedural-complexity index	52

Getting Credit	
Public credit registry operates?	Yes
Public registry coverage (borrowers/1,000 capita)	3
Public-registry index	159
Private credit-information bureau operates?	Yes
Private bureau coverage (borrowers/1,000 capita)	<1
Creditor-rights index	2

Closing a Business	
Time to go through insolvency (years)	2.6
Cost to go through insolvency (% estate)	18
Absolute priority preserved	100
Efficient outcome achieved	0
Goals-of-insolvency index	51
Court-powers index	67

Note: .. means no data available.

COLOMBIA
Latin America and Caribbean

Economy Characteristics		Starting a Business	
GNI per capita (US$)	1,830	Number of procedures	19
Population	43,035,168	Time (days)	60
Informal economy (% of income)	39.1	Cost (% of income per capita)	27.2
Legal origin	French	Minimum capital (% of income per capita)	0.0
Hiring and Firing Workers		**Enforcing a Contract**	
Flexibility-of-hiring index	33	Number of procedures	37
Conditions-of-employment index	85	Time (days)	527
Flexibility-of-firing index	60	Cost (% of income per capita)	5.9
Employment-law index	59	Procedural-complexity index	56
Getting Credit		**Closing a Business**	
Public credit registry operates?	No	Time to go through insolvency (years)	3.0
Public registry coverage (borrowers/1,000 capita)	0	Cost to go through insolvency (% estate)	1
Public-registry index	0	Absolute priority preserved	33
Private credit-information bureau operates?	Yes	Efficient outcome achieved	1
Private bureau coverage (borrowers/1,000 capita)	187	Goals-of-insolvency index	77
Creditor-rights index	0	Court-powers index	33

CONGO, DEM. REP. of
Sub-Saharan Africa

Economy Characteristics		Starting a Business	
GNI per capita (US$)	90	Number of procedures	13
Population	52,354,100	Time (days)	215
Informal economy (% of income)	..	Cost (% of income per capita)	871.9
Legal origin	French	Minimum capital (% of income per capita)	320.7
Hiring and Firing Workers		**Enforcing a Contract**	
Flexibility-of-hiring index	73	Number of procedures	55
Conditions-of-employment index	63	Time (days)	414
Flexibility-of-firing index	43	Cost (% of income per capita)	92.3
Employment-law index	60	Procedural-complexity index	54
Getting Credit		**Closing a Business**	
Public credit registry operates?	No	Time to go through insolvency (years)	No practice
Public registry coverage (borrowers/1,000 capita)	0	Cost to go through insolvency (% estate)	No practice
Public-registry index	0	Absolute priority preserved	33
Private credit-information bureau operates?	No	Efficient outcome achieved	0
Private bureau coverage (borrowers/1,000 capita)	0	Goals-of-insolvency index	8
Creditor-rights index	2	Court-powers index	33

CONGO, REP. of
Sub-Saharan Africa

Economy Characteristics		Starting a Business	
GNI per capita (US$)	700	Number of procedures	8
Population	3,103,350	Time (days)	67
Informal economy (% of income)	..	Cost (% of income per capita)	271.0
Legal origin	French	Minimum capital (% of income per capita)	205.0
Hiring and Firing Workers		**Enforcing a Contract**	
Flexibility-of-hiring index	53	Number of procedures	44
Conditions-of-employment index	78	Time (days)	500
Flexibility-of-firing index	49	Cost (% of income per capita)	51.0
Employment-law index	60	Procedural-complexity index	67
Getting Credit		**Closing a Business**	
Public credit registry operates?	Yes	Time to go through insolvency (years)	3.0
Public registry coverage (borrowers/1,000 capita)	<1	Cost to go through insolvency (% estate)	18
Public-registry index	49	Absolute priority preserved	67
Private credit-information bureau operates?	No	Efficient outcome achieved	0
Private bureau coverage (borrowers/1,000 capita)	0	Goals-of-insolvency index	42
Creditor-rights index	0	Court-powers index	100

Note: .. means no data available.

COSTA RICA
Latin America and Caribbean

Economy Characteristics		Starting a Business	
GNI per capita (US$)	4,100	Number of procedures	11
Population	3,873,000	Time (days)	80
Informal economy (% of income)	26.2	Cost (% of income per capita)	21.4
Legal origin	French	Minimum capital (% of income per capita)	0.0
Hiring and Firing Workers		**Enforcing a Contract**	
Flexibility-of-hiring index	58	Number of procedures	21
Conditions-of-employment index	83	Time (days)	370
Flexibility-of-firing index	46	Cost (% of income per capita)	22.6
Employment-law index	63	Procedural-complexity index	86
Getting Credit		**Closing a Business**	
Public credit registry operates?	Yes	Time to go through insolvency (years)	2.5
Public registry coverage (borrowers/1,000 capita)	7	Cost to go through insolvency (% estate)	18
Public-registry index	44	Absolute priority preserved	67
Private credit-information bureau operates?	Yes	Efficient outcome achieved	0
Private bureau coverage (borrowers/1,000 capita)	55	Goals-of-insolvency index	43
Creditor-rights index	1	Court-powers index	100

CÔTE D'IVOIRE
Sub-Saharan Africa

Economy Characteristics		Starting a Business	
GNI per capita (US$)	610	Number of procedures	10
Population	16,410,080	Time (days)	77
Informal economy (% of income)	39.9	Cost (% of income per capita)	143.1
Legal origin	French	Minimum capital (% of income per capita)	235.2
Hiring and Firing Workers		**Enforcing a Contract**	
Flexibility-of-hiring index	53	Number of procedures	18
Conditions-of-employment index	61	Time (days)	150
Flexibility-of-firing index	45	Cost (% of income per capita)	83.3
Employment-law index	53	Procedural-complexity index	57
Getting Credit		**Closing a Business**	
Public credit registry operates?	Yes	Time to go through insolvency (years)	2.2
Public registry coverage (borrowers/1,000 capita)	1	Cost to go through insolvency (% estate)	18
Public-registry index	22	Absolute priority preserved	67
Private credit-information bureau operates?	No	Efficient outcome achieved	0
Private bureau coverage (borrowers/1,000 capita)	0	Goals-of-insolvency index	44
Creditor-rights index	1	Court-powers index	100

CROATIA
Europe and Central Asia

Economy Characteristics		Starting a Business	
GNI per capita (US$)	4,640	Number of procedures	13
Population	4,380,780	Time (days)	50
Informal economy (% of income)	33.4	Cost (% of income per capita)	18.2
Legal origin	German	Minimum capital (% of income per capita)	50.7
Hiring and Firing Workers		**Enforcing a Contract**	
Flexibility-of-hiring index	76	Number of procedures	20
Conditions-of-employment index	89	Time (days)	330
Flexibility-of-firing index	31	Cost (% of income per capita)	6.6
Employment-law index	65	Procedural-complexity index	50
Getting Credit		**Closing a Business**	
Public credit registry operates?	No	Time to go through insolvency (years)	3.1
Public registry coverage (borrowers/1,000 capita)	0	Cost to go through insolvency (% estate)	18
Public-registry index	0	Absolute priority preserved	100
Private credit-information bureau operates?	No	Efficient outcome achieved	0
Private bureau coverage (borrowers/1,000 capita)	0	Goals-of-insolvency index	50
Creditor-rights index	3	Court-powers index	67

Note: .. means no data available.

CZECH REPUBLIC
Europe and Central Asia

Economy Characteristics		Starting a Business	
GNI per capita (US$)	5,560	Number of procedures	10
Population	10,224,000	Time (days)	88
Informal economy (% of income)	19.1	Cost (% of income per capita)	11.7
Legal origin	German	Minimum capital (% of income per capita)	110.0

Hiring and Firing Workers		Enforcing a Contract	
Flexibility-of-hiring index	17	Number of procedures	16
Conditions-of-employment index	63	Time (days)	270
Flexibility-of-firing index	27	Cost (% of income per capita)	18.5
Employment-law index	36	Procedural-complexity index	65

Getting Credit		Closing a Business	
Public credit registry operates?	Yes	Time to go through insolvency (years)	9.2
Public registry coverage (borrowers/1,000 capita)	10	Cost to go through insolvency (% estate)	38
Public-registry index	60	Absolute priority preserved	67
Private credit-information bureau operates?	Yes	Efficient outcome achieved	0
Private bureau coverage (borrowers/1,000 capita)	136	Goals-of-insolvency index	22
Creditor-rights index	3	Court-powers index	0

DENMARK
OECD: High Income

Economy Characteristics		Starting a Business	
GNI per capita (US$)	30,290	Number of procedures	4
Population	5,359,000	Time (days)	4
Informal economy (% of income)	18.2	Cost (% of income per capita)	0.0
Legal origin	Nordic	Minimum capital (% of income per capita)	52.3

Hiring and Firing Workers		Enforcing a Contract	
Flexibility-of-hiring index	33	Number of procedures	14
Conditions-of-employment index	25	Time (days)	83
Flexibility-of-firing index	17	Cost (% of income per capita)	3.8
Employment-law index	25	Procedural-complexity index	40

Getting Credit		Closing a Business	
Public credit registry operates?	No	Time to go through insolvency (years)	4.2
Public registry coverage (borrowers/1,000 capita)	0	Cost to go through insolvency (% estate)	8
Public-registry index	0	Absolute priority preserved	100
Private credit-information bureau operates?	Yes	Efficient outcome achieved	1
Private bureau coverage (borrowers/1,000 capita)	58	Goals-of-insolvency index	79
Creditor-rights index	3	Court-powers index	33

DOMINICAN REPUBLIC
Latin America and Caribbean

Economy Characteristics		Starting a Business	
GNI per capita (US$)	2,320	Number of procedures	12
Population	8,505,200	Time (days)	78
Informal economy (% of income)	32.1	Cost (% of income per capita)	48.1
Legal origin	French	Minimum capital (% of income per capita)	23.2

Hiring and Firing Workers		Enforcing a Contract	
Flexibility-of-hiring index	33	Number of procedures	19
Conditions-of-employment index	79	Time (days)	495
Flexibility-of-firing index	35	Cost (% of income per capita)	440.5
Employment-law index	49	Procedural-complexity index	69

Getting Credit		Closing a Business	
Public credit registry operates?	Yes	Time to go through insolvency (years)	3.5
Public registry coverage (borrowers/1,000 capita)	..	Cost to go through insolvency (% estate)	4
Public-registry index	42	Absolute priority preserved	0
Private credit-information bureau operates?	Yes	Efficient outcome achieved	0
Private bureau coverage (borrowers/1,000 capita)	423	Goals-of-insolvency index	37
Creditor-rights index	2	Court-powers index	67

Note: .. means no data available.

ECUADOR
Latin America and Caribbean

Economy Characteristics		Starting a Business	
GNI per capita (US$)	1,450	Number of procedures	14
Population	12,879,000	Time (days)	90
Informal economy (% of income)	34.4	Cost (% of income per capita)	63.0
Legal origin	French	Minimum capital (% of income per capita)	27.6
Hiring and Firing Workers		**Enforcing a Contract**	
Flexibility-of-hiring index	37	Number of procedures	33
Conditions-of-employment index	63	Time (days)	333
Flexibility-of-firing index	65	Cost (% of income per capita)	10.5
Employment-law index	55	Procedural-complexity index	72
Getting Credit		**Closing a Business**	
Public credit registry operates?	Yes	Time to go through insolvency (years)	3.5
Public registry coverage (borrowers/1,000 capita)	82	Cost to go through insolvency (% estate)	18
Public-registry index	55	Absolute priority preserved	0
Private credit-information bureau operates?	No	Efficient outcome achieved	0
Private bureau coverage (borrowers/1,000 capita)	0	Goals-of-insolvency index	24
Creditor-rights index	1	Court-powers index	67

EGYPT, ARAB REP. of
Middle East and North Africa

Economy Characteristics		Starting a Business	
GNI per capita (US$)	1,470	Number of procedures	13
Population	65,176,940	Time (days)	43
Informal economy (% of income)	35.1	Cost (% of income per capita)	61.2
Legal origin	French	Minimum capital (% of income per capita)	788.6
Hiring and Firing Workers		**Enforcing a Contract**	
Flexibility-of-hiring index	33	Number of procedures	19
Conditions-of-employment index	83	Time (days)	202
Flexibility-of-firing index	61	Cost (% of income per capita)	30.7
Employment-law index	59	Procedural-complexity index	50
Getting Credit		**Closing a Business**	
Public credit registry operates?	Yes	Time to go through insolvency (years)	4.3
Public registry coverage (borrowers/1,000 capita)	..	Cost to go through insolvency (% estate)	18
Public-registry index	48	Absolute priority preserved	67
Private credit-information bureau operates?	No	Efficient outcome achieved	0
Private bureau coverage (borrowers/1,000 capita)	0	Goals-of-insolvency index	39
Creditor-rights index	1	Court-powers index	67

EL SALVADOR
Latin America and Caribbean

Economy Characteristics		Starting a Business	
GNI per capita (US$)	2,080	Number of procedures	12
Population	6,400,000	Time (days)	115
Informal economy (% of income)	..	Cost (% of income per capita)	129.3
Legal origin	French	Minimum capital (% of income per capita)	549.5
Hiring and Firing Workers		**Enforcing a Contract**	
Flexibility-of-hiring index	81	Number of procedures	42
Conditions-of-employment index	75	Time (days)	240
Flexibility-of-firing index	52	Cost (% of income per capita)	7.3
Employment-law index	69	Procedural-complexity index	81
Getting Credit		**Closing a Business**	
Public credit registry operates?	Yes	Time to go through insolvency (years)	No practice
Public registry coverage (borrowers/1,000 capita)	130	Cost to go through insolvency (% estate)	No practice
Public-registry index	50	Absolute priority preserved	67
Private credit-information bureau operates?	Yes	Efficient outcome achieved	1
Private bureau coverage (borrowers/1,000 capita)	128	Goals-of-insolvency index	42
Creditor-rights index	3	Court-powers index	67

Note: .. means no data available.

ETHIOPIA
Sub-Saharan Africa

Economy Characteristics		Starting a Business	
GNI per capita (US$)	100	Number of procedures	8
Population	65,816,048	Time (days)	44
Informal economy (% of income)	40.3	Cost (% of income per capita)	421.6
Legal origin	English	Minimum capital (% of income per capita)	1756.1
Hiring and Firing Workers		**Enforcing a Contract**	
Flexibility-of-hiring index	58	Number of procedures	24
Conditions-of-employment index	67	Time (days)	895
Flexibility-of-firing index	29	Cost (% of income per capita)	34.6
Employment-law index	51	Procedural-complexity index	52
Getting Credit		**Closing a Business**	
Public credit registry operates?	No	Time to go through insolvency (years)	2.2
Public registry coverage (borrowers/1,000 capita)	0	Cost to go through insolvency (% estate)	8
Public-registry index	0	Absolute priority preserved	67
Private credit-information bureau operates?	No	Efficient outcome achieved	1
Private bureau coverage (borrowers/1,000 capita)	0	Goals-of-insolvency index	75
Creditor-rights index	3	Court-powers index	33

FINLAND
OECD: High Income

Economy Characteristics		Starting a Business	
GNI per capita (US$)	23,510	Number of procedures	4
Population	5,188,000	Time (days)	33
Informal economy (% of income)	18.3	Cost (% of income per capita)	3.1
Legal origin	Nordic	Minimum capital (% of income per capita)	32.0
Hiring and Firing Workers		**Enforcing a Contract**	
Flexibility-of-hiring index	71	Number of procedures	19
Conditions-of-employment index	43	Time (days)	240
Flexibility-of-firing index	52	Cost (% of income per capita)	15.8
Employment-law index	55	Procedural-complexity index	48
Getting Credit		**Closing a Business**	
Public credit registry operates?	No	Time to go through insolvency (years)	0.9
Public registry coverage (borrowers/1,000 capita)	0	Cost to go through insolvency (% estate)	1
Public-registry index	0	Absolute priority preserved	100
Private credit-information bureau operates?	Yes	Efficient outcome achieved	1
Private bureau coverage (borrowers/1,000 capita)	96	Goals-of-insolvency index	99
Creditor-rights index	1	Court-powers index	0

FRANCE
OECD: High Income

Economy Characteristics		Starting a Business	
GNI per capita (US$)	22,010	Number of procedures	10
Population	59,190,600	Time (days)	53
Informal economy (% of income)	15.3	Cost (% of income per capita)	3.0
Legal origin	French	Minimum capital (% of income per capita)	32.1
Hiring and Firing Workers		**Enforcing a Contract**	
Flexibility-of-hiring index	63	Number of procedures	21
Conditions-of-employment index	61	Time (days)	210
Flexibility-of-firing index	26	Cost (% of income per capita)	3.8
Employment-law index	50	Procedural-complexity index	79
Getting Credit		**Closing a Business**	
Public credit registry operates?	Yes	Time to go through insolvency (years)	2.4
Public registry coverage (borrowers/1,000 capita)	12	Cost to go through insolvency (% estate)	18
Public-registry index	53	Absolute priority preserved	67
Private credit-information bureau operates?	No	Efficient outcome achieved	0
Private bureau coverage (borrowers/1,000 capita)	0	Goals-of-insolvency index	43
Creditor-rights index	0	Court-powers index	100

Note: .. means no data available.

GEORGIA
Europe and Central Asia

Economy Characteristics		Starting a Business	
GNI per capita (US$)	650	Number of procedures	9
Population	5,224,000	Time (days)	30
Informal economy (% of income)	67.3	Cost (% of income per capita)	26.3
Legal origin	Socialist	Minimum capital (% of income per capita)	140.1
Hiring and Firing Workers		**Enforcing a Contract**	
Flexibility-of-hiring index	51	Number of procedures	17
Conditions-of-employment index	66	Time (days)	180
Flexibility-of-firing index	49	Cost (% of income per capita)	63.1
Employment-law index	55	Procedural-complexity index	48
Getting Credit		**Closing a Business**	
Public credit registry operates?	No	Time to go through insolvency (years)	3.2
Public registry coverage (borrowers/1,000 capita)	0	Cost to go through insolvency (% estate)	1
Public-registry index	0	Absolute priority preserved	100
Private credit-information bureau operates?	No	Efficient outcome achieved	0
Private bureau coverage (borrowers/1,000 capita)	0	Goals-of-insolvency index	69
Creditor-rights index	2	Court-powers index	33

GERMANY
OECD: High Income

Economy Characteristics		Starting a Business	
GNI per capita (US$)	22,670	Number of procedures	9
Population	82,333,000	Time (days)	45
Informal economy (% of income)	16.3	Cost (% of income per capita)	5.9
Legal origin	German	Minimum capital (% of income per capita)	103.8
Hiring and Firing Workers		**Enforcing a Contract**	
Flexibility-of-hiring index	63	Number of procedures	22
Conditions-of-employment index	46	Time (days)	154
Flexibility-of-firing index	45	Cost (% of income per capita)	6.0
Employment-law index	51	Procedural-complexity index	61
Getting Credit		**Closing a Business**	
Public credit registry operates?	Yes	Time to go through insolvency (years)	1.2
Public registry coverage (borrowers/1,000 capita)	5	Cost to go through insolvency (% estate)	8
Public-registry index	44	Absolute priority preserved	100
Private credit-information bureau operates?	Yes	Efficient outcome achieved	0
Private bureau coverage (borrowers/1,000 capita)	693	Goals-of-insolvency index	61
Creditor-rights index	3	Court-powers index	33

GHANA
Sub-Saharan Africa

Economy Characteristics		Starting a Business	
GNI per capita (US$)	270	Number of procedures	10
Population	19,707,740	Time (days)	84
Informal economy (% of income)	38.4	Cost (% of income per capita)	111.7
Legal origin	English	Minimum capital (% of income per capita)	1.2
Hiring and Firing Workers		**Enforcing a Contract**	
Flexibility-of-hiring index	33	Number of procedures	21
Conditions-of-employment index	56	Time (days)	90
Flexibility-of-firing index	17	Cost (% of income per capita)	23.8
Employment-law index	35	Procedural-complexity index	33
Getting Credit		**Closing a Business**	
Public credit registry operates?	No	Time to go through insolvency (years)	No practice
Public registry coverage (borrowers/1,000 capita)	0	Cost to go through insolvency (% estate)	No practice
Public-registry index	0	Absolute priority preserved	67
Private credit-information bureau operates?	Yes	Efficient outcome achieved	0
Private bureau coverage (borrowers/1,000 capita)	<1	Goals-of-insolvency index	17
Creditor-rights index	1	Court-powers index	33

Note: .. means no data available.

GREECE
OECD: High Income

Economy Characteristics		Starting a Business	
GNI per capita (US$)	11,660	Number of procedures	16
Population	10,590,870	Time (days)	45
Informal economy (% of income)	28.6	Cost (% of income per capita)	69.6
Legal origin	French	Minimum capital (% of income per capita)	145.3
Hiring and Firing Workers		**Enforcing a Contract**	
Flexibility-of-hiring index	78	Number of procedures	15
Conditions-of-employment index	81	Time (days)	315
Flexibility-of-firing index	43	Cost (% of income per capita)	8.2
Employment-law index	67	Procedural-complexity index	64
Getting Credit		**Closing a Business**	
Public credit registry operates?	No	Time to go through insolvency (years)	2.2
Public registry coverage (borrowers/1,000 capita)	0	Cost to go through insolvency (% estate)	8
Public-registry index	0	Absolute priority preserved	33
Private credit-information bureau operates?	Yes	Efficient outcome achieved	0
Private bureau coverage (borrowers/1,000 capita)	86	Goals-of-insolvency index	42
Creditor-rights index	1	Court-powers index	33

GUATEMALA
Latin America and Carribean

Economy Characteristics		Starting a Business	
GNI per capita (US$)	1,750	Number of procedures	13
Population	11,683,000	Time (days)	39
Informal economy (% of income)	51.5	Cost (% of income per capita)	66.7
Legal origin	French	Minimum capital (% of income per capita)	36.5
Hiring and Firing Workers		**Enforcing a Contract**	
Flexibility-of-hiring index	58	Number of procedures	19
Conditions-of-employment index	85	Time (days)	1460
Flexibility-of-firing index	51	Cost (% of income per capita)	20.0
Employment-law index	65	Procedural-complexity index	90
Getting Credit		**Closing a Business**	
Public credit registry operates?	No	Time to go through insolvency (years)	4.0
Public registry coverage (borrowers/1,000 capita)	0	Cost to go through insolvency (% estate)	18
Public-registry index	0	Absolute priority preserved	67
Private credit-information bureau operates?	Yes	Efficient outcome achieved	0
Private bureau coverage (borrowers/1,000 capita)	35	Goals-of-insolvency index	40
Creditor-rights index	1	Court-powers index	67

GUINEA
Sub-Saharan Africa

Economy Characteristics		Starting a Business	
GNI per capita (US$)	150	Number of procedures	13
Population	1,225,620	Time (days)	71
Informal economy (% of income)	..	Cost (% of income per capita)	229.4
Legal origin	French	Minimum capital (% of income per capita)	396.6
Hiring and Firing Workers		**Enforcing a Contract**	
Flexibility-of-hiring index	78	Number of procedures	41
Conditions-of-employment index	44	Time (days)	150
Flexibility-of-firing index	57	Cost (% of income per capita)	40.0
Employment-law index	60	Procedural-complexity index	77
Getting Credit		**Closing a Business**	
Public credit registry operates?	Yes	Time to go through insolvency (years)	No practice
Public registry coverage (borrowers/1,000 capita)	..	Cost to go through insolvency (% estate)	No practice
Public-registry index	..	Absolute priority preserved	33
Private credit-information bureau operates?	No	Efficient outcome achieved	0
Private bureau coverage (borrowers/1,000 capita)	0	Goals-of-insolvency index	8
Creditor-rights index	1	Court-powers index	100

Note: .. means no data available.

HAITI
Latin America and Caribbean

Economy Characteristics		Starting a Business	
GNI per capita (US$)	440	Number of procedures	12
Population	8,132,000	Time (days)	203
Informal economy (% of income)	..	Cost (% of income per capita)	198.9
Legal origin	French	Minimum capital (% of income per capita)	209.8
Hiring and Firing Workers		**Enforcing a Contract**	
Flexibility-of-hiring index	58	Number of procedures	41
Conditions-of-employment index	85	Time (days)	76
Flexibility-of-firing index	35	Cost (% of income per capita)	18.4
Employment-law index	60	Procedural-complexity index	69
Getting Credit		**Closing a Business**	
Public credit registry operates?	Yes	Time to go through insolvency (years)	No practice
Public registry coverage (borrowers/1,000 capita)	1	Cost to go through insolvency (% estate)	No practice
Public-registry index	59	Absolute priority preserved	67
Private credit-information bureau operates?	No	Efficient outcome achieved	1
Private bureau coverage (borrowers/1,000 capita)	0	Goals-of-insolvency index	42
Creditor-rights index	2	Court-powers index	67

HONDURAS
Latin America and Caribbean

Economy Characteristics		Starting a Business	
GNI per capita (US$)	920	Number of procedures	14
Population	6,584,730	Time (days)	80
Informal economy (% of income)	49.6	Cost (% of income per capita)	72.8
Legal origin	French	Minimum capital (% of income per capita)	165.4
Hiring and Firing Workers		**Enforcing a Contract**	
Flexibility-of-hiring index	33	Number of procedures	32
Conditions-of-employment index	87	Time (days)	225
Flexibility-of-firing index	47	Cost (% of income per capita)	6.7
Employment-law index	56	Procedural-complexity index	72
Getting Credit		**Closing a Business**	
Public credit registry operates?	Yes	Time to go through insolvency (years)	No practice
Public registry coverage (borrowers/1,000 capita)	45	Cost to go through insolvency (% estate)	No practice
Public-registry index	42	Absolute priority preserved	67
Private credit-information bureau operates?	No	Efficient outcome achieved	0
Private bureau coverage (borrowers/1,000 capita)	0	Goals-of-insolvency index	17
Creditor-rights index	2	Court-powers index	67

HONG KONG, CHINA
East Asia and Pacific

Economy Characteristics		Starting a Business	
GNI per capita (US$)	24,750	Number of procedures	5
Population	6,725,000	Time (days)	11
Informal economy (% of income)	16.6	Cost (% of income per capita)	2.3
Legal origin	English	Minimum capital (% of income per capita)	0.0
Hiring and Firing Workers		**Enforcing a Contract**	
Flexibility-of-hiring index	58	Number of procedures	17
Conditions-of-employment index	22	Time (days)	180
Flexibility-of-firing index	1	Cost (% of income per capita)	6.9
Employment-law index	27	Procedural-complexity index	50
Getting Credit		**Closing a Business**	
Public credit registry operates?	No	Time to go through insolvency (years)	1.0
Public registry coverage (borrowers/1,000 capita)	0	Cost to go through insolvency (% estate)	18
Public-registry index	0	Absolute priority preserved	33
Private credit-information bureau operates?	Yes	Efficient outcome achieved	1
Private bureau coverage (borrowers/1,000 capita)	200	Goals-of-insolvency index	63
Creditor-rights index	4	Court-powers index	67

Note: .. means no data available.

HUNGARY
Europe and Central Asia

Economy Characteristics		Starting a Business	
GNI per capita (US$)	5,280	Number of procedures	5
Population	10,187,000	Time (days)	65
Informal economy (% of income)	25.1	Cost (% of income per capita)	64.3
Legal origin	German	Minimum capital (% of income per capita)	220.3
Hiring and Firing Workers		**Enforcing a Contract**	
Flexibility-of-hiring index	46	Number of procedures	17
Conditions-of-employment index	92	Time (days)	365
Flexibility-of-firing index	23	Cost (% of income per capita)	5.4
Employment-law index	54	Procedural-complexity index	57
Getting Credit		**Closing a Business**	
Public credit registry operates?	No	Time to go through insolvency (years)	2.0
Public registry coverage (borrowers/1,000 capita)	0	Cost to go through insolvency (% estate)	38
Public-registry index	0	Absolute priority preserved	67
Private credit-information bureau operates?	Yes	Efficient outcome achieved	0
Private bureau coverage (borrowers/1,000 capita)	15	Goals-of-insolvency index	38
Creditor-rights index	2	Court-powers index	33

INDIA
South Asia

Economy Characteristics		Starting a Business	
GNI per capita (US$)	480	Number of procedures	10
Population	1,032,354,600	Time (days)	88
Informal economy (% of income)	23.1	Cost (% of income per capita)	49.8
Legal origin	English	Minimum capital (% of income per capita)	430.4
Hiring and Firing Workers		**Enforcing a Contract**	
Flexibility-of-hiring index	33	Number of procedures	22
Conditions-of-employment index	75	Time (days)	365
Flexibility-of-firing index	45	Cost (% of income per capita)	95.0
Employment-law index	51	Procedural-complexity index	50
Getting Credit		**Closing a Business**	
Public credit registry operates?	No	Time to go through insolvency (years)	11.3
Public registry coverage (borrowers/1,000 capita)	0	Cost to go through insolvency (% estate)	8
Public-registry index	0	Absolute priority preserved	33
Private credit-information bureau operates?	Developing	Efficient outcome achieved	0
Private bureau coverage (borrowers/1,000 capita)	0	Goals-of-insolvency index	21
Creditor-rights index	3	Court-powers index	33

INDONESIA
East Asia and Pacific

Economy Characteristics		Starting a Business	
GNI per capita (US$)	710	Number of procedures	11
Population	208,981,000	Time (days)	168
Informal economy (% of income)	19.4	Cost (% of income per capita)	14.5
Legal origin	French	Minimum capital (% of income per capita)	302.5
Hiring and Firing Workers		**Enforcing a Contract**	
Flexibility-of-hiring index	76	Number of procedures	29
Conditions-of-employment index	53	Time (days)	225
Flexibility-of-firing index	43	Cost (% of income per capita)	269.0
Employment-law index	57	Procedural-complexity index	67
Getting Credit		**Closing a Business**	
Public credit registry operates?	Yes	Time to go through insolvency (years)	6.0
Public registry coverage (borrowers/1,000 capita)	3	Cost to go through insolvency (% estate)	18
Public-registry index	61	Absolute priority preserved	67
Private credit-information bureau operates?	No	Efficient outcome achieved	0
Private bureau coverage (borrowers/1,000 capita)	0	Goals-of-insolvency index	35
Creditor-rights index	2	Court-powers index	100

Note: .. means no data available.

IRAN, ISLAMIC REP. of
Middle East and North Africa

Economy Characteristics		Starting a Business	
GNI per capita (US$)	1,710	Number of procedures	9
Population	64,528,160	Time (days)	48
Informal economy (% of income)	18.9	Cost (% of income per capita)	6.6
Legal origin	English	Minimum capital (% of income per capita)	7.4
Hiring and Firing Workers		**Enforcing a Contract**	
Flexibility-of-hiring index	33	Number of procedures	23
Conditions-of-employment index	77	Time (days)	150
Flexibility-of-firing index	47	Cost (% of income per capita)	5.8
Employment-law index	52	Procedural-complexity index	67
Getting Credit		**Closing a Business**	
Public credit registry operates?	Yes	Time to go through insolvency (years)	1.8
Public registry coverage (borrowers/1,000 capita)	..	Cost to go through insolvency (% estate)	8
Public-registry index	45	Absolute priority preserved	100
Private credit-information bureau operates?	No	Efficient outcome achieved	1
Private bureau coverage (borrowers/1,000 capita)	0	Goals-of-insolvency index	84
Creditor-rights index	2	Court-powers index	67

IRELAND
OECD: High Income

Economy Characteristics		Starting a Business	
GNI per capita (US$)	23,870	Number of procedures	3
Population	3,839,000	Time (days)	12
Informal economy (% of income)	15.8	Cost (% of income per capita)	10.4
Legal origin	English	Minimum capital (% of income per capita)	0.0
Hiring and Firing Workers		**Enforcing a Contract**	
Flexibility-of-hiring index	48	Number of procedures	16
Conditions-of-employment index	68	Time (days)	183
Flexibility-of-firing index	30	Cost (% of income per capita)	7.2
Employment-law index	49	Procedural-complexity index	42
Getting Credit		**Closing a Business**	
Public credit registry operates?	No	Time to go through insolvency (years)	0.4
Public registry coverage (borrowers/1,000 capita)	0	Cost to go through insolvency (% estate)	8
Public-registry index	0	Absolute priority preserved	100
Private credit-information bureau operates?	Yes	Efficient outcome achieved	1
Private bureau coverage (borrowers/1,000 capita)	730	Goals-of-insolvency index	88
Creditor-rights index	1	Court-powers index	33

ISRAEL
Middle East and North Africa

Economy Characteristics		Starting a Business	
GNI per capita (US$)	16,710	Number of procedures	5
Population	6,362,950	Time (days)	34
Informal economy (% of income)	21.9	Cost (% of income per capita)	4.7
Legal origin	English	Minimum capital (% of income per capita)	0.0
Hiring and Firing Workers		**Enforcing a Contract**	
Flexibility-of-hiring index	33	Number of procedures	19
Conditions-of-employment index	64	Time (days)	315
Flexibility-of-firing index	16	Cost (% of income per capita)	34.1
Employment-law index	38	Procedural-complexity index	51
Getting Credit		**Closing a Business**	
Public credit registry operates?	No	Time to go through insolvency (years)	4.0
Public registry coverage (borrowers/1,000 capita)	0	Cost to go through insolvency (% estate)	38
Public-registry index	0	Absolute priority preserved	100
Private credit-information bureau operates?	Yes	Efficient outcome achieved	1
Private bureau coverage (borrowers/1,000 capita)	47	Goals-of-insolvency index	67
Creditor-rights index	3	Court-powers index	67

Note: .. means no data available.

ITALY
OECD: High Income

Economy Characteristics		Starting a Business	
GNI per capita (US$)	18,960	Number of procedures	9
Population	57,948,000	Time (days)	23
Informal economy (% of income)	27	Cost (% of income per capita)	24.1
Legal origin	French	Minimum capital (% of income per capita)	49.6
Hiring and Firing Workers		**Enforcing a Contract**	
Flexibility-of-hiring index	76	Number of procedures	16
Conditions-of-employment index	62	Time (days)	645
Flexibility-of-firing index	40	Cost (% of income per capita)	3.9
Employment-law index	59	Procedural-complexity index	64
Getting Credit		**Closing a Business**	
Public credit registry operates?	Yes	Time to go through insolvency (years)	1.3
Public registry coverage (borrowers/1,000 capita)	55	Cost to go through insolvency (% estate)	18
Public-registry index	61	Absolute priority preserved	67
Private credit-information bureau operates?	Yes	Efficient outcome achieved	0
Private bureau coverage (borrowers/1,000 capita)	416	Goals-of-insolvency index	46
Creditor-rights index	1	Court-powers index	33

JAMAICA
Latin America and Caribbean

Economy Characteristics		Starting a Business	
GNI per capita (US$)	2,820	Number of procedures	7
Population	2,590,000	Time (days)	31
Informal economy (% of income)	36.4	Cost (% of income per capita)	16.2
Legal origin	English	Minimum capital (% of income per capita)	0.0
Hiring and Firing Workers		**Enforcing a Contract**	
Flexibility-of-hiring index	33	Number of procedures	14
Conditions-of-employment index	52	Time (days)	202
Flexibility-of-firing index	18	Cost (% of income per capita)	42.1
Employment-law index	34	Procedural-complexity index	38
Getting Credit		**Closing a Business**	
Public credit registry operates?	No	Time to go through insolvency (years)	1.1
Public registry coverage (borrowers/1,000 capita)	0	Cost to go through insolvency (% estate)	18
Public-registry index	0	Absolute priority preserved	33
Private credit-information bureau operates?	No	Efficient outcome achieved	1
Private bureau coverage (borrowers/1,000 capita)	0	Goals-of-insolvency index	63
Creditor-rights index	2	Court-powers index	67

JAPAN
OECD: High Income

Economy Characteristics		Starting a Business	
GNI per capita (US$)	33,550	Number of procedures	11
Population	127,034,880	Time (days)	31
Informal economy (% of income)	11.3	Cost (% of income per capita)	10.5
Legal origin	German	Minimum capital (% of income per capita)	71.3
Hiring and Firing Workers		**Enforcing a Contract**	
Flexibility-of-hiring index	39	Number of procedures	16
Conditions-of-employment index	64	Time (days)	60
Flexibility-of-firing index	9	Cost (% of income per capita)	6.4
Employment-law index	37	Procedural-complexity index	39
Getting Credit		**Closing a Business**	
Public credit registry operates?	No	Time to go through insolvency (years)	0.6
Public registry coverage (borrowers/1,000 capita)	0	Cost to go through insolvency (% estate)	4
Public-registry index	0	Absolute priority preserved	100
Private credit-information bureau operates?	Yes	Efficient outcome achieved	1
Private bureau coverage (borrowers/1,000 capita)	777	Goals-of-insolvency index	93
Creditor-rights index	2	Court-powers index	33

Note: .. means no data available.

JORDAN
Middle East and North Africa

Economy Characteristics		Starting a Business	
GNI per capita (US$)	1,760	Number of procedures	14
Population	5,030,800	Time (days)	98
Informal economy (% of income)	19.4	Cost (% of income per capita)	49.8
Legal origin	French	Minimum capital (% of income per capita)	2404.2
Hiring and Firing Workers		**Enforcing a Contract**	
Flexibility-of-hiring index	33	Number of procedures	32
Conditions-of-employment index	82	Time (days)	147
Flexibility-of-firing index	64	Cost (% of income per capita)	0.3
Employment-law index	60	Procedural-complexity index	49
Getting Credit		**Closing a Business**	
Public credit registry operates?	Yes	Time to go through insolvency (years)	4.3
Public registry coverage (borrowers/1,000 capita)	19	Cost to go through insolvency (% estate)	8
Public-registry index	47	Absolute priority preserved	33
Private credit-information bureau operates?	No	Efficient outcome achieved	0
Private bureau coverage (borrowers/1,000 capita)	0	Goals-of-insolvency index	37
Creditor-rights index	1	Court-powers index	33

KAZAKHSTAN
Europe and Central Asia

Economy Characteristics		Starting a Business	
GNI per capita (US$)	1,510	Number of procedures	10
Population	14,895,310	Time (days)	25
Informal economy (% of income)	43.2	Cost (% of income per capita)	10.1
Legal origin	Socialist	Minimum capital (% of income per capita)	35.2
Hiring and Firing Workers		**Enforcing a Contract**	
Flexibility-of-hiring index	33	Number of procedures	41
Conditions-of-employment index	89	Time (days)	120
Flexibility-of-firing index	42	Cost (% of income per capita)	7.9
Employment-law index	55	Procedural-complexity index	65
Getting Credit		**Closing a Business**	
Public credit registry operates?	No	Time to go through insolvency (years)	3.3
Public registry coverage (borrowers/1,000 capita)	0	Cost to go through insolvency (% estate)	18
Public-registry index	0	Absolute priority preserved	67
Private credit-information bureau operates?	No	Efficient outcome achieved	1
Private bureau coverage (borrowers/1,000 capita)	0	Goals-of-insolvency index	66
Creditor-rights index	2	Court-powers index	67

KENYA
Sub-Saharan Africa

Economy Characteristics		Starting a Business	
GNI per capita (US$)	360	Number of procedures	11
Population	30,735,760	Time (days)	61
Informal economy (% of income)	34.3	Cost (% of income per capita)	54.0
Legal origin	English	Minimum capital (% of income per capita)	0.0
Hiring and Firing Workers		**Enforcing a Contract**	
Flexibility-of-hiring index	33	Number of procedures	25
Conditions-of-employment index	53	Time (days)	255
Flexibility-of-firing index	16	Cost (% of income per capita)	49.5
Employment-law index	34	Procedural-complexity index	44
Getting Credit		**Closing a Business**	
Public credit registry operates?	No	Time to go through insolvency (years)	4.6
Public registry coverage (borrowers/1,000 capita)	0	Cost to go through insolvency (% estate)	18
Public-registry index	0	Absolute priority preserved	100
Private credit-information bureau operates?	Yes	Efficient outcome achieved	0
Private bureau coverage (borrowers/1,000 capita)	309	Goals-of-insolvency index	47
Creditor-rights index	4	Court-powers index	33

Note: .. means no data available.

KOREA, REP. of
OECD: High Income

Economy Characteristics		Starting a Business	
GNI per capita (US$)	9,930	Number of procedures	12
Population	47,343,000	Time (days)	33
Informal economy (% of income)	27.5	Cost (% of income per capita)	17.9
Legal origin	German	Minimum capital (% of income per capita)	402.5
Hiring and Firing Workers		**Enforcing a Contract**	
Flexibility-of-hiring index	33	Number of procedures	23
Conditions-of-employment index	88	Time (days)	75
Flexibility-of-firing index	32	Cost (% of income per capita)	4.5
Employment-law index	51	Procedural-complexity index	50
Getting Credit		**Closing a Business**	
Public credit registry operates?	No	Time to go through insolvency (years)	1.47
Public registry coverage (borrowers/1,000 capita)	0	Cost to go through insolvency (% estate)	4
Public-registry index	0	Absolute priority preserved	100
Private credit-information bureau operates?	Yes	Efficient outcome achieved	1
Private bureau coverage (borrowers/1,000 capita)	530	Goals-of-insolvency index	91
Creditor-rights index	3	Court-powers index	67

KUWAIT
Middle East and North Africa

Economy Characteristics		Starting a Business	
GNI per capita (US$)	..	Number of procedures	13
Population	2,044,270	Time (days)	34
Informal economy (% of income)	..	Cost (% of income per capita)	1.8
Legal origin	French	Minimum capital (% of income per capita)	910.6
Hiring and Firing Workers		**Enforcing a Contract**	
Flexibility-of-hiring index	33	Number of procedures	17
Conditions-of-employment index	40	Time (days)	195
Flexibility-of-firing index	50	Cost (% of income per capita)	4.4
Employment-law index	41	Procedural-complexity index	76
Getting Credit		**Closing a Business**	
Public credit registry operates?	No	Time to go through insolvency (years)	4.2
Public registry coverage (borrowers/1,000 capita)	0	Cost to go through insolvency (% estate)	1
Public-registry index	0	Absolute priority preserved	67
Private credit-information bureau operates?	Yes	Efficient outcome achieved	1
Private bureau coverage (borrowers/1,000 capita)	147	Goals-of-insolvency index	83
Creditor-rights index	2	Court-powers index	67

KYRGYZ REPUBLIC
Europe and Central Asia

Economy Characteristics		Starting a Business	
GNI per capita (US$)	290	Number of procedures	9
Population	4,955,000	Time (days)	26
Informal economy (% of income)	39.8	Cost (% of income per capita)	13.4
Legal origin	Socialist	Minimum capital (% of income per capita)	74.8
Hiring and Firing Workers		**Enforcing a Contract**	
Flexibility-of-hiring index	71	Number of procedures	44
Conditions-of-employment index	90	Time (days)	365
Flexibility-of-firing index	33	Cost (% of income per capita)	254.7
Employment-law index	64	Procedural-complexity index	48
Getting Credit		**Closing a Business**	
Public credit registry operates?	No	Time to go through insolvency (years)	4.0
Public registry coverage (borrowers/1,000 capita)	0	Cost to go through insolvency (% estate)	4
Public-registry index	0	Absolute priority preserved	100
Private credit-information bureau operates?	No	Efficient outcome achieved	0
Private bureau coverage (borrowers/1,000 capita)	0	Goals-of-insolvency index	61
Creditor-rights index	3	Court-powers index	33

Note: .. means no data available.

LAO PDR
East Asia and Pacific

Economy Characteristics		Starting a Business	
GNI per capita (US$)	310	Number of procedures	9
Population	5,403,170	Time (days)	198
Informal economy (% of income)	..	Cost (% of income per capita)	19.5
Legal origin	French	Minimum capital (% of income per capita)	150.7
Hiring and Firing Workers		**Enforcing a Contract**	
Flexibility-of-hiring index	33	Number of procedures	..
Conditions-of-employment index	87	Time (days)	..
Flexibility-of-firing index	44	Cost (% of income per capita)	..
Employment-law index	54	Procedural-complexity index	..
Getting Credit		**Closing a Business**	
Public credit registry operates?	Yes	Time to go through insolvency (years)	No practice
Public registry coverage (borrowers/1,000 capita)	..	Cost to go through insolvency (% estate)	No practice
Public-registry index	..	Absolute priority preserved	0
Private credit-information bureau operates?	No	Efficient outcome achieved	0
Private bureau coverage (borrowers/1,000 capita)	0	Goals-of-insolvency index	14
Creditor-rights index	0	Court-powers index	67

LATVIA
Europe and Central Asia

Economy Characteristics		Starting a Business	
GNI per capita (US$)	3,480	Number of procedures	7
Population	2,359,000	Time (days)	11
Informal economy (% of income)	39.9	Cost (% of income per capita)	14.7
Legal origin	German	Minimum capital (% of income per capita)	93.0
Hiring and Firing Workers		**Enforcing a Contract**	
Flexibility-of-hiring index	58	Number of procedures	19
Conditions-of-employment index	87	Time (days)	189
Flexibility-of-firing index	42	Cost (% of income per capita)	7.5
Employment-law index	62	Procedural-complexity index	56
Getting Credit		**Closing a Business**	
Public credit registry operates?	No	Time to go through insolvency (years)	1.2
Public registry coverage (borrowers/1,000 capita)	0	Cost to go through insolvency (% estate)	4
Public-registry index	0	Absolute priority preserved	100
Private credit-information bureau operates?	No	Efficient outcome achieved	1
Private bureau coverage (borrowers/1,000 capita)	0	Goals-of-insolvency index	92
Creditor-rights index	3	Court-powers index	67

LEBANON
Middle East and North Africa

Economy Characteristics		Starting a Business	
GNI per capita (US$)	3,990	Number of procedures	6
Population	4,384,680	Time (days)	46
Informal economy (% of income)	34.1	Cost (% of income per capita)	129.9
Legal origin	French	Minimum capital (% of income per capita)	83.1
Hiring and Firing Workers		**Enforcing a Contract**	
Flexibility-of-hiring index	53	Number of procedures	27
Conditions-of-employment index	50	Time (days)	721
Flexibility-of-firing index	35	Cost (% of income per capita)	54.3
Employment-law index	46	Procedural-complexity index	67
Getting Credit		**Closing a Business**	
Public credit registry operates?	No	Time to go through insolvency (years)	4.0
Public registry coverage (borrowers/1,000 capita)	0	Cost to go through insolvency (% estate)	18
Public-registry index	0	Absolute priority preserved	33
Private credit-information bureau operates?	No	Efficient outcome achieved	0
Private bureau coverage (borrowers/1,000 capita)	0	Goals-of-insolvency index	31
Creditor-rights index	4	Court-powers index	67

Note: .. means no data available.

LESOTHO
Sub-Saharan Africa

Economy Characteristics		Starting a Business	
GNI per capita (US$)	470	Number of procedures	9
Population	2,061,730	Time (days)	92
Informal economy (% of income)	..	Cost (% of income per capita)	67.4
Legal origin	English	Minimum capital (% of income per capita)	20.2
Hiring and Firing Workers		**Enforcing a Contract**	
Flexibility-of-hiring index	58	Number of procedures	..
Conditions-of-employment index	51	Time (days)	..
Flexibility-of-firing index	25	Cost (% of income per capita)	..
Employment-law index	45	Procedural-complexity index	..
Getting Credit		**Closing a Business**	
Public credit registry operates?	No	Time to go through insolvency (years)	..
Public registry coverage (borrowers/1,000 capita)	0	Cost to go through insolvency (% estate)	..
Public-registry index	0	Absolute priority preserved	..
Private credit-information bureau operates?	No	Efficient outcome achieved	..
Private bureau coverage (borrowers/1,000 capita)	0	Goals-of-insolvency index	..
Creditor-rights index	2	Court-powers index	..

LITHUANIA
Europe and Central Asia

Economy Characteristics		Starting a Business	
GNI per capita (US$)	3,660	Number of procedures	9
Population	3,482,000	Time (days)	26
Informal economy (% of income)	30.3	Cost (% of income per capita)	6.3
Legal origin	French	Minimum capital (% of income per capita)	74.4
Hiring and Firing Workers		**Enforcing a Contract**	
Flexibility-of-hiring index	71	Number of procedures	17
Conditions-of-employment index	90	Time (days)	74
Flexibility-of-firing index	31	Cost (% of income per capita)	13.0
Employment-law index	64	Procedural-complexity index	58
Getting Credit		**Closing a Business**	
Public credit registry operates?	Yes	Time to go through insolvency (years)	1.2
Public registry coverage (borrowers/1,000 capita)	7	Cost to go through insolvency (% estate)	18
Public-registry index	63	Absolute priority preserved	100
Private credit-information bureau operates?	No	Efficient outcome achieved	0
Private bureau coverage (borrowers/1,000 capita)	0	Goals-of-insolvency index	54
Creditor-rights index	2	Court-powers index	67

MACEDONIA, FYR
Europe and Central Asia

Economy Characteristics		Starting a Business	
GNI per capita (US$)	1,700	Number of procedures	13
Population	2,035,000	Time (days)	48
Informal economy (% of income)	..	Cost (% of income per capita)	13.1
Legal origin	German	Minimum capital (% of income per capita)	138.4
Hiring and Firing Workers		**Enforcing a Contract**	
Flexibility-of-hiring index	65	Number of procedures	27
Conditions-of-employment index	53	Time (days)	509
Flexibility-of-firing index	32	Cost (% of income per capita)	43.0
Employment-law index	50	Procedural-complexity index	67
Getting Credit		**Closing a Business**	
Public credit registry operates?	Yes	Time to go through insolvency (years)	3.6
Public registry coverage (borrowers/1,000 capita)	2	Cost to go through insolvency (% estate)	38
Public-registry index	42	Absolute priority preserved	67
Private credit-information bureau operates?	No	Efficient outcome achieved	0
Private bureau coverage (borrowers/1,000 capita)	0	Goals-of-insolvency index	34
Creditor-rights index	3	Court-powers index	67

Note: .. means no data available.

MADAGASCAR
Sub-Saharan Africa

Economy Characteristics		Starting a Business	
GNI per capita (US$)	240	Number of procedures	15
Population	15,975,750	Time (days)	67
Informal economy (% of income)	39.6	Cost (% of income per capita)	62.8
Legal origin	French	Minimum capital (% of income per capita)	30.5
Hiring and Firing Workers		**Enforcing a Contract**	
Flexibility-of-hiring index	48	Number of procedures	29
Conditions-of-employment index	86	Time (days)	166
Flexibility-of-firing index	49	Cost (% of income per capita)	120.2
Employment-law index	61	Procedural-complexity index	63
Getting Credit		**Closing a Business**	
Public credit registry operates?	Yes	Time to go through insolvency (years)	No practice
Public registry coverage (borrowers/1,000 capita)	2	Cost to go through insolvency (% estate)	No practice
Public-registry index	46	Absolute priority preserved	100
Private credit-information bureau operates?	No	Efficient outcome achieved	0
Private bureau coverage (borrowers/1,000 capita)	0	Goals-of-insolvency index	25
Creditor-rights index	2	Court-powers index	67

MALAWI
Sub-Saharan Africa

Economy Characteristics		Starting a Business	
GNI per capita (US$)	160	Number of procedures	11
Population	10,526,300	Time (days)	45
Informal economy (% of income)	40.3	Cost (% of income per capita)	125.4
Legal origin	English	Minimum capital (% of income per capita)	0.0
Hiring and Firing Workers		**Enforcing a Contract**	
Flexibility-of-hiring index	33	Number of procedures	16
Conditions-of-employment index	68	Time (days)	108
Flexibility-of-firing index	54	Cost (% of income per capita)	520.6
Employment-law index	52	Procedural-complexity index	48
Getting Credit		**Closing a Business**	
Public credit registry operates?	No	Time to go through insolvency (years)	2.8
Public registry coverage (borrowers/1,000 capita)	0	Cost to go through insolvency (% estate)	8
Public-registry index	0	Absolute priority preserved	33
Private credit-information bureau operates?	No	Efficient outcome achieved	0
Private bureau coverage (borrowers/1,000 capita)	0	Goals-of-insolvency index	40
Creditor-rights index	2	Court-powers index	67

MALAYSIA
East Asia and Pacific

Economy Characteristics		Starting a Business	
GNI per capita (US$)	3,540	Number of procedures	8
Population	23,802,360	Time (days)	31
Informal economy (% of income)	31.1	Cost (% of income per capita)	27.1
Legal origin	English	Minimum capital (% of income per capita)	0.0
Hiring and Firing Workers		**Enforcing a Contract**	
Flexibility-of-hiring index	33	Number of procedures	22
Conditions-of-employment index	26	Time (days)	270
Flexibility-of-firing index	15	Cost (% of income per capita)	19.4
Employment-law index	25	Procedural-complexity index	41
Getting Credit		**Closing a Business**	
Public credit registry operates?	Yes	Time to go through insolvency (years)	2.2
Public registry coverage (borrowers/1,000 capita)	105	Cost to go through insolvency (% estate)	18
Public-registry index	59	Absolute priority preserved	100
Private credit-information bureau operates?	Yes	Efficient outcome achieved	0
Private bureau coverage (borrowers/1,000 capita)	461	Goals-of-insolvency index	52
Creditor-rights index	2	Court-powers index	33

Note: .. means no data available.

MALI
Sub-Saharan Africa

Economy Characteristics		Starting a Business	
GNI per capita (US$)	240	Number of procedures	13
Population	11,094,340	Time (days)	61
Informal economy (% of income)	41	Cost (% of income per capita)	232.2
Legal origin	French	Minimum capital (% of income per capita)	597.8
Hiring and Firing Workers		**Enforcing a Contract**	
Flexibility-of-hiring index	53	Number of procedures	27
Conditions-of-employment index	86	Time (days)	150
Flexibility-of-firing index	23	Cost (% of income per capita)	7.0
Employment-law index	54	Procedural-complexity index	71
Getting Credit		**Closing a Business**	
Public credit registry operates?	Yes	Time to go through insolvency (years)	3.5
Public registry coverage (borrowers/1,000 capita)	1	Cost to go through insolvency (% estate)	18
Public-registry index	22	Absolute priority preserved	33
Private credit-information bureau operates?	No	Efficient outcome achieved	0
Private bureau coverage (borrowers/1,000 capita)	0	Goals-of-insolvency index	32
Creditor-rights index	1	Court-powers index	100

MAURITANIA
Sub-Saharan Africa

Economy Characteristics		Starting a Business	
GNI per capita (US$)	410	Number of procedures	11
Population	2,749,150	Time (days)	73
Informal economy (% of income)	..	Cost (% of income per capita)	110.2
Legal origin	French	Minimum capital (% of income per capita)	896.7
Hiring and Firing Workers		**Enforcing a Contract**	
Flexibility-of-hiring index	62	Number of procedures	..
Conditions-of-employment index	47	Time (days)	..
Flexibility-of-firing index	66	Cost (% of income per capita)	..
Employment-law index	59	Procedural-complexity index	..
Getting Credit		**Closing a Business**	
Public credit registry operates?	..	Time to go through insolvency (years)	8.0
Public registry coverage (borrowers/1,000 capita)	..	Cost to go through insolvency (% estate)	8
Public-registry index	..	Absolute priority preserved	33
Private credit-information bureau operates?	No	Efficient outcome achieved	0
Private bureau coverage (borrowers/1,000 capita)	0	Goals-of-insolvency index	28
Creditor-rights index	3	Court-powers index	67

MEXICO
Latin America and Caribbean

Economy Characteristics		Starting a Business	
GNI per capita (US$)	5,910	Number of procedures	7
Population	99,419,688	Time (days)	51
Informal economy (% of income)	30.1	Cost (% of income per capita)	18.8
Legal origin	French	Minimum capital (% of income per capita)	87.6
Hiring and Firing Workers		**Enforcing a Contract**	
Flexibility-of-hiring index	81	Number of procedures	47
Conditions-of-employment index	81	Time (days)	325
Flexibility-of-firing index	70	Cost (% of income per capita)	10.0
Employment-law index	77	Procedural-complexity index	62
Getting Credit		**Closing a Business**	
Public credit registry operates?	No	Time to go through insolvency (years)	2.0
Public registry coverage (borrowers/1,000 capita)	0	Cost to go through insolvency (% estate)	18
Public-registry index	0	Absolute priority preserved	33
Private credit-information bureau operates?	Yes	Efficient outcome achieved	1
Private bureau coverage (borrowers/1,000 capita)	382	Goals-of-insolvency index	61
Creditor-rights index	0	Court-powers index	67

Note: .. means no data available.

MOLDOVA
Europe and Central Asia

Economy Characteristics		Starting a Business	
GNI per capita (US$)	460	Number of procedures	11
Population	4,270,000	Time (days)	42
Informal economy (% of income)	45.1	Cost (% of income per capita)	26.2
Legal origin	Socialist	Minimum capital (% of income per capita)	86.3
Hiring and Firing Workers		**Enforcing a Contract**	
Flexibility-of-hiring index	71	Number of procedures	36
Conditions-of-employment index	75	Time (days)	210
Flexibility-of-firing index	54	Cost (% of income per capita)	14.2
Employment-law index	67	Procedural-complexity index	48
Getting Credit		**Closing a Business**	
Public credit registry operates?	No	Time to go through insolvency (years)	2.8
Public registry coverage (borrowers/1,000 capita)	0	Cost to go through insolvency (% estate)	8
Public-registry index	0	Absolute priority preserved	67
Private credit-information bureau operates?	No	Efficient outcome achieved	0
Private bureau coverage (borrowers/1,000 capita)	0	Goals-of-insolvency index	49
Creditor-rights index	2	Court-powers index	67

MONGOLIA
East Asia and Pacific

Economy Characteristics		Starting a Business	
GNI per capita (US$)	440	Number of procedures	8
Population	2,421,360	Time (days)	31
Informal economy (% of income)	18.4	Cost (% of income per capita)	12.0
Legal origin	Socialist	Minimum capital (% of income per capita)	2046.9
Hiring and Firing Workers		**Enforcing a Contract**	
Flexibility-of-hiring index	33	Number of procedures	26
Conditions-of-employment index	90	Time (days)	224
Flexibility-of-firing index	25	Cost (% of income per capita)	1.8
Employment-law index	50	Procedural-complexity index	71
Getting Credit		**Closing a Business**	
Public credit registry operates?	Yes	Time to go through insolvency (years)	4.0
Public registry coverage (borrowers/1,000 capita)	15	Cost to go through insolvency (% estate)	8
Public-registry index	68	Absolute priority preserved	100
Private credit-information bureau operates?	No	Efficient outcome achieved	0
Private bureau coverage (borrowers/1,000 capita)	0	Goals-of-insolvency index	54
Creditor-rights index	1	Court-powers index	67

MOROCCO
Middle East and North Africa

Economy Characteristics		Starting a Business	
GNI per capita (US$)	1,190	Number of procedures	11
Population	29,170,000	Time (days)	36
Informal economy (% of income)	36.4	Cost (% of income per capita)	19.1
Legal origin	French	Minimum capital (% of income per capita)	762.5
Hiring and Firing Workers		**Enforcing a Contract**	
Flexibility-of-hiring index	56	Number of procedures	17
Conditions-of-employment index	63	Time (days)	192
Flexibility-of-firing index	33	Cost (% of income per capita)	9.1
Employment-law index	51	Procedural-complexity index	69
Getting Credit		**Closing a Business**	
Public credit registry operates?	Yes	Time to go through insolvency (years)	1.9
Public registry coverage (borrowers/1,000 capita)	..	Cost to go through insolvency (% estate)	18
Public-registry index	33	Absolute priority preserved	33
Private credit-information bureau operates?	No	Efficient outcome achieved	0
Private bureau coverage (borrowers/1,000 capita)	0	Goals-of-insolvency index	36
Creditor-rights index	1	Court-powers index	100

Note: .. means no data available.

MOZAMBIQUE
Sub-Saharan Africa

Economy Characteristics		Starting a Business	
GNI per capita (US$)	210	Number of procedures	15
Population	18,071,160	Time (days)	153
Informal economy (% of income)	40.3	Cost (% of income per capita)	99.6
Legal origin	French	Minimum capital (% of income per capita)	30.2
Hiring and Firing Workers		**Enforcing a Contract**	
Flexibility-of-hiring index	73	Number of procedures	18
Conditions-of-employment index	85	Time (days)	540
Flexibility-of-firing index	64	Cost (% of income per capita)	9.1
Employment-law index	74	Procedural-complexity index	71
Getting Credit		**Closing a Business**	
Public credit registry operates?	Yes	Time to go through insolvency (years)	No practice
Public registry coverage (borrowers/1,000 capita)	1	Cost to go through insolvency (% estate)	No practice
Public-registry index	52	Absolute priority preserved	100
Private credit-information bureau operates?	No	Efficient outcome achieved	0
Private bureau coverage (borrowers/1,000 capita)	0	Goals-of-insolvency index	25
Creditor-rights index	2	Court-powers index	67

NAMIBIA
Sub-Saharan Africa

Economy Characteristics		Starting a Business	
GNI per capita (US$)	1,780	Number of procedures	10
Population	1,792,060	Time (days)	85
Informal economy (% of income)	..	Cost (% of income per capita)	18.7
Legal origin	English	Minimum capital (% of income per capita)	0.0
Hiring and Firing Workers		**Enforcing a Contract**	
Flexibility-of-hiring index	17	Number of procedures	..
Conditions-of-employment index	57	Time (days)	..
Flexibility-of-firing index	54	Cost (% of income per capita)	..
Employment-law index	43	Procedural-complexity index	..
Getting Credit		**Closing a Business**	
Public credit registry operates?	No	Time to go through insolvency (years)	..
Public registry coverage (borrowers/1,000 capita)	0	Cost to go through insolvency (% estate)	..
Public-registry index	0	Absolute priority preserved	..
Private credit-information bureau operates?	..	Efficient outcome achieved	..
Private bureau coverage (borrowers/1,000 capita)	..	Goals-of-insolvency index	..
Creditor-rights index	..	Court-powers index	..

NEPAL
South Asia

Economy Characteristics		Starting a Business	
GNI per capita (US$)	230	Number of procedures	8
Population	23,584,710	Time (days)	25
Informal economy (% of income)	38.4	Cost (% of income per capita)	191.0
Legal origin	English	Minimum capital (% of income per capita)	0.0
Hiring and Firing Workers		**Enforcing a Contract**	
Flexibility-of-hiring index	33	Number of procedures	24
Conditions-of-employment index	54	Time (days)	350
Flexibility-of-firing index	47	Cost (% of income per capita)	44.2
Employment-law index	45	Procedural-complexity index	63
Getting Credit		**Closing a Business**	
Public credit registry operates?	No	Time to go through insolvency (years)	5.0
Public registry coverage (borrowers/1,000 capita)	0	Cost to go through insolvency (% estate)	8
Public-registry index	0	Absolute priority preserved	33
Private credit-information bureau operates?	No	Efficient outcome achieved	0
Private bureau coverage (borrowers/1,000 capita)	0	Goals-of-insolvency index	35
Creditor-rights index	2	Court-powers index	33

Note: .. means no data available.

NETHERLANDS
OECD: High Income

Economy Characteristics		Starting a Business	
GNI per capita (US$)	23,960	Number of procedures	7
Population	16,039,000	Time (days)	11
Informal economy (% of income)	13	Cost (% of income per capita)	13.7
Legal origin	French	Minimum capital (% of income per capita)	70.7
Hiring and Firing Workers		**Enforcing a Contract**	
Flexibility-of-hiring index	51	Number of procedures	21
Conditions-of-employment index	79	Time (days)	39
Flexibility-of-firing index	33	Cost (% of income per capita)	0.5
Employment-law index	54	Procedural-complexity index	46
Getting Credit		**Closing a Business**	
Public credit registry operates?	No	Time to go through insolvency (years)	2.6
Public registry coverage (borrowers/1,000 capita)	0	Cost to go through insolvency (% estate)	1
Public-registry index	0	Absolute priority preserved	100
Private credit-information bureau operates?	Yes	Efficient outcome achieved	1
Private bureau coverage (borrowers/1,000 capita)	530	Goals-of-insolvency index	95
Creditor-rights index	3	Court-powers index	33

NEW ZEALAND
OECD: High Income

Economy Characteristics		Starting a Business	
GNI per capita (US$)	13,710	Number of procedures	3
Population	3,849,000	Time (days)	3
Informal economy (% of income)	12.7	Cost (% of income per capita)	0.2
Legal origin	English	Minimum capital (% of income per capita)	0.0
Hiring and Firing Workers		**Enforcing a Contract**	
Flexibility-of-hiring index	33	Number of procedures	19
Conditions-of-employment index	43	Time (days)	50
Flexibility-of-firing index	20	Cost (% of income per capita)	11.6
Employment-law index	32	Procedural-complexity index	31
Getting Credit		**Closing a Business**	
Public credit registry operates?	No	Time to go through insolvency (years)	2.0
Public registry coverage (borrowers/1,000 capita)	0	Cost to go through insolvency (% estate)	4
Public-registry index	0	Absolute priority preserved	100
Private credit-information bureau operates?	Yes	Efficient outcome achieved	1
Private bureau coverage (borrowers/1,000 capita)	818	Goals-of-insolvency index	90
Creditor-rights index	4	Court-powers index	0

NICARAGUA
Latin America and Caribbean

Economy Characteristics		Starting a Business	
GNI per capita (US$)	395	Number of procedures	12
Population	5,205,000	Time (days)	71
Informal economy (% of income)	45.2	Cost (% of income per capita)	337.8
Legal origin	French	Minimum capital (% of income per capita)	0.0
Hiring and Firing Workers		**Enforcing a Contract**	
Flexibility-of-hiring index	33	Number of procedures	17
Conditions-of-employment index	90	Time (days)	125
Flexibility-of-firing index	58	Cost (% of income per capita)	17.7
Employment-law index	61	Procedural-complexity index	79
Getting Credit		**Closing a Business**	
Public credit registry operates?	Yes	Time to go through insolvency (years)	2.3
Public registry coverage (borrowers/1,000 capita)	50	Cost to go through insolvency (% estate)	8
Public-registry index	45	Absolute priority preserved	100
Private credit-information bureau operates?	No	Efficient outcome achieved	0
Private bureau coverage (borrowers/1,000 capita)	0	Goals-of-insolvency index	58
Creditor-rights index	4	Court-powers index	67

Note: .. means no data available.

NIGER
Sub-Saharan Africa

Economy Characteristics		Starting a Business	
GNI per capita (US$)	170	Number of procedures	11
Population	11,184,130	Time (days)	27
Informal economy (% of income)	41.9	Cost (% of income per capita)	446.6
Legal origin	French	Minimum capital (% of income per capita)	844.0

Hiring and Firing Workers		Enforcing a Contract	
Flexibility-of-hiring index	53	Number of procedures	29
Conditions-of-employment index	89	Time (days)	365
Flexibility-of-firing index	34	Cost (% of income per capita)	57.1
Employment-law index	59	Procedural-complexity index	63

Getting Credit		Closing a Business	
Public credit registry operates?	Yes	Time to go through insolvency (years)	5.0
Public registry coverage (borrowers/1,000 capita)	1	Cost to go through insolvency (% estate)	18
Public-registry index	22	Absolute priority preserved	67
Private credit-information bureau operates?	No	Efficient outcome achieved	0
Private bureau coverage (borrowers/1,000 capita)	0	Goals-of-insolvency index	37
Creditor-rights index	1	Court-powers index	100

NIGERIA
Sub-Saharan Africa

Economy Characteristics		Starting a Business	
GNI per capita (US$)	290	Number of procedures	10
Population	129,874,976	Time (days)	44
Informal economy (% of income)	57.9	Cost (% of income per capita)	92.3
Legal origin	English	Minimum capital (% of income per capita)	28.6

Hiring and Firing Workers		Enforcing a Contract	
Flexibility-of-hiring index	17	Number of procedures	23
Conditions-of-employment index	76	Time (days)	730
Flexibility-of-firing index	36	Cost (% of income per capita)	6.6
Employment-law index	43	Procedural-complexity index	52

Getting Credit		Closing a Business	
Public credit registry operates?	Yes	Time to go through insolvency (years)	1.6
Public registry coverage (borrowers/1,000 capita)	<1	Cost to go through insolvency (% estate)	18
Public-registry index	55	Absolute priority preserved	67
Private credit-information bureau operates?	No	Efficient outcome achieved	0
Private bureau coverage (borrowers/1,000 capita)	0	Goals-of-insolvency index	45
Creditor-rights index	4	Court-powers index	67

NORWAY
OECD: High Income

Economy Characteristics		Starting a Business	
GNI per capita (US$)	37,850	Number of procedures	4
Population	4,513,000	Time (days)	24
Informal economy (% of income)	4,519,398	Cost (% of income per capita)	3.9
Legal origin	Nordic	Minimum capital (% of income per capita)	33.1

Hiring and Firing Workers		Enforcing a Contract	
Flexibility-of-hiring index	58	Number of procedures	12
Conditions-of-employment index	39	Time (days)	87
Flexibility-of-firing index	25	Cost (% of income per capita)	10.4
Employment-law index	41	Procedural-complexity index	48

Getting Credit		Closing a Business	
Public credit registry operates?	No	Time to go through insolvency (years)	0.9
Public registry coverage (borrowers/1,000 capita)	0	Cost to go through insolvency (% estate)	1
Public-registry index	0	Absolute priority preserved	100
Private credit-information bureau operates?	Yes	Efficient outcome achieved	1
Private bureau coverage (borrowers/1,000 capita)	945	Goals-of-insolvency index	99
Creditor-rights index	2	Court-powers index	67

Note: .. means no data available.

OMAN
Middle East and North Africa

Economy Characteristics		Starting a Business	
GNI per capita (US$)	..	Number of procedures	9
Population	2,478,000	Time (days)	34
Informal economy (% of income)	..	Cost (% of income per capita)	5.3
Legal origin	French	Minimum capital (% of income per capita)	720.9
Hiring and Firing Workers		**Enforcing a Contract**	
Flexibility-of-hiring index	58	Number of procedures	54
Conditions-of-employment index	78	Time (days)	250
Flexibility-of-firing index	25	Cost (% of income per capita)	4.8
Employment-law index	54	Procedural-complexity index	51
Getting Credit		**Closing a Business**	
Public credit registry operates?	No	Time to go through insolvency (years)	7.0
Public registry coverage (borrowers/1,000 capita)	0	Cost to go through insolvency (% estate)	4
Public-registry index	0	Absolute priority preserved	0
Private credit-information bureau operates?	No	Efficient outcome achieved	0
Private bureau coverage (borrowers/1,000 capita)	0	Goals-of-insolvency index	29
Creditor-rights index	0	Court-powers index	67

PAKISTAN
South Asia

Economy Characteristics		Starting a Business	
GNI per capita (US$)	410	Number of procedures	10
Population	141,450,144	Time (days)	22
Informal economy (% of income)	36.8	Cost (% of income per capita)	46.8
Legal origin	English	Minimum capital (% of income per capita)	0.0
Hiring and Firing Workers		**Enforcing a Contract**	
Flexibility-of-hiring index	65	Number of procedures	30
Conditions-of-employment index	75	Time (days)	365
Flexibility-of-firing index	33	Cost (% of income per capita)	45.8
Employment-law index	58	Procedural-complexity index	53
Getting Credit		**Closing a Business**	
Public credit registry operates?	Yes	Time to go through insolvency (years)	2.8
Public registry coverage (borrowers/1,000 capita)	1	Cost to go through insolvency (% estate)	4
Public-registry index	42	Absolute priority preserved	100
Private credit-information bureau operates?	Yes	Efficient outcome achieved	0
Private bureau coverage (borrowers/1,000 capita)	<1	Goals-of-insolvency index	63
Creditor-rights index	1	Court-powers index	33

PANAMA
Latin America and Caribbean

Economy Characteristics		Starting a Business	
GNI per capita (US$)	4,020	Number of procedures	7
Population	2,897,000	Time (days)	19
Informal economy (% of income)	64.1	Cost (% of income per capita)	26.3
Legal origin	French	Minimum capital (% of income per capita)	0.0
Hiring and Firing Workers		**Enforcing a Contract**	
Flexibility-of-hiring index	81	Number of procedures	44
Conditions-of-employment index	87	Time (days)	197
Flexibility-of-firing index	68	Cost (% of income per capita)	20.0
Employment-law index	79	Procedural-complexity index	82
Getting Credit		**Closing a Business**	
Public credit registry operates?	No	Time to go through insolvency (years)	6.5
Public registry coverage (borrowers/1,000 capita)	0	Cost to go through insolvency (% estate)	38
Public-registry index	0	Absolute priority preserved	100
Private credit-information bureau operates?	Yes	Efficient outcome achieved	0
Private bureau coverage (borrowers/1,000 capita)	302	Goals-of-insolvency index	36
Creditor-rights index	4	Court-powers index	33

Note: .. means no data available.

PAPUA NEW GUINEA
East Asia and Pacific

Economy Characteristics		Starting a Business	
GNI per capita (US$)	530	Number of procedures	7
Population	5,252,530	Time (days)	69
Informal economy (% of income)	..	Cost (% of income per capita)	26.4
Legal origin	English	Minimum capital (% of income per capita)	0.0
Hiring and Firing Workers		**Enforcing a Contract**	
Flexibility-of-hiring index	17	Number of procedures	22
Conditions-of-employment index	57	Time (days)	270
Flexibility-of-firing index	4	Cost (% of income per capita)	41.1
Employment-law index	26	Procedural-complexity index	45
Getting Credit		**Closing a Business**	
Public credit registry operates?	No	Time to go through insolvency (years)	..
Public registry coverage (borrowers/1,000 capita)	0	Cost to go through insolvency (% estate)	..
Public-registry index	0	Absolute priority preserved	..
Private credit-information bureau operates?	No	Efficient outcome achieved	..
Private bureau coverage (borrowers/1,000 capita)	0	Goals-of-insolvency index	..
Creditor-rights index	2	Court-powers index	..

PARAGUAY
Latin America and Caribbean

Economy Characteristics		Starting a Business	
GNI per capita (US$)	1,170	Number of procedures	18
Population	5,390,000	Time (days)	73
Informal economy (% of income)	..	Cost (% of income per capita)	160.9
Legal origin	French	Minimum capital (% of income per capita)	0.0
Hiring and Firing Workers		**Enforcing a Contract**	
Flexibility-of-hiring index	58	Number of procedures	46
Conditions-of-employment index	90	Time (days)	188
Flexibility-of-firing index	71	Cost (% of income per capita)	34.0
Employment-law index	73	Procedural-complexity index	67
Getting Credit		**Closing a Business**	
Public credit registry operates?	..	Time to go through insolvency (years)	3.9
Public registry coverage (borrowers/1,000 capita)	..	Cost to go through insolvency (% estate)	8
Public-registry index	..	Absolute priority preserved	67
Private credit-information bureau operates?	..	Efficient outcome achieved	0
Private bureau coverage (borrowers/1,000 capita)	..	Goals-of-insolvency index	46
Creditor-rights index	2	Court-powers index	67

PERU
Latin America and Caribbean

Economy Characteristics		Starting a Business	
GNI per capita (US$)	2,050	Number of procedures	9
Population	26,347,000	Time (days)	100
Informal economy (% of income)	59.9	Cost (% of income per capita)	24.9
Legal origin	French	Minimum capital (% of income per capita)	0.0
Hiring and Firing Workers		**Enforcing a Contract**	
Flexibility-of-hiring index	71	Number of procedures	35
Conditions-of-employment index	81	Time (days)	441
Flexibility-of-firing index	69	Cost (% of income per capita)	29.7
Employment-law index	73	Procedural-complexity index	82
Getting Credit		**Closing a Business**	
Public credit registry operates?	Yes	Time to go through insolvency (years)	2.1
Public registry coverage (borrowers/1,000 capita)	92	Cost to go through insolvency (% estate)	8
Public-registry index	54	Absolute priority preserved	33
Private credit-information bureau operates?	Yes	Efficient outcome achieved	1
Private bureau coverage (borrowers/1,000 capita)	185	Goals-of-insolvency index	67
Creditor-rights index	0	Court-powers index	33

Note: .. means no data available.

PHILIPPINES
East Asia and Pacific

Economy Characteristics		Starting a Business	
GNI per capita (US$)	1,020	Number of procedures	11
Population	78,317,032	Time (days)	59
Informal economy (% of income)	43.4	Cost (% of income per capita)	24.4
Legal origin	French	Minimum capital (% of income per capita)	9.5
Hiring and Firing Workers		**Enforcing a Contract**	
Flexibility-of-hiring index	58	Number of procedures	28
Conditions-of-employment index	73	Time (days)	164
Flexibility-of-firing index	50	Cost (% of income per capita)	103.7
Employment-law index	60	Procedural-complexity index	75
Getting Credit		**Closing a Business**	
Public credit registry operates?	No	Time to go through insolvency (years)	5.7
Public registry coverage (borrowers/1,000 capita)	0	Cost to go through insolvency (% estate)	38
Public-registry index	0	Absolute priority preserved	100
Private credit-information bureau operates?	Yes	Efficient outcome achieved	0
Private bureau coverage (borrowers/1,000 capita)	22	Goals-of-insolvency index	38
Creditor-rights index	1	Court-powers index	100

POLAND
Europe and Central Asia

Economy Characteristics		Starting a Business	
GNI per capita (US$)	4,570	Number of procedures	12
Population	38,641,000	Time (days)	31
Informal economy (% of income)	27.6	Cost (% of income per capita)	20.3
Legal origin	German	Minimum capital (% of income per capita)	21.4
Hiring and Firing Workers		**Enforcing a Contract**	
Flexibility-of-hiring index	33	Number of procedures	18
Conditions-of-employment index	92	Time (days)	1000
Flexibility-of-firing index	39	Cost (% of income per capita)	11.2
Employment-law index	55	Procedural-complexity index	65
Getting Credit		**Closing a Business**	
Public credit registry operates?	No	Time to go through insolvency (years)	1.5
Public registry coverage (borrowers/1,000 capita)	0	Cost to go through insolvency (% estate)	18
Public-registry index	0	Absolute priority preserved	67
Private credit-information bureau operates?	Yes	Efficient outcome achieved	1
Private bureau coverage (borrowers/1,000 capita)	543	Goals-of-insolvency index	70
Creditor-rights index	2	Court-powers index	67

PORTUGAL
OECD: High Income

Economy Characteristics		Starting a Business	
GNI per capita (US$)	10,840	Number of procedures	11
Population	10,024,000	Time (days)	95
Informal economy (% of income)	22.6	Cost (% of income per capita)	12.5
Legal origin	French	Minimum capital (% of income per capita)	43.4
Hiring and Firing Workers		**Enforcing a Contract**	
Flexibility-of-hiring index	76	Number of procedures	22
Conditions-of-employment index	88	Time (days)	420
Flexibility-of-firing index	73	Cost (% of income per capita)	4.9
Employment-law index	79	Procedural-complexity index	54
Getting Credit		**Closing a Business**	
Public credit registry operates?	Yes	Time to go through insolvency (years)	2.6
Public registry coverage (borrowers/1,000 capita)	496	Cost to go through insolvency (% estate)	8
Public-registry index	61	Absolute priority preserved	33
Private credit-information bureau operates?	Yes	Efficient outcome achieved	1
Private bureau coverage (borrowers/1,000 capita)	24	Goals-of-insolvency index	66
Creditor-rights index	1	Court-powers index	33

Note: .. means no data available.

PUERTO RICO
Latin America and Caribbean

Economy Characteristics		Starting a Business	
GNI per capita (US$)	..	Number of procedures	6
Population	3,840,000	Time (days)	6
Informal economy (% of income)	..	Cost (% of income per capita)	2.8
Legal origin	French	Minimum capital (% of income per capita)	0.0
Hiring and Firing Workers		**Enforcing a Contract**	
Flexibility-of-hiring index	33	Number of procedures	55
Conditions-of-employment index	67	Time (days)	365
Flexibility-of-firing index	24	Cost (% of income per capita)	20.9
Employment-law index	41	Procedural-complexity index	52
Getting Credit		**Closing a Business**	
Public credit registry operates?	No	Time to go through insolvency (years)	3.8
Public registry coverage (borrowers/1,000 capita)	0	Cost to go through insolvency (% estate)	8
Public-registry index	0	Absolute priority preserved	67
Private credit-information bureau operates?	..	Efficient outcome achieved	1
Private bureau coverage (borrowers/1,000 capita)	..	Goals-of-insolvency index	71
Creditor-rights index	1	Court-powers index	33

ROMANIA
Europe and Central Asia

Economy Characteristics		Starting a Business	
GNI per capita (US$)	1,850	Number of procedures	6
Population	22,408,000	Time (days)	27
Informal economy (% of income)	34.4	Cost (% of income per capita)	11.7
Legal origin	French	Minimum capital (% of income per capita)	3.3
Hiring and Firing Workers		**Enforcing a Contract**	
Flexibility-of-hiring index	48	Number of procedures	28
Conditions-of-employment index	85	Time (days)	225
Flexibility-of-firing index	29	Cost (% of income per capita)	13.1
Employment-law index	54	Procedural-complexity index	60
Getting Credit		**Closing a Business**	
Public credit registry operates?	Yes	Time to go through insolvency (years)	3.2
Public registry coverage (borrowers/1,000 capita)	1	Cost to go through insolvency (% estate)	8
Public-registry index	59	Absolute priority preserved	33
Private credit-information bureau operates?	No	Efficient outcome achieved	0
Private bureau coverage (borrowers/1,000 capita)	0	Goals-of-insolvency index	39
Creditor-rights index	0	Court-powers index	33

RUSSIAN FEDERATION
Europe and Central Asia

Economy Characteristics		Starting a Business	
GNI per capita (US$)	2,140	Number of procedures	12
Population	144,752,000	Time (days)	29
Informal economy (% of income)	46.1	Cost (% of income per capita)	9.3
Legal origin	Socialist	Minimum capital (% of income per capita)	29.8
Hiring and Firing Workers		**Enforcing a Contract**	
Flexibility-of-hiring index	33	Number of procedures	16
Conditions-of-employment index	77	Time (days)	160
Flexibility-of-firing index	71	Cost (% of income per capita)	20.2
Employment-law index	61	Procedural-complexity index	48
Getting Credit		**Closing a Business**	
Public credit registry operates?	No	Time to go through insolvency (years)	1.53
Public registry coverage (borrowers/1,000 capita)	0	Cost to go through insolvency (% estate)	4
Public-registry index	0	Absolute priority preserved	67
Private credit-information bureau operates?	No	Efficient outcome achieved	0
Private bureau coverage (borrowers/1,000 capita)	0	Goals-of-insolvency index	58
Creditor-rights index	2	Court-powers index	67

Note: .. means no data available.

RWANDA
Sub-Saharan Africa

Economy Characteristics	
GNI per capita (US$)	230
Population	7,933,000
Informal economy (% of income)	..
Legal origin	French

Hiring and Firing Workers	
Flexibility-of-hiring index	53
Conditions-of-employment index	94
Flexibility-of-firing index	32
Employment-law index	60

Getting Credit	
Public credit registry operates?	Yes
Public registry coverage (borrowers/1,000 capita)	<1
Public-registry index	57
Private credit-information bureau operates?	No
Private bureau coverage (borrowers/1,000 capita)	0
Creditor-rights index	1

Starting a Business	
Number of procedures	9
Time (days)	43
Cost (% of income per capita)	232.3
Minimum capital (% of income per capita)	457.3

Enforcing a Contract	
Number of procedures	..
Time (days)	..
Cost (% of income per capita)	..
Procedural-complexity index	..

Closing a Business	
Time to go through insolvency (years)	No practice
Cost to go through insolvency (% estate)	No practice
Absolute priority preserved	33
Efficient outcome achieved	0
Goals-of-insolvency index	8
Court-powers index	33

SAUDI ARABIA
Middle East and North Africa

Economy Characteristics	
GNI per capita (US$)	7,065
Population	21,408,470
Informal economy (% of income)	18.4
Legal origin	English

Hiring and Firing Workers	
Flexibility-of-hiring index	33
Conditions-of-employment index	58
Flexibility-of-firing index	16
Employment-law index	36

Getting Credit	
Public credit registry operates?	Yes
Public registry coverage (borrowers/1,000 capita)	<1
Public-registry index	42
Private credit-information bureau operates?	Developing
Private bureau coverage (borrowers/1,000 capita)	0
Creditor-rights index	2

Starting a Business	
Number of procedures	14
Time (days)	95
Cost (% of income per capita)	130.5
Minimum capital (% of income per capita)	1610.5

Enforcing a Contract	
Number of procedures	19
Time (days)	195
Cost (% of income per capita)	..
Procedural-complexity index	50

Closing a Business	
Time to go through insolvency (years)	3.0
Cost to go through insolvency (% estate)	18
Absolute priority preserved	100
Efficient outcome achieved	0
Goals-of-insolvency index	50
Court-powers index	33

SENEGAL
Sub-Saharan Africa

Economy Characteristics	
GNI per capita (US$)	470
Population	9,767,780
Informal economy (% of income)	43.2
Legal origin	French

Hiring and Firing Workers	
Flexibility-of-hiring index	48
Conditions-of-employment index	83
Flexibility-of-firing index	30
Employment-law index	54

Getting Credit	
Public credit registry operates?	Yes
Public registry coverage (borrowers/1,000 capita)	2
Public-registry index	22
Private credit-information bureau operates?	No
Private bureau coverage (borrowers/1,000 capita)	0
Creditor-rights index	1

Starting a Business	
Number of procedures	9
Time (days)	58
Cost (% of income per capita)	123.6
Minimum capital (% of income per capita)	296.1

Enforcing a Contract	
Number of procedures	30
Time (days)	335
Cost (% of income per capita)	48.6
Procedural-complexity index	75

Closing a Business	
Time to go through insolvency (years)	3.0
Cost to go through insolvency (% estate)	8
Absolute priority preserved	67
Efficient outcome achieved	1
Goals-of-insolvency index	73
Court-powers index	100

Note: .. means no data available.

SERBIA AND MONTENEGRO
Europe and Central Asia

Economy Characteristics

Economy Characteristics	
GNI per capita (US$)	1,400
Population	10,651,000
Informal economy (% of income)	29.1
Legal origin	German

Starting a Business

Starting a Business	
Number of procedures	10
Time (days)	44
Cost (% of income per capita)	13.3
Minimum capital (% of income per capita)	5.5

Hiring and Firing Workers

Hiring and Firing Workers	
Flexibility-of-hiring index	51
Conditions-of-employment index	88
Flexibility-of-firing index	29
Employment-law index	56

Enforcing a Contract

Enforcing a Contract	
Number of procedures	40
Time (days)	1028
Cost (% of income per capita)	20.0
Procedural-complexity index	61

Getting Credit

Getting Credit	
Public credit registry operates?	Yes
Public registry coverage (borrowers/1,000 capita)	<1
Public-registry index	33
Private credit-information bureau operates?	No
Private bureau coverage (borrowers/1,000 capita)	0
Creditor-rights index	2

Closing a Business

Closing a Business	
Time to go through insolvency (years)	7.3
Cost to go through insolvency (% estate)	38
Absolute priority preserved	33
Efficient outcome achieved	1
Goals-of-insolvency index	42
Court-powers index	67

SIERRA LEONE
Sub-Saharan Africa

Economy Characteristics

Economy Characteristics	
GNI per capita (US$)	140
Population	5,133,380
Informal economy (% of income)	..
Legal origin	English

Starting a Business

Starting a Business	
Number of procedures	9
Time (days)	26
Cost (% of income per capita)	1297.6
Minimum capital (% of income per capita)	0.0

Hiring and Firing Workers

Hiring and Firing Workers	
Flexibility-of-hiring index	56
Conditions-of-employment index	84
Flexibility-of-firing index	62
Employment-law index	67

Enforcing a Contract

Enforcing a Contract	
Number of procedures	48
Time (days)	114
Cost (% of income per capita)	8.3
Procedural-complexity index	29

Getting Credit

Getting Credit	
Public credit registry operates?	No
Public registry coverage (borrowers/1,000 capita)	0
Public-registry index	0
Private credit-information bureau operates?	No
Private bureau coverage (borrowers/1,000 capita)	0
Creditor-rights index	2

Closing a Business

Closing a Business	
Time to go through insolvency (years)	2.5
Cost to go through insolvency (% estate)	38
Absolute priority preserved	0
Efficient outcome achieved	0
Goals-of-insolvency index	20
Court-powers index	33

SINGAPORE
East Asia and Pacific

Economy Characteristics

Economy Characteristics	
GNI per capita (US$)	20,690
Population	4,131,000
Informal economy (% of income)	13.1
Legal origin	English

Starting a Business

Starting a Business	
Number of procedures	7
Time (days)	8
Cost (% of income per capita)	1.2
Minimum capital (% of income per capita)	0.0

Hiring and Firing Workers

Hiring and Firing Workers	
Flexibility-of-hiring index	33
Conditions-of-employment index	26
Flexibility-of-firing index	1
Employment-law index	20

Enforcing a Contract

Enforcing a Contract	
Number of procedures	23
Time (days)	50
Cost (% of income per capita)	14.4
Procedural-complexity index	49

Getting Credit

Getting Credit	
Public credit registry operates?	No
Public registry coverage (borrowers/1,000 capita)	0
Public-registry index	0
Private credit-information bureau operates?	Developing
Private bureau coverage (borrowers/1,000 capita)	0
Creditor-rights index	3

Closing a Business

Closing a Business	
Time to go through insolvency (years)	0.7
Cost to go through insolvency (% estate)	1
Absolute priority preserved	100
Efficient outcome achieved	1
Goals-of-insolvency index	99
Court-powers index	33

Note: .. means no data available.

SLOVAK REPUBLIC
Europe and Central Asia

Economy Characteristics		Starting a Business	
GNI per capita (US$)	3,950	Number of procedures	10
Population	5,404,000	Time (days)	98
Informal economy (% of income)	18.9	Cost (% of income per capita)	10.2
Legal origin	German	Minimum capital (% of income per capita)	111.8
Hiring and Firing Workers		**Enforcing a Contract**	
Flexibility-of-hiring index	34	Number of procedures	26
Conditions-of-employment index	89	Time (days)	420
Flexibility-of-firing index	60	Cost (% of income per capita)	13.3
Employment-law index	61	Procedural-complexity index	40
Getting Credit		**Closing a Business**	
Public credit registry operates?	Yes	Time to go through insolvency (years)	4.8
Public registry coverage (borrowers/1,000 capita)	2	Cost to go through insolvency (% estate)	18
Public-registry index	48	Absolute priority preserved	100
Private credit-information bureau operates?	No	Efficient outcome achieved	1
Private bureau coverage (borrowers/1,000 capita)	0	Goals-of-insolvency index	71
Creditor-rights index	2	Court-powers index	67

SLOVENIA
Europe and Central Asia

Economy Characteristics		Starting a Business	
GNI per capita (US$)	9,810	Number of procedures	10
Population	1,992,000	Time (days)	61
Informal economy (% of income)	27.1	Cost (% of income per capita)	15.5
Legal origin	German	Minimum capital (% of income per capita)	89.1
Hiring and Firing Workers		**Enforcing a Contract**	
Flexibility-of-hiring index	53	Number of procedures	22
Conditions-of-employment index	84	Time (days)	1003
Flexibility-of-firing index	41	Cost (% of income per capita)	3.6
Employment-law index	59	Procedural-complexity index	65
Getting Credit		**Closing a Business**	
Public credit registry operates?	Yes	Time to go through insolvency (years)	3.7
Public registry coverage (borrowers/1,000 capita)	14	Cost to go through insolvency (% estate)	18
Public-registry index	60	Absolute priority preserved	67
Private credit-information bureau operates?	No	Efficient outcome achieved	0
Private bureau coverage (borrowers/1,000 capita)	0	Goals-of-insolvency index	41
Creditor-rights index	3	Court-powers index	67

SOUTH AFRICA
Sub-Saharan Africa

Economy Characteristics		Starting a Business	
GNI per capita (US$)	2,600	Number of procedures	9
Population	43,240,000	Time (days)	38
Informal economy (% of income)	28.4	Cost (% of income per capita)	8.7
Legal origin	English	Minimum capital (% of income per capita)	0.0
Hiring and Firing Workers		**Enforcing a Contract**	
Flexibility-of-hiring index	42	Number of procedures	26
Conditions-of-employment index	36	Time (days)	207
Flexibility-of-firing index	30	Cost (% of income per capita)	16.7
Employment-law index	36	Procedural-complexity index	56
Getting Credit		**Closing a Business**	
Public credit registry operates?	No	Time to go through insolvency (years)	2.0
Public registry coverage (borrowers/1,000 capita)	0	Cost to go through insolvency (% estate)	18
Public-registry index	0	Absolute priority preserved	100
Private credit-information bureau operates?	Yes	Efficient outcome achieved	0
Private bureau coverage (borrowers/1,000 capita)	469	Goals-of-insolvency index	53
Creditor-rights index	3	Court-powers index	67

Note: .. means no data available.

SPAIN
OECD: High Income

Economy Characteristics		Starting a Business	
GNI per capita (US$)	14,430	Number of procedures	11
Population	41,117,000	Time (days)	115
Informal economy (% of income)	22.6	Cost (% of income per capita)	16.4
Legal origin	French	Minimum capital (% of income per capita)	19.6
Hiring and Firing Workers		**Enforcing a Contract**	
Flexibility-of-hiring index	76	Number of procedures	20
Conditions-of-employment index	88	Time (days)	147
Flexibility-of-firing index	45	Cost (% of income per capita)	10.7
Employment-law index	70	Procedural-complexity index	83
Getting Credit		**Closing a Business**	
Public credit registry operates?	Yes	Time to go through insolvency (years)	1.53
Public registry coverage (borrowers/1,000 capita)	305	Cost to go through insolvency (% estate)	8
Public-registry index	64	Absolute priority preserved	33
Private credit-information bureau operates?	Yes	Efficient outcome achieved	1
Private bureau coverage (borrowers/1,000 capita)	0	Goals-of-insolvency index	0
Creditor-rights index	0	Court-powers index	0

SRI LANKA
South Asia

Economy Characteristics		Starting a Business	
GNI per capita (US$)	840	Number of procedures	8
Population	18,732,000	Time (days)	58
Informal economy (% of income)	44.6	Cost (% of income per capita)	18.3
Legal origin	English	Minimum capital (% of income per capita)	0.0
Hiring and Firing Workers		**Enforcing a Contract**	
Flexibility-of-hiring index	33	Number of procedures	17
Conditions-of-employment index	52	Time (days)	440
Flexibility-of-firing index	40	Cost (% of income per capita)	7.6
Employment-law index	42	Procedural-complexity index	59
Getting Credit		**Closing a Business**	
Public credit registry operates?	No	Time to go through insolvency (years)	2.3
Public registry coverage (borrowers/1,000 capita)	0	Cost to go through insolvency (% estate)	18
Public-registry index	0	Absolute priority preserved	33
Private credit-information bureau operates?	Yes	Efficient outcome achieved	0
Private bureau coverage (borrowers/1,000 capita)	9	Goals-of-insolvency index	35
Creditor-rights index	2	Court-powers index	67

SWEDEN
OECD: High Income

Economy Characteristics		Starting a Business	
GNI per capita (US$)	24,820	Number of procedures	3
Population	8,894,000	Time (days)	16
Informal economy (% of income)	19.1	Cost (% of income per capita)	0.8
Legal origin	Nordic	Minimum capital (% of income per capita)	41.4
Hiring and Firing Workers		**Enforcing a Contract**	
Flexibility-of-hiring index	56	Number of procedures	21
Conditions-of-employment index	39	Time (days)	190
Flexibility-of-firing index	31	Cost (% of income per capita)	7.6
Employment-law index	42	Procedural-complexity index	44
Getting Credit		**Closing a Business**	
Public credit registry operates?	No	Time to go through insolvency (years)	2
Public registry coverage (borrowers/1,000 capita)	0	Cost to go through insolvency (% estate)	8
Public-registry index	0	Absolute priority preserved	100
Private credit-information bureau operates?	Yes	Efficient outcome achieved	1
Private bureau coverage (borrowers/1,000 capita)	489	Goals-of-insolvency index	84
Creditor-rights index	1	Court-powers index	33

Note: .. means no data available.

SWITZERLAND
OECD: High Income

Economy Characteristics		Starting a Business	
GNI per capita (US$)	37,930	Number of procedures	6
Population	7,231,000	Time (days)	20
Informal economy (% of income)	8.8	Cost (% of income per capita)	8.5
Legal origin	German	Minimum capital (% of income per capita)	33.8
Hiring and Firing Workers		**Enforcing a Contract**	
Flexibility-of-hiring index	33	Number of procedures	14
Conditions-of-employment index	53	Time (days)	224
Flexibility-of-firing index	23	Cost (% of income per capita)	3.9
Employment-law index	36	Procedural-complexity index	44
Getting Credit		**Closing a Business**	
Public credit registry operates?	No	Time to go through insolvency (years)	4.6
Public registry coverage (borrowers/1,000 capita)	0	Cost to go through insolvency (% estate)	4
Public-registry index	0	Absolute priority preserved	100
Private credit-information bureau operates?	Yes	Efficient outcome achieved	0
Private bureau coverage (borrowers/1,000 capita)	178	Goals-of-insolvency index	59
Creditor-rights index	1	Court-powers index	67

SYRIAN ARAB REPUBLIC
Middle East and North Africa

Economy Characteristics		Starting a Business	
GNI per capita (US$)	1,130	Number of procedures	10
Population	16,593,210	Time (days)	42
Informal economy (% of income)	19.3	Cost (% of income per capita)	16.7
Legal origin	French	Minimum capital (% of income per capita)	5627.2
Hiring and Firing Workers		**Enforcing a Contract**	
Flexibility-of-hiring index	33	Number of procedures	36
Conditions-of-employment index	79	Time (days)	596
Flexibility-of-firing index	22	Cost (% of income per capita)	31.3
Employment-law index	45	Procedural-complexity index	69
Getting Credit		**Closing a Business**	
Public credit registry operates?	No	Time to go through insolvency (years)	4.1
Public registry coverage (borrowers/1,000 capita)	0	Cost to go through insolvency (% estate)	8
Public-registry index	0	Absolute priority preserved	33
Private credit-information bureau operates?	No	Efficient outcome achieved	0
Private bureau coverage (borrowers/1,000 capita)	0	Goals-of-insolvency index	37
Creditor-rights index	3	Court-powers index	67

TAIWAN, CHINA
East Asia and Pacific

Economy Characteristics		Starting a Business	
GNI per capita (US$)	13,300	Number of procedures	8
Population	22,342,000	Time (days)	48
Informal economy (% of income)	19.6	Cost (% of income per capita)	6.1
Legal origin	German	Minimum capital (% of income per capita)	217.4
Hiring and Firing Workers		**Enforcing a Contract**	
Flexibility-of-hiring index	81	Number of procedures	15
Conditions-of-employment index	59	Time (days)	210
Flexibility-of-firing index	32	Cost (% of income per capita)	0.5
Employment-law index	57	Procedural-complexity index	37
Getting Credit		**Closing a Business**	
Public credit registry operates?	Yes	Time to go through insolvency (years)	0.8
Public registry coverage (borrowers/1,000 capita)	27	Cost to go through insolvency (% estate)	4
Public-registry index	70	Absolute priority preserved	0
Private credit-information bureau operates?	Yes	Efficient outcome achieved	1
Private bureau coverage (borrowers/1,000 capita)	..	Goals-of-insolvency index	68
Creditor-rights index	1	Court-powers index	100

Note: .. means no data available.

TANZANIA
Sub-Saharan Africa

Economy Characteristics		Starting a Business	
GNI per capita (US$)	280	Number of procedures	13
Population	34,449,620	Time (days)	35
Informal economy (% of income)	58.3	Cost (% of income per capita)	199.0
Legal origin	English	Minimum capital (% of income per capita)	0.0
Hiring and Firing Workers		**Enforcing a Contract**	
Flexibility-of-hiring index	57	Number of procedures	14
Conditions-of-employment index	77	Time (days)	127
Flexibility-of-firing index	49	Cost (% of income per capita)	3.8
Employment-law index	61	Procedural-complexity index	62
Getting Credit		**Closing a Business**	
Public credit registry operates?	No	Time to go through insolvency (years)	3.0
Public registry coverage (borrowers/1,000 capita)	0	Cost to go through insolvency (% estate)	8
Public-registry index	0	Absolute priority preserved	33
Private credit-information bureau operates?	No	Efficient outcome achieved	1
Private bureau coverage (borrowers/1,000 capita)	0	Goals-of-insolvency index	65
Creditor-rights index	2	Court-powers index	67

THAILAND
East Asia and Pacific

Economy Characteristics		Starting a Business	
GNI per capita (US$)	1,980	Number of procedures	9
Population	61,183,900	Time (days)	42
Informal economy (% of income)	52.6	Cost (% of income per capita)	7.3
Legal origin	English	Minimum capital (% of income per capita)	0.0
Hiring and Firing Workers		**Enforcing a Contract**	
Flexibility-of-hiring index	78	Number of procedures	19
Conditions-of-employment index	73	Time (days)	210
Flexibility-of-firing index	30	Cost (% of income per capita)	29.6
Employment-law index	61	Procedural-complexity index	53
Getting Credit		**Closing a Business**	
Public credit registry operates?	No	Time to go through insolvency (years)	2.6
Public registry coverage (borrowers/1,000 capita)	0	Cost to go through insolvency (% estate)	38
Public-registry index	0	Absolute priority preserved	67
Private credit-information bureau operates?	Yes	Efficient outcome achieved	1
Private bureau coverage (borrowers/1,000 capita)	98	Goals-of-insolvency index	62
Creditor-rights index	3	Court-powers index	33

TOGO
Sub-Saharan Africa

Economy Characteristics		Starting a Business	
GNI per capita (US$)	270	Number of procedures	14
Population	4,653,400	Time (days)	63
Informal economy (% of income)	..	Cost (% of income per capita)	281.4
Legal origin	French	Minimum capital (% of income per capita)	531.4
Hiring and Firing Workers		**Enforcing a Contract**	
Flexibility-of-hiring index	53	Number of procedures	43
Conditions-of-employment index	80	Time (days)	503
Flexibility-of-firing index	36	Cost (% of income per capita)	21.4
Employment-law index	57	Procedural-complexity index	63
Getting Credit		**Closing a Business**	
Public credit registry operates?	Yes	Time to go through insolvency (years)	No practice
Public registry coverage (borrowers/1,000 capita)	1	Cost to go through insolvency (% estate)	No practice
Public-registry index	22	Absolute priority preserved	33
Private credit-information bureau operates?	No	Efficient outcome achieved	0
Private bureau coverage (borrowers/1,000 capita)	0	Goals-of-insolvency index	8
Creditor-rights index	2	Court-powers index	100

Note: .. means no data available.

TUNISIA
Middle East and North Africa

Economy Characteristics		Starting a Business	
GNI per capita (US$)	2,000	Number of procedures	10
Population	9,673,600	Time (days)	46
Informal economy (% of income)	38.4	Cost (% of income per capita)	16.4
Legal origin	French	Minimum capital (% of income per capita)	351.7
Hiring and Firing Workers		**Enforcing a Contract**	
Flexibility-of-hiring index	73	Number of procedures	14
Conditions-of-employment index	53	Time (days)	7
Flexibility-of-firing index	44	Cost (% of income per capita)	4.1
Employment-law index	57	Procedural-complexity index	60
Getting Credit		**Closing a Business**	
Public credit registry operates?	Yes	Time to go through insolvency (years)	2.5
Public registry coverage (borrowers/1,000 capita)	4	Cost to go through insolvency (% estate)	8
Public-registry index	48	Absolute priority preserved	67
Private credit-information bureau operates?	No	Efficient outcome achieved	0
Private bureau coverage (borrowers/1,000 capita)	0	Goals-of-insolvency index	50
Creditor-rights index	0	Court-powers index	67

TURKEY
Europe and Central Asia

Economy Characteristics		Starting a Business	
GNI per capita (US$)	2,500	Number of procedures	13
Population	68,529,000	Time (days)	38
Informal economy (% of income)	32.1	Cost (% of income per capita)	37.1
Legal origin	French	Minimum capital (% of income per capita)	13.2
Hiring and Firing Workers		**Enforcing a Contract**	
Flexibility-of-hiring index	58	Number of procedures	18
Conditions-of-employment index	91	Time (days)	105
Flexibility-of-firing index	17	Cost (% of income per capita)	5.4
Employment-law index	55	Procedural-complexity index	38
Getting Credit		**Closing a Business**	
Public credit registry operates?	Yes	Time to go through insolvency (years)	1.8
Public registry coverage (borrowers/1,000 capita)	7	Cost to go through insolvency (% estate)	8
Public-registry index	44	Absolute priority preserved	67
Private credit-information bureau operates?	Yes	Efficient outcome achieved	0
Private bureau coverage (borrowers/1,000 capita)	204	Goals-of-insolvency index	51
Creditor-rights index	2	Court-powers index	67

UGANDA
Sub-Saharan Africa

Economy Characteristics		Starting a Business	
GNI per capita (US$)	250	Number of procedures	17
Population	22,788,000	Time (days)	36
Informal economy (% of income)	43.1	Cost (% of income per capita)	135.1
Legal origin	English	Minimum capital (% of income per capita)	0.0
Hiring and Firing Workers		**Enforcing a Contract**	
Flexibility-of-hiring index	33	Number of procedures	16
Conditions-of-employment index	44	Time (days)	99
Flexibility-of-firing index	50	Cost (% of income per capita)	10.0
Employment-law index	42	Procedural-complexity index	40
Getting Credit		**Closing a Business**	
Public credit registry operates?	No	Time to go through insolvency (years)	2.0
Public registry coverage (borrowers/1,000 capita)	0	Cost to go through insolvency (% estate)	38
Public-registry index	0	Absolute priority preserved	33
Private credit-information bureau operates?	No	Efficient outcome achieved	1
Private bureau coverage (borrowers/1,000 capita)	0	Goals-of-insolvency index	55
Creditor-rights index	2	Court-powers index	67

Note: .. means no data available.

UKRAINE
Europe and Central Asia

Economy Characteristics		Starting a Business	
GNI per capita (US$)	770	Number of procedures	14
Population	49,093,000	Time (days)	40
Informal economy (% of income)	52.2	Cost (% of income per capita)	27.3
Legal origin	Socialist	Minimum capital (% of income per capita)	450.8
Hiring and Firing Workers		**Enforcing a Contract**	
Flexibility-of-hiring index	58	Number of procedures	20
Conditions-of-employment index	93	Time (days)	224
Flexibility-of-firing index	69	Cost (% of income per capita)	11.0
Employment-law index	73	Procedural-complexity index	51
Getting Credit		**Closing a Business**	
Public credit registry operates?	No	Time to go through insolvency (years)	2.97
Public registry coverage (borrowers/1,000 capita)	0	Cost to go through insolvency (% estate)	18
Public-registry index	0	Absolute priority preserved	67
Private credit-information bureau operates?	No	Efficient outcome achieved	0
Private bureau coverage (borrowers/1,000 capita)	0	Goals-of-insolvency index	42
Creditor-rights index	2	Court-powers index	33

UNITED ARAB EMIRATES
Middle East and North Africa

Economy Characteristics		Starting a Business	
GNI per capita (US$)	20,218	Number of procedures	10
Population	2,976,290	Time (days)	29
Informal economy (% of income)	26.4	Cost (% of income per capita)	24.5
Legal origin	English	Minimum capital (% of income per capita)	404.0
Hiring and Firing Workers		**Enforcing a Contract**	
Flexibility-of-hiring index	33	Number of procedures	27
Conditions-of-employment index	66	Time (days)	559
Flexibility-of-firing index	37	Cost (% of income per capita)	10.6
Employment-law index	45	Procedural-complexity index	56
Getting Credit		**Closing a Business**	
Public credit registry operates?	Yes	Time to go through insolvency (years)	5
Public registry coverage (borrowers/1,000 capita)	12	Cost to go through insolvency (% estate)	38
Public-registry index	44	Absolute priority preserved	33
Private credit-information bureau operates?	No	Efficient outcome achieved	0
Private bureau coverage (borrowers/1,000 capita)	0	Goals-of-insolvency index	23
Creditor-rights index	2	Court-powers index	33

UNITED KINGDOM
OECD: High Income

Economy Characteristics		Starting a Business	
GNI per capita (US$)	25,250	Number of procedures	6
Population	58,800,000	Time (days)	18
Informal economy (% of income)	12.6	Cost (% of income per capita)	1.0
Legal origin	English	Minimum capital (% of income per capita)	0.0
Hiring and Firing Workers		**Enforcing a Contract**	
Flexibility-of-hiring index	33	Number of procedures	12
Conditions-of-employment index	42	Time (days)	101
Flexibility-of-firing index	9	Cost (% of income per capita)	0.5
Employment-law index	28	Procedural-complexity index	36
Getting Credit		**Closing a Business**	
Public credit registry operates?	No	Time to go through insolvency (years)	1.0
Public registry coverage (borrowers/1,000 capita)	0	Cost to go through insolvency (% estate)	8
Public-registry index	0	Absolute priority preserved	100
Private credit-information bureau operates?	Yes	Efficient outcome achieved	1
Private bureau coverage (borrowers/1,000 capita)	652	Goals-of-insolvency index	86
Creditor-rights index	4	Court-powers index	0

Note: .. means no data available.

UNITED STATES
OECD: High Income

Economy Characteristics		Starting a Business	
GNI per capita (US$)	35,060	Number of procedures	5
Population	285,318,016	Time (days)	4
Informal economy (% of income)	8.8	Cost (% of income per capita)	0.6
Legal origin	English	Minimum capital (% of income per capita)	0.0
Hiring and Firing Workers		**Enforcing a Contract**	
Flexibility-of-hiring index	33	Number of procedures	17
Conditions-of-employment index	29	Time (days)	365
Flexibility-of-firing index	5	Cost (% of income per capita)	0.4
Employment-law index	22	Procedural-complexity index	46
Getting Credit		**Closing a Business**	
Public credit registry operates?	No	Time to go through insolvency (years)	3.0
Public registry coverage (borrowers/1,000 capita)	0	Cost to go through insolvency (% estate)	4
Public-registry index	0	Absolute priority preserved	100
Private credit-information bureau operates?	Yes	Efficient outcome achieved	1
Private bureau coverage (borrowers/1,000 capita)	810	Goals-of-insolvency index	88
Creditor-rights index	1	Court-powers index	33

URUGUAY
Latin America and Caribbean

Economy Characteristics		Starting a Business	
GNI per capita (US$)	4,370	Number of procedures	10
Population	3,361,000	Time (days)	27
Informal economy (% of income)	51.1	Cost (% of income per capita)	46.7
Legal origin	French	Minimum capital (% of income per capita)	699.0
Hiring and Firing Workers		**Enforcing a Contract**	
Flexibility-of-hiring index	58	Number of procedures	38
Conditions-of-employment index	56	Time (days)	360
Flexibility-of-firing index	3	Cost (% of income per capita)	13.7
Employment-law index	39	Procedural-complexity index	55
Getting Credit		**Closing a Business**	
Public credit registry operates?	Yes	Time to go through insolvency (years)	4.0
Public registry coverage (borrowers/1,000 capita)	49	Cost to go through insolvency (% estate)	8
Public-registry index	57	Absolute priority preserved	100
Private credit-information bureau operates?	Yes	Efficient outcome achieved	0
Private bureau coverage (borrowers/1,000 capita)	479	Goals-of-insolvency index	54
Creditor-rights index	3	Court-powers index	67

UZBEKISTAN
Europe and Central Asia

Economy Characteristics		Starting a Business	
GNI per capita (US$)	450	Number of procedures	9
Population	25,068,000	Time (days)	33
Informal economy (% of income)	34.1	Cost (% of income per capita)	16.0
Legal origin	Socialist	Minimum capital (% of income per capita)	64.3
Hiring and Firing Workers		**Enforcing a Contract**	
Flexibility-of-hiring index	46	Number of procedures	34
Conditions-of-employment index	69	Time (days)	258
Flexibility-of-firing index	50	Cost (% of income per capita)	2.1
Employment-law index	55	Procedural-complexity index	57
Getting Credit		**Closing a Business**	
Public credit registry operates?	No	Time to go through insolvency (years)	3.25
Public registry coverage (borrowers/1,000 capita)	0	Cost to go through insolvency (% estate)	4
Public-registry index	0	Absolute priority preserved	33
Private credit-information bureau operates?	No	Efficient outcome achieved	0
Private bureau coverage (borrowers/1,000 capita)	0	Goals-of-insolvency index	46
Creditor-rights index	2	Court-powers index	67

Note: .. means no data available.

VENEZUELA, RB
Latin America and Caribbean

Economy Characteristics		Starting a Business	
GNI per capita (US$)	4,090	Number of procedures	14
Population	24,632,000	Time (days)	119
Informal economy (% of income)	33.6	Cost (% of income per capita)	19.3
Legal origin	French	Minimum capital (% of income per capita)	0.0
Hiring and Firing Workers		**Enforcing a Contract**	
Flexibility-of-hiring index	78	Number of procedures	41
Conditions-of-employment index	88	Time (days)	360
Flexibility-of-firing index	60	Cost (% of income per capita)	46.9
Employment-law index	75	Procedural-complexity index	81
Getting Credit		**Closing a Business**	
Public credit registry operates?	Yes	Time to go through insolvency (years)	4.0
Public registry coverage (borrowers/1,000 capita)	97	Cost to go through insolvency (% estate)	38
Public-registry index	46	Absolute priority preserved	100
Private credit-information bureau operates?	No	Efficient outcome achieved	1
Private bureau coverage (borrowers/1,000 capita)	0	Goals-of-insolvency index	67
Creditor-rights index	2	Court-powers index	67

VIETNAM
East Asia and Pacific

Economy Characteristics		Starting a Business	
GNI per capita (US$)	430	Number of procedures	11
Population	79,526,048	Time (days)	63
Informal economy (% of income)	15.6	Cost (% of income per capita)	29.9
Legal origin	French	Minimum capital (% of income per capita)	0.0
Hiring and Firing Workers		**Enforcing a Contract**	
Flexibility-of-hiring index	43	Number of procedures	28
Conditions-of-employment index	77	Time (days)	120
Flexibility-of-firing index	48	Cost (% of income per capita)	8.5
Employment-law index	56	Procedural-complexity index	46
Getting Credit		**Closing a Business**	
Public credit registry operates?	Yes	Time to go through insolvency (years)	No practice
Public registry coverage (borrowers/1,000 capita)	2	Cost to go through insolvency (% estate)	No practice
Public-registry index	67	Absolute priority preserved	33
Private credit-information bureau operates?	No	Efficient outcome achieved	1
Private bureau coverage (borrowers/1,000 capita)	0	Goals-of-insolvency index	33
Creditor-rights index	0	Court-powers index	67

YEMEN, REP. of
Middle East and North Africa

Economy Characteristics		Starting a Business	
GNI per capita (US$)	490	Number of procedures	13
Population	18,045,750	Time (days)	96
Informal economy (% of income)	27.4	Cost (% of income per capita)	264.1
Legal origin	English	Minimum capital (% of income per capita)	1716.9
Hiring and Firing Workers		**Enforcing a Contract**	
Flexibility-of-hiring index	33	Number of procedures	27
Conditions-of-employment index	66	Time (days)	240
Flexibility-of-firing index	28	Cost (% of income per capita)	0.5
Employment-law index	43	Procedural-complexity index	60
Getting Credit		**Closing a Business**	
Public credit registry operates?	Yes	Time to go through insolvency (years)	2.4
Public registry coverage (borrowers/1,000 capita)	7	Cost to go through insolvency (% estate)	4
Public-registry index	38	Absolute priority preserved	33
Private credit-information bureau operates?	No	Efficient outcome achieved	0
Private bureau coverage (borrowers/1,000 capita)	0	Goals-of-insolvency index	47
Creditor-rights index	0	Court-powers index	33

Note: .. means no data available.

ZAMBIA
Sub-Saharan Africa

Economy Characteristics		Starting a Business	
GNI per capita (US$)	330	Number of procedures	6
Population	10,282,500	Time (days)	40
Informal economy (% of income)	48.9	Cost (% of income per capita)	24.1
Legal origin	English	Minimum capital (% of income per capita)	137.8
Hiring and Firing Workers		**Enforcing a Contract**	
Flexibility-of-hiring index	33	Number of procedures	16
Conditions-of-employment index	64	Time (days)	188
Flexibility-of-firing index	40	Cost (% of income per capita)	15.8
Employment-law index	46	Procedural-complexity index	32
Getting Credit		**Closing a Business**	
Public credit registry operates?	No	Time to go through insolvency (years)	3.7
Public registry coverage (borrowers/1,000 capita)	0	Cost to go through insolvency (% estate)	8
Public-registry index	0	Absolute priority preserved	100
Private credit-information bureau operates?	No	Efficient outcome achieved	0
Private bureau coverage (borrowers/1,000 capita)	0	Goals-of-insolvency index	55
Creditor-rights index	1	Court-powers index	33

ZIMBABWE
Sub-Saharan Africa

Economy Characteristics		Starting a Business	
GNI per capita (US$)	463	Number of procedures	10
Population	12,820,650	Time (days)	122
Informal economy (% of income)	59.4	Cost (% of income per capita)	285.3
Legal origin	English	Minimum capital (% of income per capita)	0.0
Hiring and Firing Workers		**Enforcing a Contract**	
Flexibility-of-hiring index	33	Number of procedures	13
Conditions-of-employment index	22	Time (days)	197
Flexibility-of-firing index	26	Cost (% of income per capita)	39.5
Employment-law index	27	Procedural-complexity index	50
Getting Credit		**Closing a Business**	
Public credit registry operates?	No	Time to go through insolvency (years)	2.3
Public registry coverage (borrowers/1,000 capita)	0	Cost to go through insolvency (% estate)	18
Public-registry index	0	Absolute priority preserved	100
Private credit-information bureau operates?	No	Efficient outcome achieved	0
Private bureau coverage (borrowers/1,000 capita)	0	Goals-of-insolvency index	52
Creditor-rights index	4	Court-powers index	67

Note: .. means no data available.

We would like to thank the following organizations and individuals who have generously contributed to the data collection of the Doing Business project. Contact details of local partners are available on the Doing Business website at http://rru.worldbank.org/doingbusiness/.

Global Contributors
Baker & McKenzie
Dun and Bradstreet International
International Bar Association
Lex Mundi Association of law firms
TransUnion International

Albania
Artur Asllani *Studio Legale Tonucci*
Yair Baranes *USAID*
Genc Boga *Boga & Associates*
Vilsa Dado *Kalo & Associates*
Shpati Hoxha *Boga & Associates*
Sonila Ibrahimi *Boga & Associates*
Perparim Kalo *Kalo & Associates*
Miranda Ramaj *Bank of Albania*

Algeria
Branka Achari-Djokic *Bank of Algeria*
Mamoun Aidoud *Aidoud Law Firm*
Amine Hadad *Ghellal & Mekerba*
Mustapha Hamdane *Cabinet d Avocats Mustapha Hamdane*
Samir Hamouda *Cabinet d Avocats Samir Hamouda*
Yamina Kebir *Yemina Kebir Law Offices*
Said Maherzi *Bank of Algeria*

Angola
Fátima Freitas *Fátima Freitas Law Firm*
Leão Peres *National Bank of Angola*

Argentina
Juan Arocena *Allende & Brea Law Firm*
Vanesa Balda *Manoff & Feilbogen Law Firm*
Oscar Del Rio *Central Bank of Argentina*
Bernardo Horacio Fernandez *Central Bank of Argentina*
Alejandro Fiuza *Marval O'Farrell & Mairal*
Nicolas Garcia Pinto *Baker & McKenzie*
Martin Lanfranco *Marval O'Farrell & Mairal*
Roberto Laterza *Organización Veraz*
Patricia Lopez Aufranc *Marval O'Farrell & Mairal*
Maria Lujan Bianchi *Brons & Salas Law Firm*
Eugenio Maurette *Abeledo Gottheil Abogados*
Sean McCormick *Llerena & Asociados Abogados*
Olga Muino *Centro de Estudios Bonaerenses*
Miguel Murray *Estudio Juridico Borda*
Alfredo O'Farrell *Marval O'Farrell & Mairal*
Juan Manuel Alvarez Prado *Alvarez Prado & Asociados*
Jorge Raul Postiglione *Brons & Salas Law Firm*
Liliana Segade *Quattrini Laprida & Asociados*
Alfredo Vicens *Organización Veraz*
Octavio Miguel Zenarruza *Alvarez Prado & Asociados*

Armenia
Karen Hambardzoumyan *Central Bank of Armenia*
Alan Kuchukyan *KPMG Armenia*
Suren Melikyan *KPMG Armenia*
Tom Samuelian *Arlex International*
Artur Tunyan *Tunyan & Associates*

Australia
Brett Cook *Allens Arthur Robinson*
David Cross *Allens Arthur Robinson*
Christopher Davie *Clayton Utz*
Paul James *Clayton Utz*
Sonya Karo *ASIC*
Timothy L'Estrange *Allens Arthur Robinson*
Judy Lau *Australian Prudential Regulation Authority*
John Lobban *Blake Dawson Waldron*
Helen MacKay *Allens Arthur Robinson*
Tim O'Doherty *Baker & McKenzie*
Michael O'Donnell *Thomson Playford*
Michael Quintan *Allens Arthur Robinson*
Andrew Smith *Mallesons Stephen Jaques*
Reinhard Toegl *Dr. Reinhard Toegl Law Offices*
Jane Wilson *Baycorp Advantage*

Austria
Johannes Barbist *Binder Grösswang Rechtsanwälte*
Walter Bornett *Austrian Institute for SME Research*

Tibor Fabian *Binder Grösswang Rechtsanwälte*
Julian Feichtinger *Cerha Hempel & Spiegelfeld*
Bernhard Gumpoldsberger *Saxinger Chalupsky Weber & Partners*
Harald Heschl *Kreditschutzverband von 1870*
Sylvia Hofinger *Vienna City Government*
Alexander Klauser *Brauneis, Klauser & Prändl*
Robert Kovacs *Coface Intercredit*
Christian Lettmayr *Austrian Institute for SME Research*
Irene Mandl *Austrian Institute for SME Research*
Leopold Mayer *Dun and Bradstreet Information Services*
Wolfgang Messeritsch *National Bank of Austria*
Norbert Scherbaum *Scherbaum/Seebacher Rechtsanwälte*
Benedikt Spiegelfeld *Cerha Hempel & Spiegelfeld*
Dagmar Straka *National Bank of Austria*
Reinhard Töegl *Reinhard Töegl Law Offices*

Azerbaijan
Ofelia Abdulaeva *Salans*
Nazli Ahmadova *Baku Law Centre*
Elgar Alekperov *Baku Law Centre*
Aykhan Asadov *Baker & McKenzie*
Rufat Aslanov *National Bank of Azerbaijan*
Alum Bati *Salans Hertzfeld & Heilbronn Law Firm*
Christine Ferguson *Baker Botts*
Farhad Hajizade *Salans Hertzfeld & Heilbronn Law Firm*
Daniel Matthews *Baker & McKenzie*
Kanan Safarov *Ledingham Chalmers*
Safkhan Shahmammadli *Baker Botts*
Michael Walsh *Ledingham Chalmers*

Bangladesh
Jasim Uddin Ahmad *Bank of Bangladesh*
Halim Bepari *Supreme Court of Bangladesh*
Shirin Chaudhury *The Law Associates*
A.B.M. Badrud Doulah *Doulah & Doulah Advocates Attorneys & Notaries*
Shamsud Doulah *Doulah & Doulah Advocates*
Aneek Haque *Huq & Company*
Raquibul Haque *Miah Advocates & Attorneys Law Firm*
Mirza Quamrul Hasan *Advisers' Legal Alliance*
Khondker Shamsuddin Mahmood *Advisers' Legal Alliance*
Amir-Ul Islam *The Law Associates*

Belarus
Vladimir Biruk *Belarusian Community of Specialists in Crisis Management*
Svetlana Dashuk *Vlasova & Partners*
Alexandr Dovgenko *Incorporation Lawyer*
Yuri Krasnov *National Bank of the Republic of Belarus*
Igor Likhogrud *National Bank of the Republic of Belarus*
Vassili Salei *Borovtsov & Salei*
Vitaliy Sevrukevich *DICSA International Group of Lawyers*
Vassili Voloshinets *Incorporation Lawyer*

Belgium
Pamela Cordova *Loyens*
Ludo Cornelis *Eubelius Attorneys*
Dirk De Backer *Allen & Overy*
Marc Dechevre *Union Professionnelle du Crédit*
Luc Demeyere *Allen & Overy*
Pieter De Koster *Allen & Overy*
Joan Dubaere *Peeters Advocaten-Avocats*
Alain Francois *Eubelius Attorneys*
Ignace Maes *Baker & McKenzie*
Andre Moreau *National Bank of Belgium*
Didier Muraille *National Bank of Belgium*
Leo Peeters *Peeters Advocaten-Avocats*
Hans Seeldrayers *Eubelius Attorneys*
Rudy Trogh *National Bank of Belgium*
Jan Van Celst *Allen & Overy*

Benin
Rafikou Alabi *Cabinet Rafikou Alabi*
Vilevo Biova Devo *Centrale des Risques de l'Union Monetaire Ouest Africaine*
Evelyne Mandessi Bell *Mandessi Bell Law Firm*
Jacques Migan *Jacques Migan Law Firm*
Edgar-Yves Monnou *Cabinet Edgar-Yves Monnou*
Francois Nare *Centrale des Risques de l'Union Monetaire Ouest Africaine*

Bolivia
Carolina Aguirre Urioste *Bufete Aguirre*
Fernando Aguirre *Bufete Aguirre*
Carlos Ferreira *C.R. & F. Rojas Abogados*
Primitivo Gutiérrez *Guevara & Gutierrez*
Enrique Hurtado *Superintendency of Banks and Financial Entities*
Ricardo Indacochea *San Martín Indacochea & Asociados*
Manfredo Kempff *C.R. & F. Rojas Abogados*
Fernando Rojas *C.R. & F. Rojas Abogados*
Pablo Rojas *C.R & F. Rojas Abogados*
Sergio Salazar-Machicado *Salazar, Salazar & Asociados*

Bosnia and Herzegovina
Yair Baranes *USAID*
Adnan Hrenovica *LRC Credit Bureau*
Nikola Jankovic *Lansky & Partner Attorneys*
Kerim Karabdic *Advokati Salih & Kerim Karabdic*
Vesna Mrkovic *Lansky & Partner Attorneys*
Ibrahim Polimac *Agency for Banking of Federation of Bosnia and Herzegovina*

Botswana
Neill Armstrong *Armstrongs Attorneys*
John Carr-Hartley *Armstrongs Attorneys*
Topiwa Chilume *Armstrongs Attorneys Notaries & Conveyancers*
Edward Fashole Luke II *Luke & Associates*
Vincent Galeromeloe *Information Trust Corporation*
Kwadwo Osei-Ofei *Armstrongs Attorneys*
Moses Pelaelo *Bank of Botswana*
Virgil Vergeer *Collins Newman & Co*

Brazil
Adriana Baroni Santi *Ulhôa Canto Rezende e Guerra-Advogados*
Thomas Benes Felsberg *Felsberg e Associados*

Heloisa Bonciani Nader di Cunto *Duarte Garcia Caselli Guimarães e Terra Advogados*
Altamiro Boscoli *Demarest e Almeida Advogados*
Ulhôa Canto *Ulhôa Canto Rezende e Guerra-Advogados*
Gustavo Castro *Viseu Castro Cunha e Oricchio Advogados*
Paulo Sérgio Cavalheiro *Central Bank of Brazil*
Pedro Vitor Araujo da Costa *Escritorio de Advocacia Gouvêa Vieira*
Silvia Poggi de Carvalho *Duarte Garcia Caselli Guimarães e Terra Advogados*
Aloysio Meirelles de Miranda *Ulhôa Canto Rezende e Guerra-Advogados*
Silvio de Salvo *Venosa Demarest e Almeida Advogados*
Duarte Garcia *Duarte Garcia Caselli Guimarães e Terra Advogados*
Regina Gasulla Bouza *Goulart Penteado, Iervolino e Lefosse – Advogados*
Renato Giovanni Filho *Ulhôa Canto Rezende e Guerra-Advogados*
Caio Julius *Bolina Lazzareschi Advogados*
Maria Fernanda Lopes Ferraz Tella *Felsberg and Associados*
José Augusto Martins *Baker & McKenzie*
André Megale *Goulart Penteado, Iervolino e Lefosse – Advogados*
Fabiano Milani *Goulart Penteado, Iervolino e Lefosse – Advogados*
Andrea Oricchio *Kirsh Viseu Castro Cunha e Oricchio Advogados*
Valéria Salomão *Central Bank of Brazil*

Bulgaria
Svetlin Adrianov *Legal InterConsult - Penkov Markov and Partners Law Office*
Borislav Boyanov *Borislav Boyanov & Co*
George Dimitrov *O.R.A.C. Dimitrov Petrov & Co*
Vasil Iliev *Consult*
Georgi Kitanov *Totev Partners*
S. Kyutchukov *Djingov Gouginski Kyutchukov & Velichkov*
Yordan Manahilov *Bulgarian National Bank*
Stoyan Manolov *Bulgarian National Bank*
Vladimir Penkov *Legal InterConsult - Penkov Markov and Partners Law Office*
Kamelia Popova *Coface Intercredit Bulgaria*
Irina Tsvetkova *Landwell Bulgaria*

Burkina Faso
Bernardin Dabire *Dabire Sorgho & Toe*
Vilevo Biova Devo *Centrale des Risques de l'Union Monetaire Ouest Africaine*
Frank Didier Toe *Dabire Sorgho & Toe*
Barthélémy Kere *Cabinet d'Avocats Barthélémy Kere*
Evelyne Mandessi Bell *Mandessi Bell Law Firm*
Francois Nare *Centrale des Risques de l'Union Monetaire Ouest Africaine*
Marie-Antoinette Sorgho-Sery *Dabire Sorgho & Toe*
Oumarou Ouedraogo *Ohada Legis*
Dieudonné Bonkoungou *Ohada Legis*

Burundi
Tharcisse Ntakiyica *Cabinet Tharcisse Ntakiyica*
Yves Ntivumbura *Central Bank of Burundi*

Cambodia
Phyroath Heng *IMC Consulting*
Tim Smyth *IMC Consulting*

Cameroon
D. Etah Akoh *Etah-Nan & C. Société d'Avocats Barristers & Solicitors*
David Boyo *Henri Job & Partners*
Emmanuel Ekobo *Cabinet Ekobo*
Isabelle Fomukong *Cabinet Fomukong*
Tahir Souleyman Haggar *La Commission Bancaire de l'Afrique Centrale*
Paul Jing *Henri Job & Partners*
Gaston Kenfack *Ministry of Justice*
Kumfa Jude Kwenyui *Juris Consul Law Firm*
Daniel Mwambo Ndeley *Juris Consul Law Firm*
Evelyne Mandessi Bell *Mandessi Bell Law Firm*
Mwambo Litombe Ndeley *Juris Consul Law Firm*
Rafael Tung Nsue *La Commission Bancaire de l'Afrique Centrale*
Henri Pierre Job *Henri Job & Partners*

Canada
Arthur Adams *Southern Ontario Credit Bureau*
Eldon Bennett *Aird & Berlis*
David Bish *Goodmans*
Jay Carfagnini *Goodmans*
Thomas Cumming *Gowling Lafleur Henderson*
Yoine Goldstein *Goldstein Flanz & Fishman*
Leonid Gorelik *Baker & McKenzie*
Karen Grant *TransUnion*
Charles Johnston *Superintendency of Financial Institutions*
Charles Magerman *Baker & McKenzie*
Patrick McCarthy *Borden Ladner Gervais*
Tim Paleczny *Government of Ontario*
Debbie Ranger *Canada Customs and Revenue Agency*
Jeff Rosekat *Baker & McKenzie*
Paul Schabas *Blake Cassels & Graydon*
Leneo Sdao *Baker & McKenzie*
Jason Vonderhaar *TransUnion*
Jonathan Wigley *Baker & McKenzie*
Christopher William *Besant Cassels Brock*

Central African Republic
Tahir Souleyman Haggar *La Commission Bancaire de l'Afrique Centrale*
Rafael Tung Nsue *La Commission Bancaire de l'Afrique Centrale*
Nicolas Tiangaye *Nicolas Tiangaye Law Firm*

Chad
Thomas Dingamgoto *Cabinet Dingamgoto et Associes*
Tahir Souleyman Haggar *Centrale Recapitulative des Risques*
Rafael Tung Nsue *La Commission Bancaire de l'Afrique Centrale*

Chile
Cristian Araya *Alcaino Rodriguez & Sahli Law Firm*
Manuel Blanco *Blanco & Cia Abogados*
Jimena Bronfman *Guerrero Olivos Novoa y Errázuriz*

Jaime Cordova *Superintendencia de Bancos y Institutciones Financieras Chile*
Rodrigo Cuchacovich *Baker & McKenzie*
Martín del Río *Vial y Palma Abogados*
Ricardo Escobar *Carey y Cia Law Firm*
Cristian Eyzaguirre *Claro & Cia*
Roberto Guerrero *Guerrero Olivos Novoa y Errázuriz*
Luis Gutierrez *Puga & Ortiz*
Silvio Figari Napoli *Databusiness*
Andrés Jana *Linetzky Alvarez Hinzpeter Jana & Valle*
Cesar Jimenez Ortiz *Superintendencia de Bancos y Instituciones Financieras Chile*
León Larrain *Baker & McKenzie*
Claudio Ortiz Tello *Boletin Comercial*
Felipe Ossa *Claro & Cia*
Juan Eduardo Palma *Vial y Palma Abogados*
Daniela Peña Fergadiott *Barros Court Correa y Cia. Abogados*
Sebastián Yunge *Guerrero Olivos Novoa y Errázuriz*

China
Brian Barron *Baker & McKenzie*
Charles Booth *University of Hong Kong*
Barry Cheng *Baker & McKenzie*
Bob Kwauk *Blake Cassels & Graydon*
Joseph Lam *Deacons*
Wang Li *De Heng Law Offices*
Yang Ling *Huaxia International Business Credit Consulting*
Jerry Liu *Huaxia International Business Credit Consulting*
Linfei Liu *Jun He Law Offices*
Chen Min *Blake Cassels & Graydon*
Li Wang *DeHeng Law Office*
Xiaochuan Yang *PricewaterhouseCoopers New York Office*
Jin Zhong *Jun He Law Offices*
Zhang Zihong *People's Bank of China*

Colombia
Dario Cárdenas Navas *Cárdenas & Cárdenas*
Jorge Lara *Baker & McKenzie*
José Antonio Lloreda *José Lloreda Camacho & Co*
Juan Manuel *Villaveces Hollmann DataCrédito*
Luis E. Nieto *Arrieta Mantilla & Asociados*
Juanita Olaya Garcia *National Department of Planning*
Ricardo León Otero *Superintendencia Bancaria de Colombia*
Daniel Posse *Posse Herrera & Ruiz*
Zuli Rodríguez *Legal Department División de Personas Jurídicas*
Bernardo Salazar *Brigard & Urrutia*
Paula Samper Salazar *Gomez Pinzon Linares Samper Suarez Villamil*
Carlos Urrutia-Holguin *Brigard & Urrutia*
Juan Manuel Villaveces Hollmann *Computec*

Congo, Dem. Rep. of
Louman Mpoy *Cabinet Louman Mpoy*

Congo, Rep. of
Tahir Souleyman Haggar *Centrale Recapitulative des Risques*
Jean Petro *Cabinet d'Avocats Jean Petro*
Rafael Tung Nsue *La Commission Bancaire de l'Afrique Centrale*

Costa Rica
Bernardo Alfaro Araya *Superintendencia General de Entidades Bancarias de Costa Rica*
Kathya Araya *Facio & Cañas*
Freddy Fachler *Pacheco Coto*
Alfredo Fournier *Beeche Fournier Asociados*
Manuel Gonzalez Sanz *Facio & Cañas*
Tomás Guardia *Facio & Cañas*
Fernando Mora *Rojas Mora Bolanos y Asociados*
Rodrigo Oreamuno *Facio & Cañas*
Mario Quintana *Asesores Juridicos Asociados Doninelli & Quintana*
Luis Monge Sancho *Teletec*

Côte d'Ivoire
Jean-Charles Daguin *Fidafrica Member of PricewaterhouseCoopers*
Vilevo Biova Devo *Centrale des Risques de l'Union Monetaire Ouest Africaine*
Karim Fadika *Fadika-Delafosse-Kacoutie-Anthony*
Colette Kacoutie *Fadika-Delafosse-Kacoutie-Anthony*
Evelyne Mandessi Bell *Mandessi Bell Law Firm*
Ghislaine Moise-Bazie *SCPA Konate Moise-Bazie & Koyo*
Francois Nare *Centrale des Risques de l'Union Monetaire Ouest Africaine*
Georges N'Goan *N'Goan Asman & Associes*
Dominique Taty *Fidafrica Member of PricewaterhouseCoopers*

Croatia
Mladen Duliba *Croatian National Bank*
Marijan Hanzekovic *Hanzekovic & Radakovic*
Zdenko Haramija *Koprer & Haramija*
Irina Jelcic *Hanzekovic & Radakovic*
Sanja Juric *Juric Law Offices*
Davor Juros *Coface Intercredit Croatia*
Vanja Kalogera *Croatian Investment Promotion Agency*
Jerina Malesevic *Koprer & Haramija*
Ana Mataga *Croatian National Bank*
Tin Matić *Matić Law Office*
Zeljko Pazur *Ministry of Finance*
Vlado Sevsek *Vlado Sevsek & Zeljka Brlecic*
Lidija Stopfer *Vukmir Law Office*
Jane Tait *PricewaterhouseCoopers*
Hrvoje Vukic *Vukic Jelušic Sulina Stankovic Jurcan & Jabuka*
Eugen Zadravec *Eugen Zadravec Law Firm*

Czech Republic
Vladimir Ambruz *Ambruz & Dark advokati v.o.s.*
Libor Basl *Baker & McKenzie*
Jiri Bobek *Squire Sanders & Dempsey*
Jiri Cerny *Peterka Leuchterova & Partners*
Tomas Denmark *Czech Banking Credit Bureau*
Andrea Korpasova *Baker & McKenzie*
Petr Kucera *Aspekt Kilcullen*
Jan Molik *Judr Jan Molik Advokat*
Jarmila Musilova *Czech National Bank*
Ivo Nesrovnal *Gleiss Lutz Advokati*
Petr Ríha *Procházka Randl Kubr*

Roman Studnicny *Coface Intercredit Czechia*
Ružena Trojánková *Linklaters & Alliance*
Katerina Trojanova *Czech Banking Credit Bureau*
Ludek Vrána *Linklaters & Alliance*

Denmark
Jens Arnesen *Eversheds*
Susanne Baekvig *Danish Commerce and Companies Agency*
Claus Bennetsen *Accura*
Ole Borch *Bech-Bruun Dragsted*
N.V. Falling Olsen *Poul Schmith Kammeradvokaten*
Ulrik Frirs *Danish Commerce and Companies Agency*
Steen Halmind *Bech-Bruun Dragsted*
Jørn Skovslund Hansen *RKI Kredit Information*
Mikkel Hesselgren *Gorrissen Federspiel Kierkegaard Law Firm*
Jørgen Jepsen *Kromann Reumert*
Jørgen Kjærgaard Madsen *Kromann Reumert*
Poul Meisler *Danish Commerce and Companies Agency*
Pia Møller *Danish Financial Authority*
Thomas Nielsen *Hjejle Gersted & Mogensen*
Kurt Skovlund *Kromann Reumert*

Dominican Republic
Jesus Almanzar *Rojas De Marchena Kaluche & Asociados*
Flavia Baez de George *Pellerano & Herrera*
Luis Heredia Bonetti *Russin Vecchi & Heredia Bonetti*
Ana Isabel Caceres *Troncoso & Caceres*
Franklin Guilamo *De Marchena Kaluche & Asociados*
Hipolito Herrera *V. Pellerano & Herrera*
Porfirio Lopez *Data-Credito*
Xavier Marra *Dhimes & Marra*
Roberto Payano *Superintendencia de Bancos de la Republica Domenicana*
Luis Pellerano *Pellerano & Herrera*
Marcelino San Miguel *CICLA*
Juan Suero *Aaron Suero & Pedersini*
Manuel Tapia *Dr. Ramon Tapia Espinal & Asociados*

Ecuador
Xavier Amador Pino *Estudio Juridico Amador*
Miguel Macías Carmigniani *Macias Hurtado & Macias*
Jose Rumazo Arcos *Perez Bustamante & Ponce Abogados*
Francisco Boloña *Morales Boloña Abogados*
Lucía Cordero-Ledergerber *Falconi Puig Abogados*
Antonio Donoso Naranjo *Superintendencia de Bancos e Seguros, Republica del Ecuador*
Luis Eduardo Garcia *Asesoria Legal Légalas*
Jacob Hidrowoh *Perez Bustamante & Ponce Abogados*
Sebastián Pérez-Arteta *Asesoria Legal Légalas*
Falconi Puig *Falconi Puig Abogados*
Hernan Santacruz *Perez Bustamante & Ponce Abogados*
Santiago Terán Muñoz *Estudio Jurídico Moeller & Cia*
Guillermo Torres *Infaes*

Egypt, Arab Rep. of
Amal Afifi *Dawood Denton Wilde Sapte*
Mohamed Ajsa *Central Bank of Egypt*
Rania Bata *Sarwat A. Shahid Law Firm*
Ashraf Elibrachy *Ibrachy & Dermarkar*
Diaa El-Din Abd Rabou *Central Bank of Egypt*
Sarwat Abd El-Shahid *Sarwat A. Shahid Law Firm Gotshal & Manges*
Ahmed Farid Mohamed El-Sherbiny *Ahmed El-Sherbiny Law Firm*
Samiha Fawzy *The Egyptian Center for Economic Studies*
Karim Adel Kamel *Adel Kamel Law Office*
Mohamed Kamel *Kamel Law Firm*
Katerina Miltiadou *Mecos*
Attef Mohmed *Alfeky Soliman & Partners Advocates*
Mahmoud Shedid *Shalakany Law Office*
Ragy Soliman *Ibrachy & Dermarkar*
Mohamad Talaat *Baker & McKenzie*
Mona Zulficar *Shalakany Law Office*

El Salvador
Francisco Armando *Arias Rivera F.A. Arias & Muñoz*
Roberta Gallardo *F.A. Arias & Muñoz*
Marcela Mancia *F.A. Arias & Muñoz*
Astrud Melendez *Asociacion Protectora de Creditos de el Salvador*
Hilda Morena Segovia *Superintendencia del Sistema Financiero, El Salvador*
Celina Padilla *F.A. Arias & Muñoz*
Jose Roberto *Romero Pineda & Asociados*
Roxana Romero *Romero Pineda & Asociados*

Ethiopia
Teshome Gabre-Mariam Bokan *Teshome Gabre-Mariam*
Debebe Legesse *Debebe Legesse Law Firm*
Lakew Lemma *National Bank of Ethiopia*
Tameru Wondm Agegnehu *Tameru Wondm Agegnehu Law Offices*

Finland
Ahti Auikolinen *Ministry of Labor*
Timo Esko *Esko Timo & Uoti Sami*
Berndt Heikel *Hannes Snellman*
Pekka Jaatinen *Castren & Snellman*
Bernt Juthstrom *Roschier-Holmberg & Waselius*
Kaija Kilappa *Financial Supervision Authority*
Gisela Knuts *Roschier-Holmberg & Waselius*
Patrik Lindfors *Hannes Snellman Attorneys at Law*
Tomas Lindholm *Roschier-Holmberg & Waselius*
Samu Palkonen *Roschier-Holmberg & Waselius*
Mikko Parjanne *Suomen Asiakastieto Oy Finska*
Bekka Rasane *Employment and Economic Development Center*
Mikko Reinikainen *PricewaterhouseCoopers*
Sakari E Sorri *Bützow Nordia*
Sarah Tähkälä *Hannes Snellman*
Sami Tuominen *PricewaterhouseCoopers*
Eeva Vahtera *Ministry of Labor*
Helena Viita *Roschier-Holmberg & Waselius*
Gunnar Westerlund *Roschier-Holmberg & Waselius*

France
Antoine Azam-Darley *Azam-Darley & Associes*
Laurent Barbara *Baker & McKenzie*
Nicolas Barberis *Ashurst Morris Crisp*
Louis Bernard Buchman *Caubet Chouchana Meyer*
Stéphanie Chatelon *Deloitte & Touche Juridique et Fiscal*
John Crothers *Gide Loyrette Nouel*
Bertrand Debosque *Bignon Lebray Delsol & Associes*
Olivier Jaudoin *Banque de France*
Antoine Maffei *De Pardieu Brocas Maffei & Leygonie*
Philippe Prevost *Banque de France*
Alexia Simon *Azam-Darley & Associes*
Laurent Valadoux *Banque de France*
Philippe Xavier-Bender *Gide Loyrette Nouel*

Georgia
Irakli Adeishvili *Okraliashvili & Partners*
Giorgi Begiashvili *Begiashvili & Co.*
Lado Chanturia *Supreme Court of Georgia*
Murtaz Kikoria *National Bank of Georgia*
Victor Kipiani *Mgaloblishvili Kipiani Dzidziguri*
Dimitri Kitoshvili *Okraliashvili & Partners*
Rainer Magold *Baker & McKenzie*
Archil Melikadze *Center for Enterprises Restructuring and Management Analysis*
Avto Namicheishvili *Begiashvili & Co.*
Vakhtang Shepardnadze *Mgaloblishvili Kipiani Dzidziguri*

Germany
Wulf Bach *Schufa*
Jennifer Bierly-Seipp *Gassner Stockmann & Kollegen*
Hans-Joachim Dohr *Federal Financial Supervisory Authority*
Ute Foshag *Hogan & Hartson Raue*
Klaus Günther *Oppenhoff & Rädler-Linklaters & Alliance*
Manfred Heinrich *Deutsche Bundesbank*
Peter Hoegen *Allen & Overy*
Christof Kautzsch *Haarmann Hemmelrath*
Joerg Rossen *Creditreform*
Ingrid Seitz *Deutsche Bundesbank*
Holger Thomas *SJ Berwin Knopf Tulloch Steininger*
Frank Vogel *SJ Berwin Knopf Tulloch Steininger*

Ghana
Reginald Bannerman *Bruce-Lyle Bannerman & Thompson*
Stella Bentsi-Enchill *Lexconsult & Co*
William Fugar *Fugar & Co Legal Practitioners and Notaries Public*
David Hesse *Hesse & Larsey Law Firm*
Kenneth Laryea *Laryea Laryea & Co PC*
D.A.K Mensah *Central Databank*
Sam Okudzeto *Sam Okudzeto & Associates*
Lawrence Otto *Fugar and Company*
Jacob Saah *PricewaterhouseCoopers*
V.J. Dela Selormey *Bank of Ghana*

Greece
Themis Antoniou *Bank of Greece*
Georgios Bazinas *Anagnostopoulos Bazinas Fifis Counsellor & Attorneys at Law*
Ioanna Bokorou *Kyriakides-Georgopoulos Law Firm*
Angeliki Delicostopoulou *A & A Delicostopoulou*
Stefanoyannis Economou *Law Offices Economou and Associates*
John Kyriakides *Kyriakides-Geogropoulos Law Firm*
Konstantinos Mellios *Sarantitis & Partners*
Effie Mitsopoulou *Kyriakides-Geogropoulos Law Firm*
Dimitris Paraskevas *Elias Sp. Paraskevas*
Kleanthis Roussos *Roussos Law Firm*
Victoria Zachopoulou *Tiresias*

Guatemala
Juan Luis Aguilar *Salguero Aguilar & Zarceño*
Alfonso Carrillo *Carrillo & Asociados*
Rodimiro Castaneda *Superintendencia de Bancos Guatemala*
Anabella Chaclan *Arenales & Skinner-Klée*
Guillermo Contreras *Bancared ORBE*
Juan Diaz Lopez *Superintendencia de Bancos Guatemala*
Gabriela Maria Franco *TransUnion*
Rodolfo Fuentes *Protectora de Credito Comercial*
Eduardo Mayora *Dawe Mayora & Mayora*
Alfredo Rodríguez-Mahuad *Rodríguez Archila Castellanos Solares & Aguilar*
Luis Turk Mejia *Superintendencia de Bancos Guatemala*

Guinea
Boubacar Barry *Boubacar Barry Law Firm*

Haiti
Yves Joseph *Bank of the Republic of Haiti*
Louis Gary Lissade *Cabinet Lissade*
Salim Succar *Cabinet Lissade*

Honduras
Tania Casco *Bufete Casco & Asociados*
Jorge Omar Casco *Bufete Casco & Asociados*
Estela Chavez *TransUnion*
León Gómez *B & B Abogados*
Laureano Gutierrez *Bufete Gutierrez Falla*
F. Dario Lobo *Bufete Gutierrez Falla*
Armida Maria Lopez de Arguello *ACZLAW Bufete Internacional de Abogados*
Ulises Mejía *B & B Abogados*
Ana Cristina de Pereira *Comisión Nacional de Bancos y Seguros, Honduras*
Jose Ramon Paz *J.R. Paz & Asociados*
Jose Rafael Rivera Ferrari *J.R. Paz & Asociados*
Rene Lopez Rodezno *Lopez Rodezno & Asociados*
Roberto Zacarias Jr. *Zacarias Aguilar & Asociados*
Violeta Zuniga de Godoy *Comision Nacional de Bancos y Seguros, Honduras*

Hong Kong, China
Andrew Baggio *Baker & McKenzie*
Brian Barron *Baker & McKenzie*
Charles Booth *University of Hong Kong*
Teresa Ma *Linklaters*

Rupert Nicholl *Johnson Stokes & Master*
Richard Tollan *Johnson Stokes & Master*
Jim Hy Wong *Hong Kong Monetary Authority*
Alex Yuen *TransUnion*
Shirley Yuen *TransUnion*

Hungary
Csendes Agnes *Dessewffy Bellák & Partners Law Office*
Barbara Bognar *Hungarian Financial Supervisory Authority*
Tunde Ezsias *Coface Intercredit Hungary*
Gábor Felsen *Köves Clifford Chance Pünder*
Gabor Horvàth *Oppenheim ès Tàrsai Freshfields Bruckhaus Deringer*
Andrea Jardi *Nemeth Haarmann Hemmelrath*
Istvan Nagy *Creditreform Interinfo*
Péter Nógrádi *Nógrádi Law Office*
Klara Oppenheim *Oppenheim ès Tàrsai Freshfields Bruckhaus Deringer*
Ádám Pethő *Interbank Informatics Services*
Konrád Siegler *Baker & McKenzie*
Benedek Sipöcz *Dewey Ballantine*
Gábor Spitz *Haarmann Hemmelrath & Partner*
Ágnes Szent-Ivány *Sándor Szegedi Szent-Ivány*
Erica Voros *Hungarian Financial Supervisory Authority*

India
R. Amurty *Commerce & Co Agency*
Freyan Desai *Kachwaha & Partners*
Rajkumar Dubey *Singhania & Co*
R.J. Gagrat *Gagrat & Co-Advocates & Solicitors*
Vishal Gandhi *Nishith Desai Associates*
Trupti Garach *Brand Farrar Buxbaum LLP*
Ravi Kulkarni *Little & Co*
N. Marwah *Commerce & Co Agency*
Stephen Mathias *Kochhar & Co Bangalore*
Shri Vijay Mathur *Ministry of Finance Department of Revenue Central Board of Direct Taxes*
Dara Mehta *Little & Co*
Ganpat Raj Mehta *India Law Info*
S. K. Mitra *Indian Investment Center*
Ajit Mittal *Reserve Bank of India*
Ravi Nath *Rajinder Narain & Co*
G. S. Ram *Ministry of Labor*
K.K. Ramani *Laws4India*
Abhishek Saket *Singhania & Co*
D.C. Singhania *Singhania & Co*
Suhas Srinivasiah *Kochhar & Co Bangalore*
K. Suresh *Startupbazaar*

Indonesia
Eman Achmad *Lubis Santosa & Maulana*
Abrahem Adrinaaz *PricewaterhouseCoopers*
Andu Ambuml *Investment Coordinating Board*
H.M.U. Fachri Asaari *Warens & Achyar*
Theodoor Bakker *Ali Budiardjo Nugroho Reksodiputro Counsellors at Law*
Steven Bloom *KPMG*
Danmawan Dgayusmam *Investment Coordinating Board*
Erwandi Hendarta *Baker & McKenzie*
Ali Imron Murim *Central Bank of Indonesia*
Darrell Johnson *SSEK Indonesian Legal Consultants*
Timbul Thomas Lubis *Lubis Ganie Surowidjojo*
Bill Macdonald *PricewaterhouseCoopers*
Ferry Madian *Nugroho Reksodiputro*
Yoga Mulya *Baker & McKenzie*
Luhut Pangaribuan *Luhut M.P. Pangaribuan & Partners*
Basuui Sidharta *KPMG*
Ernst Tehuteru *Ali Budiardjo Nugroho Reksodiputro Counsellors at Law*

Iran, Islamic Rep. of
Alexander Aghayan *Alexander Aghayan & Associates*
Behrooz Akhlaghi *Dr. Behrooz Akhlaghi & Associates*
Reza Askari *Foreign Legal Affairs Group*
B.F. Zarin-Ghalam *Banking Information Department*
Katerina Miltiadou *Mecos*
Parviz Savrai *Dr. Parviz Savrai and Associates*
M. Shahabi *Tavakoli & Shahabi Attorneys and Counselors at Law*
B.F. Zarin-Ghalam *Central Bank of the Islamic Republic of Iran*

Ireland
Andrew Bates *Dillon Eustace*
Declan Black *Mason Hayes & Curran*
Tanya Colbert *Mason Hayes & Curran*
Anthony Collins *Eugene F. Collins Solicitors*
Kathryn Copeland *Central Bank of Ireland*
John Doyle *Dillon Eustace*
Melissa Jennings *Arthur Cox*
William Johnston *Arthur Cox*
N. McDonald *Companies Registration Office*
Michael Meghen *Arthur Cox*
David O'Donohoe *Arthur Cox*
Barry O'Neill *Eugene F. Collins Solicitors*
Dermot Rowe *Dublin Corporation*
Maurice Phelan *Mason Hayes & Curran*
Seamus Tighearnaigh *Irish Credit Bureau*
Deirdre Ward *Company Formations International*

Israel
Eli Arbel *Bank of Israel*
Paul Baris *Yigal Arnon & Co.*
Gil Birger *Embassy of Israel in Washington, DC*
Sabina Blank *Small Business Authority of Israel*
Amihud Doron *A. Doron & Co.*
David Drutman *Amihud Doron & Co., Law Offices*
Alex Hertman *S. Horowitz & Co.*
Zvi Howard *Nixon Elchanan Landau Law Offices*
Pinchas Katz *Bank of Israel*
Gideon Koren *Ben Zvi Koren*
Michelle Liberman *S. Horowitz & Co.*
Jakob Melcer *E.S. Shimron I. Molho Persky & Co.*
Vazana Mordechai *Ministry of Finance*
VIVID Management Systems
Stel Pinhasov *Embassy of Israel in Washington, DC*

Eliot Sacks *Herzog Fox & Neeman*
Yaacov Salomon *Lipschutz & Co.*
Asaf Samuel *Lipschutz & Co.*
Ron Storch *Global Credit Services*
Dror Vigdor *Yigal Arnon & Co.*

Italy
Maria Pia Ascenzo *Bank of Italy*
Giuseppe Alemani *Mallet-Prevost Colt & Mosle*
Gian Bruno Bruni *Bruni Gramellini e Associati*
Lisa Curran *Allen & Overy*
Federico Dettori *Gianni Origoni Grippo & Partners*
Giuseppe Godano *Bank of Italy*
Enrico Lodi *CRIF*
Giuseppe Lombardi *Pedersoli Lombardi e Associati*
Stefano Macchi di Cellere *Studio Legale Macchi di Cellere e Gangemi*
Alberto Maria Fornari *Baker & McKenzie Giuseppe Alemani Curtis Mallet-Prevost Colt & Mosle*
Fabrizio Mariotti *Studio Legale Beltramo*
Ida Marotta *The Brosio Casati e Associati*
Francesco Pensato *Franzosi Dal Negro*
Andrea Rescigno *White & Case – Varrenti e Associati*
Nerio Saguatti *Consorzio per la Tutela del Credito*
Pensato Setti *Studio Legale Macchi di Cellere e Gangemi*
Vittorio Tadei *Chiomenti Studio Legale*
Fabio Tortora *Experian Credit Bureau*

Jamaica
Rosslyn Combie Sykes *Nunes Scholefield Deleon & Co.*
Dave Garcia *Myers Fletcher & Gordon*
Gayon Hosin *Bank of Jamaica*
Anthony Jenkinson *Nunes Scholefield DeLeon & Co.*
Derek Jones *Myers Fletcher & Gordon*
Rattray Misheca *Seymour Myers Fletcher & Gordon*
Alfred Rattray *Myers Fletcher & Gordon*
O. J. Rattray *Patterson & Rattray*

Japan
Shinichiro Abe *Credit Information Center Corp*
Naoki Eguchi *Baker & McKenzie*
Tamotsu Hatasawa *Hatasawa & Wakai Law Firm*
Osamu Kawakami *Japan Information Center Corp*
Nobuaki Matsuoka *Yamaguchi International*
Toshio Miyatake *Law Firm Adachi Henderson Miyatake & Fujita*
Satoshi Ogishi *Nishimura & Partners*
Yuji Onuki *Asahi Law Offices*
Jeremy Pitts *Baker & McKenzie*
Setsuko Sato *CCB*
Tomoe Sato *Credit Information Center Corp*
Gaku Suzuki *Asahi Koma Law Offices*
Shinjiro Takagi *Industrial Revitalization Corporation of Japan*
Tadeshi Yokoyama *Financial Services Agency*

Jordan
Sami Al-Louzi *Ali Sharif Zu'Bi & Sharif Ali Zu'Bi*
Nelly Batchoun *Central Bank of Jordan*
Francis Bawab *PricewaterhouseCoopers*
Micheal Dabit *Micheal Dabit & Associates Attorneys at Law*
Salahel Dine *Al-Bashir International Business Legal Associates*
Yousef Khalilieh *Rajai Dajani & Associates Law Office*
Michel Mazto *Ministry of Finance*
Katerina Miltiadou *Mecos*
Shadi Zghoul *DaJani & Associates*
Ali Sharif Zu'bi *Ali Sharif Zu'Bi & Sharif Ali Zu'Bi*

Kazakhstan
Ahmetzhan Abdulaev *Grata Law Firm*
Madiar Balken *Graduate Law Academy Adilet*
John W. Barnum *McGuireWoods, Kazakhstan*
Yuri Bassin *Aequitas*
Yuri A. Bolotov *Michael Wilson & Partners*
Olga Chentsova *Salans*
Mariya Gekko *Baker & McKenzie*
Eric Imashev *McGuireWoods, Kazakhstan*
Kuliash Muratovna Iliasova *Scientific Research Institute for Private Law, Humanities and Law University*
Snezhana V. Popova *McGuireWoods, Kazakhstan*
Jazykbaeva Raushan *Aequitas*
Richard Remias *McGuireWoods, Kazakhstan*
Marla Valdez *Denton Wilde Sapte Law Firm*
Valerie Zhakenov *Zhakenov and Partners, in affiliation with White Savelieva*
Rima Zhakupova *Salans*

Kenya
K.S. Anjarwalla *Kapila Anjarwalla & Khanna Advocates*
Bill Deverell *Kaplan & Stratton*
W.S. Deverell *Kaplan & Stratton*
Oliver Fowler *Kaplan & Stratton*
Fiona Fox *PricewaterhouseCoopers*
Sheetal Kapila *Kapila Anjarwalla & Khanna*
Hamish Keith *Daly & Figgis Advocates*
John Murugu *Central Bank of Kenya*
Wanjiru Nduati *Kaplan & Stratton*
Conrad Nyakuri *PricewaterhouseCoopers*
Fred Ochieng *Kaplan & Stratton*
Richard Omwela *Hamilton Harrison & Mathews Law Firm*
Sonal Sejpal *Kapila Anjarwalla & Khanna Advocates*

Korea, Rep. of
Duck-Soon Chang *First Law Offices of Korea*
Eui Jong Chung *Kim & Lee*
Ju Myung Hwang *Hwang Mok Park & Jin*
James (Ik-Soo) Jeon *Sojong Partners*
Daniel Y. Kim *Sojong Partners*
Gahng Hee Lee *Ministry of Labor*
K. C. Lee *Korea Trade-Investment Promotion Agency*
Dong Chin Lim *Chung & Suh Attorneys at Law*
Sharon Noh *Korea Information Services*
Paul Stephan Penczner *Lee International IP & Law Group*
Kyung-Han Sohn *Aram International Law Offices*
Sung-il Yang *Ministry of Health and Welfare.*

Kuwait
Walid Abd Elrahim Ahmed *Abdullah Kh. Al-Ayoub & Associates*
Abdullah Kh. Al-Ayoub *Abdullah Kh. Al-Ayoub & Associates*
Mishare M. Al-Ghazali *Mishare M. Al-Ghazali & Partners*
Ruba El- Habel *Abdullah Kh. Al-Ayoub & Associates*
Jasmin Kohina *Abdullah Kh. Al-Ayoub & Associates*

Kyrgyz Republic
Julia Bulatova *Law Firm Partner*
Gulnara Kalikova *Chadbourne & Parke*
Natalia Sidorovna Galiampova *Third Arbitrage Court*
Nurlanbek Tynaev *National Bank of the Kyrgyz Republic*
Emil Oskonbale *Sphynx Consult*
Mirgul Smanalieva *Law Firm Partner*
Larisa Tashtemirovna Zhanibekova *Larisa Tashtemirovna Zhanibekova Law Firm*

Lao PDR
Edwards Nicholas *DFDL*
Isabelle Robineau *DFDL*
Louis-Martin Desautels *DFDL*

Latvia
Irina Ivanova *Financial and Capital Markets Commission*
Dace Jenava *A. Jenava Birojs*
Filip Klavins *Klavins Slaidins & Loze*
Valters Kronbergs *Kronbergs Law Office*
Monika Kuprijanova *Council of Sworn Notaries of Latvia*
Juris Puce *Creditreform Latvija*
Anita Tamberga-Salmane *Klavins Slaidins & Loze*
Ugis Treilons *Klavins Slaidins & Loze*
Ziedonis Udris *CB&M Law Firm*
Asnata Venckava *IGK-System*
Romualds Vonsovics *Lejins Torgans & Vonsovics Ziedonis Udris Skudra & Udris*

Lebanon
Antoine Abboud *Law Office of A. Abboud & Associates*
Walid Alamuddin *Banking Control Commission of Lebanon*
Ramy Aoun *Badri and Salim El Meouchi Law Firm*
Raymond Azar *Raymond Azar Law Offices*
Randa Bahsoun *PricewaterhouseCoopers*
Raymonde Eid *Badri and Salim El Meouchi Law Firm*
Sadim El Meouchi *Badri and Salim El Meouchi Law Firm*
Ramzi George *PricewaterhouseCoopers*
Nabil Mallat *Hyam Mallat Law Offices*
Yara Maroun *The Law Offices of Tyan & Zgheib*
Katerina Miltiadou *Mecos*
Fadi Moghaizel *Moghaizel Law Offices*
Chandra Muki *PricewaterhouseCoopers*
Walid Nasser *Walid Nasser & Associates*
Nada Abu Samra *Badri and Salim El Meouchi Law Firm*
Nady Tyan *The Law Offices of Tyan & Zgheib*

Lesotho
Stefan Carl Buys *Du Preez Liebetrau & Co.*
Arshad Farouk *Du Preez Liebetrau & Co.*
Margarete Higgs *Du Preez Liebetrau & Co.*

Lithuania
Renata Berzanskiene *Sorainen Law Offices*
Tomas Davidonis *Sorainen Law Offices*
Dalia Foigt *Regija Law Firm*
Kornelija Francuzeviciute *Bank of Lithuania*
Rolandas Galvenas *Lideika Petrauskas Valiunas ir Partneriai*
Marius Jakulis *Jason AAA Law Firm*
Mindaugas Kiškis *Lideika Petrauskas Valiunas ir Partneriai*
Jurate Kugyte *Lideika Petrauskas Valiunas ir Partneriai*
Marius Navickas *Foresta Business Law Group*
Ramunas Petravicius *Lideika Petrauskas Valiunas ir Partneriai*
Kazimieras Ramonas *Bank of Lithuania*
Laimonas Skibarka *Lideika Petrauskas Valiunas ir Partneriai*
Marius Urbelis *Sorainen Law Offices*
Victor Vaitkevicius *Kredoline*
Rolandas Valiunas *Lideika Petrauskas Valiunas ir Partneriai*

Macedonia, FYR
Zlatko Antevski *Lawyers Antevski*
Dragana Vukobrat *National Bank of the Republic of Macedonia*

Madagascar
Raphaël Jakoba *MCI Law Firm*
Hanta Radilofe *Cabinet Félicien Radilofe*
Theodore Ramangalahy *Commission de Supervision Bancaire et Financiere*
Henri Bernard Razakariasa *Banque Centrale de Madagascar*

Malawi
Robert Atherstone *Stumbles Sacranie Gow & Co.*
Roseline Gramani *Savjani & Co.*
S. E. Jussab *Sacranie Gow & Co.*
Shabir Latif *Sacranie Gow & Co.*
W. R. Milonde *Reserve Bank of Malawi*
Ben Ndau *Savjani & Associates Law Firm*
D.A. Ravel *Wilson & Morgan*
Loganath Sabapathy *Logan Sabapathy & Co.*

Malaysia
Sbdul Rahim Ali *Registrar of Companies*
Francis Chan *Basis Corporation*
H. Y. Chong *Azman Davidson & Co.*
Wong Chong *Wah Skrine*
J. Wilfred Durai *Azlan Zain, Zain & Co.*
Chin Sok Ee *Bank Negara Malaysia*
Wan Hashim *Malaysian Industrial Development Authority*
Mohammad Haszri *Abu Hassan Azmi & Associates*
Ar Karunakaran *The Malaysian Industrial Development Authority*
Christopher Lee *Baker & McKenzie*
Azmi Mohd *Ali Azmi & Associates*
Rajendra Navaratnam *Azman Davidson & Co.*
Loganath Sabapathy *Registrar of Companies*
Francis Tan *Azman Davidson & Co.*
Chung Tze Keog *CTOS Sdn Bhd*
J. Wilfred *Durai Zain & Co.*
Azlan Zain *Zain & Co.*

Mali
Vilevo Biova Devo *Centrale des Risques de l'Union Monetaire Ouest Africaine*
Seydou Ibrahim Maiga *Cabinet d'Avocats Seydou Ibrahim Maiga*
Francois Nare *Centrale des Risques de l'Union Monetaire Ouest Africaine*

Mauritania
A. S. Bouhoubeyni *Cabinet Bouhoubeyni*
Ould Bouhoubeyni *Ahmed Salem Ould Bouhoubeyni Ahmed Salem Law Firm*

Mexico
Gerardo Carreto-Chávez *Barrera Siqueiros y Torres Landa Attorneys at Law*
María Casas *Baker & McKenzie*
Carlos Grimm *Baker & McKenzie*
Eduardo Heftye *López Velarde Heftye y Soria*
Bill Kryzda *Goodrich Riquelme Y Asociados*
Jorge Leon-Orantes *Goodrich Riquelme Y Asociados*
Eduardo Llamosa *Profancresa*
Enrique Nort *Comision Nacional Bancaria y de Valores*
Pablo Perezalonso *Ritch Heather y Mueller*
Jose Luis Quiroz *Mateos Winstead y Rivera*
Rafael Ramirez Arroyo *Martínez Algaba Estrella De Haro y Galvan-Duque*
Juan Manuel Rincón *Franck Galicia y Robles*
Arturo Saavedra Rodríguez *Rodríguez Vega Rubio Y Asociados*
Martinez Arrieta Rodríguez *Vega Rubio Y Asociados*
Carlos Sanchez-Mejorada *Sanchez-Mejorada y Pasquel*
Juan Francisco Torres-Landa *R. Barrera Siqueiros y Torres Landa SC Attorneys at Law*

Moldova
David Brodsky *Brodsky Uskov Looper Reed & Partners*
Procop Buruiana *Buruiana & Partners*
Stela Cibotari *National Bank of Moldova*
Victoria Ciofu *National Bank of Moldova*
Iurie Lungu *Levintsa & Associates*
Victor Levintsa *Levintsa & Associates*
Irina Moghiliova *Brodsky Uskov Looper Reed & Partners*
Alexander Turcan *Turcan & Turcan*

Mongolia
Bayarmaa Badarch *Lynch & Mahoney*
Batzaya Bodikhuu *Anderson & Anderson Mongolia*
David Buxbaum *Anderson & Anderson Mongolia*
L. Chimgee *Bank of Mongolia*
Maurice Lynch *Lynch & Mahoney*
Daniel Mahoney *Lynch & Mahoney*
Ulziideleg Taivan *Credit Information Bureau*

Morocco
Myriam Bennani *Hajji & Associés Association d'Avocats*
Richard Cantin *Cabinet Naciri & Associés*
Frédéric Elbar *C.M.S. Bureau Francis Lefebvre Maroc*
Amin Hajji *Amin Hajji Law Offices*
Azzedine Kettani *Kettani Law Firm*
Nadia Kettani *Kettani Law Firm*
Ahmed Lahrache *Bank Al-Maghrib*
Hicham Naciri *Cabinet Naciri & Associés*
Mehdi Salmouni-Zerhouni *Hajji & Associés Association d'Avocats*

Mozambique
Alexandra Carvalho *Vasconcelos Porto & Asociados*
Carlos de Sousa e Brito *Carlos de Sousa e Brito & Associados*
Antonio de Vasconcelos Porto *Vasconcelos Porto & Asociados*
Aquiles Dimene *Vasconcelos Porto & Asociados*
Rita Furtado *H.Gamito, Cuito, Goncalves Pereira, Castelo Branco & Associado*
Joao Martins *PricewaterhouseCoopers*
Carol Christie Smit *American Embassy in Maputo*
Bonifácia Mario Suege *Bank of Mozambique*
Eric Whitaker *American Embassy in Maputo*

Namibia
Hanno Bossau *Lorentz & Bone*
Natasha Cochrane *P.F. Koep & Co.*
Peter *P.F. Koep & Co* Frank Koep *P.F. Koep & Co.*
Richard Mueller P.F. Koep & Co.
Phillip Mwangala *Bank of Namibia*
Deon Obbes *Lorentz & Bone*
Marius van Breda *Information Trust Corporation*

Nepal
Indra Lohani *Dhruba Bar Singh Thapa & Associates*
Surendra Man Pradhan *Nepal Rastra Bank*
Kusum Shrestha *Kusum Law Firm*
Sudheer Shrethha *Kusum Law Firm*
Sajjan Thapa *Dhruba Bar Singh Thapa & Associates*
Bharat Rej Upreti *Pioneer Law Associates*

Netherlands
Rob Abendroth *Allen & Overy*
Casper Banz *Houthoff Buruma*
Michiel Gorsira *Simmons & Simmons*
Glenn Haulussy *Haulussy Advokaten*
M. de Kogel *De Netherlandshe Bank*
R. Koster *Chamber of Commerce Amsterdam*
Joop Lobstein *Stichting Bureau Krediet Registratie*
L. Moll *Chamber of Commerce*
Piet Schroeder *Baker & McKenzie*
Jaap-Jan Trommel *NautaDutilh Attorneys*
Peter Wakkie *De Brauw Blackstone Westbroek*
Marcel Willems *Kennedy Van der Laan*

New Zealand
Tim Buckley *Chapman Tripp*
Niels Campbell *Bell Gully*
Margaret Griffin *Reserve Bank of New Zealand*
Paul Heath *High Court of New Zealand*
Janine Jackson *Baycorp Advantage*
Kirri Lynn *Companies' Office*
Laurence Mayne *Russell McVeagh*

Lee-Ann McArthur *NZ Companies Office*
Richard Peach *Baycorp Advantage*
Nicola Penman-Chambers *Simpson Grierson*
Charlotte Rose *Simpson Grierson*
Douglas Seymour Alderslade *Chapman Tripp*
Peter Sheerin *Baycorp Advantage*
Arthur Young *Chapman Tripp*

Nicaragua
Roberto Arguello *F.A. Arias & Muñoz Law Firm*
Carlos Bonilla *Superintendencia de Bancos y de Otras Instituciones Financieras*
Jose Evenor Taboada *Taboada & Asociados*
María José Guerrero *F.A. Arias & Muñoz Law Firm*
Pedro Muñoz *F.A. Arias & Muñoz Law Firm*
Ana Rizo *F.A. Arias & Muñoz Law Firm*
Oscar Silva *Delaney & Associates*

Niger
Vilevo Biova Devo *Centrale des Risques de l'Union Monetaire Ouest Africaine*
Samna Daouda *Ohada Legis*
Aïssatou Djibo *Maitre Djibo Aissatou*
Bernar-Oliver Kouaovi *Cabinet Kouaovi*
Francois Narem *Centrale des Risques de l'Union Monetaire Ouest Africaine*

Nigeria
Lara Ademola *Lara Ademola & Co.*
John Adetiba *PricewaterhouseCoopers*
Daniel Agbor *Udo Udoma & Belo-Osagie*
Oluseyi Abiodun Akinwunmi *Akinwunmi & Busari*
Samuel Etuk *Etuk & Urua*
Anse Ezetha *Chief Law Agu Ezetah & Co.*
Mohammed Ibrahim *Embassy of Nigeria in Washington, DC*
O. I. Imala *Central Bank of Nigeria*
Evelyne Mandessi Bell *Mandessi Bell Law Firm*
Ndubisi Chuks Nwasike *Chuks Nwasike Solicitor*
Chike Obianwu *Udo Udoma & Belo-Osagie*
Uzoma Ogbonna *Chief Law Agu Ezetah & Co.*
Joy Okeaya-Inneh *Chief Rotimi Williams' Chambers*

Norway
Edgar Barsgoe *Ministry of Labor and Government Administration*
Morten Beck *Advokatfirmaet PricewaterhouseCoopers*
Frode Berntsen *Advokatfirmaet PricewaterhouseCoopers*
Paul Buche *Tax Law Department*
Lars Carlsson *Creditinform*
Finn Erik *Engzelius Thommessen Greve Lund*
Stein Fagerhaug *Thommessen Greve Lund*
Claus Flinder *Simonsen Føyen Advokatfirma*
Hans Haugstad *Thommessen Greve Lund*
Aase Aa. Lundgaard *Deloitte Touche Tohmatsu*
Glenn McKenzie *Brønnøysund Register Centre*
Guri Midttun *Norwegian Trade Council*
Christian Mueller *Thommessen Greve Lund*
Finn Rime *Rime & Co. Advokatfirma*
Vegard Sivertsen *Deloitte & Touche, Norway*
Lisbeth Strand *The Banking, Insurance and Securities Commission of Norway*
Anne Thorsheim *Oslo Business*
Elste Torsvik *Ministry of Labor and Government Administration*
Sverre Tyrhaug *Thommessen Greve Lund*
Preben Willoch *Advokatfirmaet PricewaterhouseCoopers*

Oman
Mansoor Jamal *Malik Al Alawi Mansoor Jamal & Co.*

Pakistan
Masood Khan Afridi *Afridi & Angell & Khan*
M. Bilal Aftab *News-VIS Credit Information Services*
Shamim Ahmed *Securities and Exchange Commission*
Salman Aslam Butt *Cornelius Lane & Mufti*
Mohammad Azam Chaudhry *Azam Chaudhry Law Associates*
Syed Ahmad Hassan *Shah Afridi & Angell & Khan*
Ishrat Husain *State Bank of Pakistan*
Kairas Kabraji *Kabraji & Talibuddin*
Muhammad Khalid Javed *Board of Investment, Pakistan*
Muhammad Akram Khan *Board of Investment, Pakistan*
Sikandar Hassan Khan *Cornelius Lane & Mufti*
Rashad Miyan *Board of Investment, Pakistan*
Babar Mufti *International Credit Information*
Amna Piracha *International Credit Information*
Talat Rasheed *Board of Investment, Pakistan*
Muhammad Saleem *Credit Information Bureau*
Haider Shamsi *Haider Shamsi and Co.*

Panama
Leonor Alvarado *Alvarado Ledezma & De Sanctis*
Ebrahim Asvat *Patton Moreno & Asvat*
Eric Britton *Infante Garrido & Garrido Abogados*
Delia Cardenas *Superintendencia de Bancos de Panama*
Julio Cesar Contreras III *Arosemena Noriega & Contreras*
Jorge Garrido *M. Infante Garrido & Garrido*
Francisco Pérez Ferreira *Patton Moreno & Asvat*
Lizbeth Ramsey *Asociación Panameña de Crédito*
Analita Romero *KPMG*
Juan Tejada Mora *Icaza Gonzalez-Ruiz & Aleman*

Papua New Guinea
Kirsten Kobus *Allens Arthur Robinson*
Vincent Bull *Allens Arthur Robinson*
Rio Fiocco *Posman Kua Aisi Lawyers*
Richard Flynn *Blake Dawson Waldron*

Paraguay
Hugo Berkemeyer *Berkemeyer Attorneys and Counselors*
Luis Breuer *Berkemeyer Attorneys and Counselors*
Esteban Burt *Peroni Sosa Tellechea Burt & Narvaja*

Peru
Marco Antonio Alarcón Piana *Estudio Luis Echecopar Garcia*
Luis Felipe Arizmendi Echecopar *Superintendencia de Bancos y Seguros del Peru*

Guilhermo Alceu Auler *Muniz Forsyth Ramirez Perez-Taiman & Luna Victoria*
Luís Fuentes *Barrios Fuentes Urquiaga*
Manuel Olaechea *Du Bois*
Alonso Rey Bustamante *Payet Rey Cauvi Abogados*
Ricardo Silva *Muniz Law Firm*
Manuel Villa-García *Estudio Olaechea*
Gino Zolezzi *Certicom*

Philippines
Marissa Acain *PhilBizInfo*
Theresa Ballelos *Baker & McKenzie*
Manuel Batallones *BAP Credit Bureau*
Angelica Cayas *Board of Investment*
Kenneth Chua *Castillo Laman Tan Pantaleon & San Jose*
Emerico De Guzman *Angara Abello Concepcion Regala & Cruz*
Benjamin Dela Cruz *Board of Investments*
Mila Digan *Board of Investments*
Nestor Espenilla *Central Bank of the Philippines*
Gilberto Gallos *Abello Concepción Regala & Cruz*
Andres Gatmaitan *Sycip Salazar Hernandez & Gatmaitan*
Tadeo Hilado *Abello Concepcion Regala & Cruz*
Natividad Kwan *Baker & McKenzie*
Romeo Mendoza *Romulo Mabanta Buenaventura Sayoc & de Los Angeles*
Yolanda Mendoza-Eleazar *Castillo Laman Tan Pantaleon & San Jose.*
Efren Lee No *Investment Management Department*
Nicanor Padilla *Siguion Reyna, Montecillo & Ongsiako Law Offices*
Polo Pantaleón *Castillo Laman Tan Pantaleon & San Jose*
Emmanuel Paras *Cecile M.E. Caro*
Teodoro Regala *Angara Abello Concepcion Regala & Cruz*
Ricardo Romulo *Romulo Mabanta Buenaventura Sayoc & de Los Angeles*
Roger Sapanta *Board of Investments*
Tess Sianghio-Baac *Abello Concepcion Regala & Cruz*
Cirilo T Tolosa *Sycip Salazar Hernandez & Gatmaitan*

Poland
Tomasz Brudkowski *Kochanski Brudkowski & Partners*
Renata Cichocka *Haarmann Hemmelrath*
Slawomir Domzal *Biuro Informacji Kredytowej*
Maciej Duszczyk *Biuro Informacji Kredytowej*
Paweł Ignatjew *Baker & McKenzie*
Iwona Janeczek *Commercial Debtor Register/KSV Information Services*
Tomasz Kanski *Soltysiñski Kawecki & Szlezak*
Katarzyna Kompowska *Coface Intercredit Poland*
Petr Kucera *Aspekt Kilcullen*
Wojciech Kwasniak *National Bank of Poland*
Bartlomiej Raczkowski *Soltysiñski Kawecki & Szlezak*
Jean Rossi *Gide Loyrette Nouel Polska*
Tomasz Stawecki *Baker & McKenzie*
Przemyslaw Pietrzak *Nörr Stiefenhofer Lutz*
Robert Siuchmo *Biuro Informacji Kredytowej*
Anna Talar Jeschke *Haarmann Hemmelrath*
Tomasz Turek *Nikiel & Zacharzewski*
Tomasz Wardynski *Wardynski & Partners*
Robert Windmill *Haarmann Hemmelrath*
Steven Wood *TGC Polska Law Firm*

Portugal
Fernando Resina Da Silva *Vieira de Almeida & Associados*
João Cadete de Matos *Banco de Portugal*
Cristina Dein *Jalles Advogados*
Rosemary de Rougemont *Neville de Rougemont & Associados*
Carlos de Sousa e Brito *Carlos de Sousa e Brito & Asociados*
Paulo Lowndes *Marques Abreu & Marques Vinhas e Associados*
Fernando Marta *Credinformacoes*
Inês Batalha Mendes *Abreu Cardigos & Asociados*
Miguel de Avillez Pereira *Abreu Cardigos & Asociados*
Vicky Rodriguez *Neville de Rougemont & Asociados Sociedade de Advogados*
Ana Isabel Vieira *Banco de Portugal*

Puerto Rico
Vicente Antonetti *Goldman Antonetti & Cordova*
Marcelo Lopez *Goldman Antonetti & Cordova*

Romania
Philip Ankel *Moore Vartires & Associates SCPA*
Tiberiu Csaki *Altheimer & Gray Moore*
Teodor Gigea *Coface Intercredit Romania*
Veronica Gruzsnicki *Babiuc Sulica & Associates*
Andrea Ionescu *Altheimer & Gray Moore*
Corina Gabriela Ionescu *Nestor Nestor Diculescu Kingston Petersen*
Nicoleta Kalman *Nicoleta Kalman Law Office*
Daniel Lungu *Racoti Predoiu & Partners*
Elena Mirea *Delos Creditinfo*
Ion I. Nestor *Nestor Nestor Diculescu Kingston Petersen*
Theodor Nicolescu *Theodor Nicolescu Law Office*
David Stabb *Sinclair Roche & Temperley*
Arin Octav Stanescu *National Association of Practitioners in Reorganization and Winding Up*
Paraschiva Suica-Neagu *Nestor Nestor Diculescu Kingston Petersen*
Valeria Tomesou *Credit reform Romania*
Catalin Tripon *Babiuc Sulica & Associates*
Florentin Tuca *Musat & Asociatii*
Petre Tulin *National Bank of Romania*
Perry Zizzi *Moore Vartires & Associates SCPA*

Russian Federation
Irina Astrakhan *PricewaterhouseCoopers*
Peter Barenboim *Moscow Interbank Currency Exchange*
Christian Becker *Haarmann Hemmelrath & Partner*
Maria Blagowolina *Haarmann Hemmelrath & Partner*
Vladimir Dragunov *Baker & McKenzie*
Igor Gorchakov *Baker & McKenzie*
John Hammond *CMS Cameron McKenna*
David Lasfargue *Gide Loyrette Nouel*
Sergei Lazarev *Russin & Vecchi*

Ludmila Malykhina *CMS Cameron McKenna*
Alexey Simanovskiy *Bank of Russia*
Vladislav Talantsev *Russin & Vecchi*

Rwanda
Jean Haguma *Haguma & Associes*
Angelique Kantengwa *National Bank of Rwanda*

Saudi Arabia
Fahd Al-Mufarrij *Saudi Arabian Monetary Agency*
Mujahid Al-Sawwaf *Law Offices of Dr. Mujahid M. Al-Sawwaf*
Mohammed Jaber *Nader Nader Law*
Hassan Mahassni *Law Offices of Hassan Mahassni*
Francois Majdy *Kasseem Al-Fallaj Law Firm*
Katerina Miltiadou *Mecos*
Akram Mohamed *Nader Nader Law*
Sameh Toban *Toban Law Firm*
Ebaish Zebar *Law Firm of Salah Al-Hejailany*

Senegal
Vilevo Biova Devo *Centrale des Risques de l'Union Monetaire Ouest Africaine*
Aboubacar Fall *Fall Associates Law Offices*
Cheikh Fall *Cheikh Fall Law Offices*
Mame Adama Gueye *SCP Mame Adama Gueye & Associes*
Mamadou Mbaye *SCP Mame Adama Gueye & Associés*
Ibrahima Mbodj *Etude Maitre Ibrahima Mbodj*
Francois Nare *Centrale des Risques de l'Union Monetaire Ouest Africaine*
François Sarr *François Sarr & Associes*
Mamadou Seck *SCP Sow Seck*

Serbia and Montenegro
Miroslav Basic *Studio Legale Sutti*
Yorgos Chairetis *IKRP Rokas & Partners*
Ilija Drazic *Drazic Lazarevic & Beatovic*
Kerim Karabdic *Advokati Salih & Kerim Karabdic*
Dubravka Kosic *Kosic & Sutti*
Nikola Kosic *Agency Sportnet DiN*
Mirko Lovric *National Bank of Serbia and Montenegro*
Neli Markovic *Credit Information System*
Milos Zivkovic *Zivkovic & Samardzic Law Office*

Sierra Leone
Emmanuel Roberts *Roberts & Partners*

Singapore
Leslie Chew *SC Khattar Wong & Partners*
Tan Peng Chin *Tan Peng Chin*
Cheah Swee Gim *Kelvin Chia Partnership*
Deborah Evaline *Barker Khattar Wong & Partners*
Ng Wai King *Venture Law*
Tham Yew Kong *Monetary Authority of Singapore*
Angela Lim *Baker & McKenzie*
Daphne Teo *Monetary Authority of Singapore*
Lincoln Teo *Credit Bureau Singapore*
Lee Kuan Wei *Venture Law*
Jennifer Yeo *Yeo-Leong & Peh*
Samuel Yuen *David Lim & Partners*

Slovak Republic
Martin Bednár *HMG & Partners*
Katarina Cechova *Advokátska kancelária*
Milan Horvath *National Bank of Slovakia*
Tomáš Kamenec *Dedák & Partners*
Renátus Kollár *Allen & Overy*
Petr Kucera *Aspekt Kilcullen s*
Vladimir Malik *Coface Intercredit Slovakia*
Čechová Rakovský *Advokátska kancelária*
Zuzana Valerova *PricewaterhouseCoopers*

Slovenia
Crtomir Borec *Deloitte & Touche*
Stane Berlec *Trade and Investment Promotion Office*
Simon Bracun *Colja Rojs & Partnerji*
Petra Drobne *Small Business Development Center*
Joze Golobic *Small Business Development Center*
Vilma Hanzel *Bank of Slovenia*
Sreco Jadek *Odvetniska Fisarna Jadek & Pensa*
Andrej Jarkovič Selih *Selih Janezic & Jarkovic*
Denis Kostrevc *Deloitte & Touche*
Gerald Lambert *Deloitte & Touche*
Klemen Sesok *Deloitte & Touche*
Irena Skocir *Coface Intercredit Slovenija*
Barbara Smolnikar *SKB Banka DD*

South Africa
Marianne Brown *Institute for Public Finance and Auditing*
Peter Eugene *Whelan Bowman Gilfillan Findlay & Tait*
Mike Forsyth *Austen Smith Attorneys*
David Garegae *Greater Pretoria Metropolitan Council*
Tim Gordon-Grant *Bowman Gilfillan*
Desere Jordaan *LT Attorneys Notaries & Conveyancers*
Renee Kruger *Webber Wentzel Bowens*
Francis Manickum *Department of Trade and Industry*
Andrew Muir *Austen Smith Attorneys*
Johan Neser *Cliffe Dekker*
Laurence Pereira *Vorster Pereira*
Joe Pietersen *South Africa Reserve Bank*
Hugo Stark *South Africa Reserve Bank*
Jacques Van Wyk *Cliffe Dekker*
Greg Ward *TransUnion ITC*
David Watkins *Bowman Gilfillan*
Phillip Webster *LeBoeuf Lamb Greene & MacRae*
Ralph Zulman *Supreme Court of Appeal of South Africa*

Spain
Agustí Bou Maqueda *Jausas, Nadal & Vidal*
Ariadna Cambronero *Uría & Menéndez*
Soledad Cruces de Abia *Bank of Spain*
Sergio del Bosque *Uría & Menéndez*
Anselmo Diaz Fernández *Bank of Spain*
Alejandro Ferreres *Uría & Menéndez*
Ana Just *Iuris Valls Abogados*

Alfonso Pedrajas *Mullerat*
Arturo Rainer Pan *Echecopar Abogados Law Firm*
Eduardo Rodriguez *Rovira Uria & Menendez*
Maria Gracia Rubio *Baker & McKenzie*
Rafael Sebastián *Uría & Menéndez*
Miguel Torres *The Bufete Mullerat Law Firm*
Carlos Valls *Iuris Valls Abogados*
Carlos Viladás *Jené Uría & Menéndez*

Sri Lanka
Asanka Abeysekera *Tichurelvam Associates*
N. P. H. Amarasena *Credit Information Bureau of Sri Lanka*
Savantha De Saram *D. L. & F. De Saram*
Sharmela De Silva *Tichurelvam Associates*
Desmond Fernando *Fernando & Co.*
T.G. Gooneratne *Julius & Creasy Solicitors Attorneys at Law*
Ananda Lecamwasam *PricewaterhouseCoopers*
Ramani Muttetuwegama *Tichurelvam Associates*
Kandiah Neelakandan *Kandiah Neelakandan Law Firm*
Aruni Rajakariar *National Development Bank*
P. Samarasiri *Central Bank of Sri Lanka*
R. Senathi *Rajah Julius & Creasy Solicitors Attorneys at Law*
Niranjan Sinnethamby *Tiruchelvam Associates*
Neelan Tiruchelvam *Tiruchelvam Associates*
John Wilson Jr. *John Wilson Partners*

Sweden
Mats Berter *Magnusson Wahlin Qvist Stanbrook Advokatbyra*
Tommy Bisander *UC AB*
Vibekke Eliasson *Finansinspektionen*
Jörgen Estving *Magnusson Wahlin Qvist Stanbrook Advokatbyrå*
Elisabet Fura-Sandstrom *Advokatfirman Vinge & KB*
Leif Gustafsson *Baker & McKenzie*
Eric Halvarsson *Hammarskiöld & Co.*
Peder Hammarskiöld *Hammarskiöld & Co.*
Paula Hammarstrom *Andersson Magnusson Wahlin Qvist Stanbrook Advokatbyra*
Stefan Holmberg *Gärde Wesslau*
John Henwood *Robinson Bertram*
Margret Inger *Finansinspektionen*
Mattias Larsson *Advokatfirman Cederquist KB*
Knox Nxumalo *Robinson Bertram*
Lars Nylund *Advokatfirman Fylgia*
Cecilia Rembert *Invest in Sweden Agency*
Martin Wallin *Linklaters Lagerlöf*

Switzerland
Peter R. Altenburger *Altenburger & Partners*
Karl Arnold *Pestalozzi Lachenal Patry*
Vischer Frédéric Bétrisey *Baker & McKenzie*
Christian Etter *Swiss Embassy in Washington, DC*
Rolf Gertsch *Swiss Federal Banking Commission*
Erwin Griesshammer *Vischer*
Hans R. Hintermeister *ZEK Switzerland*
Iur. Yvonne Hintermeister *Handelsregisteramt des Kantons Zurich*
Andrea Molino *Spiess Brunoni Pedrazzini Molino*
Guy-Philippe Rubeli *Pestalozzi Lachenal Patry*
Kurt Spinnler *Swiss Federal Banking Commission*

Syrian Arab Republic
Kanaan Al-Ahmar *Al-Ahmar & Partners*
Hani Bitar *Syrian Arab Consultants Law Office*
Riad Daoudi *Syrian Arab Consultants Law Office*
Antoun Joubran *Syrian Arab Consultants Law Office*
Muhammed Jumma *Bank of Syria*
Fadi Kardous *Kardous Law Office*
Katerina Miltiadou *Mecos*
Moussa Mittry *Louka & Mitry*
Gabriel Oussi *Syrian Arab Consultants Law Office*

Taiwan, China
Jack J. T. Huang *Jones Day*
Serina Chung *Jones Day*
Julie Chu *Jones Day*
Angela Wu *Yangming Partners*
Mark Ohlson *Yangming Partners*
Edgar Chen *Tsar & Tsai Law Firm*
John Chen *Formosa Transnational Attorneys at Law*
Helen Chou *Russin & Vecchi LLC*
Patrick Pai-Chiang Chu *Lee and Li*
Joyce Fan *Lee and Li*
James Hwang *Tsar & Tsai Law Firm*
Edward Lai *Central Bank of China*
Bee Leay Teo *Baker & McKenzie*
Justin Liang *Baker & McKenzie*
Jeffrey Lin *Joint Credit Information Center*
Jennifer Lin *Tsar & Tsai Law Firm*
Jen Kong Loh *Alliance International Law Offices*
Thomas McGowan *Russin & Vecchi LLC*
Shiau Pan Yang *Lee and Li*

Tanzania
Naimi Dyer *Mkono & Co. Law Firm*
Ademba Gomba *Gomba & Co. Advocates*
A. K. Kameja *Kameja & Nguluma Advocates*
Wilbert Kapinga *Mkono & Co. Law Firm*
Pauline Kasonda *Mkono & Co. Law Firm*
Ishengoma Masha *Mujulizi & Magai Advocates*
L.H. Mkila *Bank of Tanzania*
Nimrod Mkono *Mkono & Co. Law Firm*
Charles Rwechungura *Maajar Rwechungura & Kameja*
Maajar Rwechungura *Kameja & Nguluma Advocates*
Constantine Rweyemamu *Mutalemwa Masha Mujulizi & Magai Advocates*
Henry Sato *Massaba Kameja & Nguluma Advocates*
Leopold Thomas *Kagula Kalunga & Company*

Thailand
Rujira Bunnag *Marut Bunnag International Law Office*
Vira Kammee *International Legal Counsellors Thailand*
Khun Kanok *Thailand-US Business Council*
Komkrit Kietduriyakul *Baker & McKenzie*
Dej-Udom Krairit *Dej-Udom & Associates*
K. Kunjara *Thai Credit Bureau*
David Lyman *Tilleke & Gibbins International*
Steven Miller *Johnson Stokes & Master*

Cynthia Pornavalai *Tilleke & Gibbins International*
Nuttida Samalapa *Baker & McKenzie*
Anongporn Thanachaiary *Tilleke & Gibbins International*
Boonchai Thaveekittikul Boonchai *Arthur Andersen*
Harold Vickery Jr. *Vickery & Worachai*
Pimvimol Vipamaneerut *Tilleke & Gibbins International*
Prapakorn Wannakano *Bank of Thailand*

Togo
Jean-Marie Adenka *Cabinet Adenka*
Vilevo Biova Devo *Centrale des Risques de l'Union Monetaire Ouest Africaine*
Francois Nare *Centrale des Risques de l'Union Monetaire Ouest Africaine*

Tunisia
Badreddine Barkia *Central Bank of Tunisia*
Bouaziz Belaiba *Yasmina Sorenco*
Adly Bellagha *Adly Bellagha & Associates*
Lamine Bellagha *Adly Bellagha and Associates*
Celine Dupont *Ferchiou & Associates Meziou Knani*
Salaheddine Caid Essebsi *The Salaheddine Caid Essebsi & Associates*
Faiza Feki *Central Bank of Tunisia*
Noureddine Ferchiou *Ferchiou & Associates Meziou Knani*
Elyès Ben Mansour *Gide Loyrette Nouel Tunisie*
Faouzi Mili *Mili and Associates*
Ilhem Ouanes *Tekaya Ferchiou & Associes*
Kamel Ben Salah *Gide Loyrette Nouel Tunisie*

Turkey
Burcu Acarturk *Pekin & Pekin*
I. Hakki Arslan *Central Bank of the Republic of Turkey*
Erol Bircanoglu Jr. *Bircanoglu Law Firm*
Ibrahim Canakci *Banking Regulation and Supervision Agency*
Mesut Cakmak *Cakmak Ortak Avukat Burosu*
Zeynep Cakmak *Cakmak Ortak Avukat Burosu*
Fadlullah Cerrahoglu *Mehmet Can Ekzen*
Kazim Derman *KKB Kredi Kayit Burosu*
Semiha Gorgulu *Yamaner & Yamaner*
Ali Gozutok *Pekin & Pekin*
Selen Gures *Law Offices of M. Fadlullah Cerrahoglu*
Fahri Okumus *Central Bank of the Republic of Turkey*
Sebnem Onder *Cakmak Ortak Avukat Burosu*
Eser Ozer *Anorbis Uluslararasi Bilgi Merkezi*
Ahmed Pekin *Pekin & Pekin*
Y. Selim Sariibrahimoglu *DTB Dis Ticaret Bilgi Merkezi*
Yesim Sezgingil *DTB Dis Ticaret Bilgi Merkezi*
Paul Sheridan *Denton Wilde Sapte & Guner*
Selcuk Tayfun Ok *Chamber of Commerce*
Aysegül Yalçinmani *Law Offices of M. Fadlullah Cerrahoglu*
Mehtap Yildirim-Ozturk *Cakmak Ortak Avukat Burosu*

Uganda
Justine Bagyenda *Bank of Uganda*
Moses Jurua Adriko *Adriko & Karugaba Advocates*
Oscar Kambona *Kampala Associated Advocates*
Masembe Kanyerezi *Mugerwa & Masembe*
Sim Katende *Katende Sempebwa & Co. Advocates*
David Mpanga *Mugerwa & Masembe Advocates*
Gabriel Mpubani *Gabriel Mpubani Law Offices*
Rose Namarome *Odere & Nalyanya Law Firm*
Charles Odere *Odere & Nalyanya Law Firm*
Justin Semuyaba *Semuyaba Iga & Co. Advocates*
Alan Shonubi *Shonubi Musoke & Co.*

Ukraine
Valeria Kazadarova *Baker & McKenzie*
James T. Hitch *Baker & McKenzie*
Olyana Rudyakova *Baker & McKenzie*
Oleg Alyoshin *Vasil Kisil & Partners*
Natalia Artemova *Grischenko & Partners*
Daniel Bilak *Jurvneshservice Attorneys & Counsels*
Serhiy Chorny *Baker & McKenzie*
Olexandr Fedoriv *Credit Rating Agency SlavRating*
Anna Globina *Altheimer & Gray*
Yaroslav Gregirchak *Magister & Partners*
James Hitch III *Baker & McKenzie*
Ruslan Israpilov *Grischenko & Partners*
Aleksandr Kireyev *National Bank of Ukraine*
Sergei Konnov *Konnov Law Offices*
Svetlana Kustova *Konnov Law Offices*
Olexander Martinenko *Scott and Martinenko Law Firm*
Andrii Palianytsia *LCPS*
Markian Silecky *Silecky Law Firm*
Mykola Stetsenko *Scott and Martinenko Law Firm*
Sergei Voitovich *Grischenko & Partners*
Alexander Yefimov *Alexander Yefomiv Law Offices*
Oleg Zinkevych *Kravets & Levenets*

United Arab Emirates
Murad Abida *Hadef Al Dhahiri & Associates*
Bashir Ahmed *Afridi & Angell*
Saeed Abdulla Al Hamiz *Central Bank of the United Arab Emirates*
Habib Al Mulla *Habib Al Mulla & Co.*
Hassen Ferris *Afridi & Angell*
Nabil Issa *Afridi & Angell*
Katerina Miltiadou *Mecos*
Stephen Rodd *Bryan Cave*
Jonathan Silver *Clyde & Co.*

United Kingdom
Kenneth Baird *Freshfields Bruckhaus Deringer*
Richard Boulton *Financial Services Authority*
Greg Boyd *Baker & McKenzie*
Richard Clark *Slaughter & May*
John Hadlow *Experian*
Andrew Haywood *Attorney at Law*
Michael Prior *Shawn Coulson International Lawyers*
Milton Psyllides *Eversheds Law Firm*
Kathy Smith *Slaughter & May*
Michael Steiner *Denton Wilde Sapte*
Philip Wood *Allen & Overy*
John Young *Eversheds Low Firm*

United States
David Adkins *Federal Reserve Board*
Richard Broude *Law Offices of Richard F. Broude*
Peter Chaffetz *Clifford Chance*
Larry Haas *Baker & McKenzie*
Charles Kerr *Morrison & Foerster*
Erik Lindauer *Sullivan & Cromwell*
Stephen Raslavich *United States Bankruptcy Court*
Richard Spillenkothen *Federal Reserve Board*

Uruguay
Maria Elena Abo *Muxi & Asociados*
Conrado Hughes Delgado *Hughes & Hughes*
Noelia Eiras *Hughes & Hughes*
Daniel Ferrere *Ferrere Lamaison*
Diego Galante *Galante & Martins*
Manuel González Rocco *Banco Central del Uruguay*
Rosario Garat *Superintendencia de Instituciones de Intermediación Financiera*
Marcela Hughes *Hughes & Hughes*
Mercedes Jimenez de Arrechaga *Guyer & Regules*
Estudio Jurídico *Muxí & Asociados*
Elbio Kuster *Bado Kuster Zerbino & Rachetti*
Jose Lorieto *Clearing de informes*
Matilde Milicevic *Clearing de Informes*
Alejandro Miller *Artola Guyer & Regules*
Ricardo Olivera *Olivera & Delpiazzo*
Veronica Raffo *Ferrere Lamaison*
Bruno Santin *Estudio Jurídico Muxí & Asociados*
Alvaro Tarabal *Guyer & Regules*

Uzbekistan
Sanjarbek Abdukhalilov *Denton Wilde Sapte*
Sanjar Abduhalilov *Denton Wilde Sapte*
Daniel Ferrere *Ferrere Lamaison*
R. Gulyamov *Central Bank of the Republic of Uzbekistan*
Thomas Johnson *Denton Wilde Sapte*
Tatiana Lopaeva *Tashkent City Economic Court*
Veronica Raffo *Ferrere Lamaison*
Vakhid Saparov *Baker & McKenzie*
Sofiya Shaikhrazieva *Denton Wilde Sapte*
Umarov Abdurakhim Vakhidovich *Uzbek Association of Banks*
Marla Valdez *Denton Wilde Sapte*

Venezuela, RB
Carolina Armada *ITP Consulting*
Gertrudiz Bonilla *Romero-Muci & Asociados*
Carlos Dominguez *Hoet Pelaez Castillo & Duque*
Rossanna D'Onza *Baker & McKenzie*
Gustavo Muci *Romero-Muci & Asociados*
Irving Ochoa *Superintendencia de Bancos y Otras Instituciones Financieras*
Fernando Pelaez-Pier *Hoet Pelaez Castillo & Duque*
Carlos Plaza *Baker & McKenzie*
Victor Sanchez *Leal Bentata Abogados*
Patricia Wallis *ITP Consulting*

Vietnam
Fred Burke *Baker & McKenzie*
Uan Pham Cong *State Bank of Vietnam*
Florent Fassier *Gide Loyrette Nouel*
Nguyen Viet Ha *Russin & Vecchi*
Ngo Thanh Hang *PricewaterhouseCoopers*
John Hickin *Johnson Stokes & Master*
Richard Irwin *PricewaterhouseCoopers*
Nguyen Hoang Kim Oanh *Baker & McKenzie*
Ian Lewis *Johnson Stokes & Master*
Han Mahn Tien *Concetti Consulting*
John Malcolm Hickin *Johnson Stokes & Master*
Pham Nghiem Xuan Bac *Vision & Associates Investment & Management Consultants*
Tran Thi Thanh Ha *Baker & McKenzie*
Giles Thomas Cooper *Baker & McKenzie*

Yemen, Rep. of
Sheikh Khalid Abdullah *Law Offices of Sheikh Tariq Abdullah*
Adel Adham *Adham & Associates*
Anwar Adham *Adham & Associates*
Jamal Adimi *Jamal Adimi Law Offices*
Abdalla Al-Meqbeli *Abdalla Al-Meqbeli & Associates*
Abdula Al-Olofi *Central Bank of Yemen*
Katerina Miltiadou *Mecos*
Honorable Mohamed Jaffer Kassim *Ministry of Justice*

Zambia
Moses Chatulika *Bank of Zambia*
Mwelwa Chibesakunda *Corpus Globe Advocates*
Elias Chipimo *Corpus Globe Advocates*
Abdul Dudhia *Musa Dudhia & Co.*
Pixie Linda Mwila Kasonde-Yangailo *PricewaterhouseCoopers*
N.K. Mubonda *Dhkemp & Co. Law Firm*
Morris Mulomba *Bank of Zambia*
Kanti Patel *Christopher Russell Cook & Co.*
Solly Patel *Christopher Russell Cook & Co.*

Zimbabwe
Roger Chadwick *Scanlen & Holderness*
Innocent Chagonda *Atherstone & Cook*
Lindsay Cook *Atherstone & Cook*
C.L. Dhliwayo *Reserve Bank of Zimbabwe*
Stephen Gwasira *Reserve Bank of Zimbabwe*
Brenda Wood Kahari *B.W. Kahari Law Offices*
Peter Lloyd *Gill Godlonton & Gerrans*
Piniel Mkushi *Sawyer & Mkushi*
Sternford Moyo *Scanlen & Holderness*
N.K. Mubonda *D.H. Kemp and Company*
Kanti Patel *Christopher Russell Cook & Co.*
Alwyn Pichanick *Wintertons Law Firm*
Yuezhen Wei *PricewaterhouseCoopers*